Water Governance in Motion

Towards Socially and Environmentally Sustainable Water Laws

Edited by

P. Cullet • A. Gowlland-Gualtieri
R. Madhav • U. Ramanathan

SWISS NATIONAL SCIENCE FOUNDATION

International Environmental
Law Research Centre

FOUNDATION BOOKS

Delhi • Bangalore • Mumbai • Kolkata • Chennai • Hyderabad • Pune

Published by
Cambridge University Press India Pvt. Ltd.
under the imprint of Foundation Books
Cambridge House, 4381/4 Ansari Road, Daryaganj, **New Delhi** 110 002

Cambridge University Press India Pvt. Ltd.
C-22, C-Block, Brigade M.M., K.R. Road, Jayanagar, **Bangalore** 560 070
Plot No. 80, Service Industries, Shirvane, Sector-1, Nerul,
Navi Mumbai 400 706
10 Raja Subodh Mullick Square, 2nd Floor, **Kolkata** 700 013
21/1 (New No. 49), 1st Floor, Model School Road, Thousand Lights,
Chennai 600 006
House No. 3-5-874/6/4, (Near Apollo Hospital), Hyderguda,
Hyderabad 500 029
Agarwal Pride, 'A' Wing, 1308 Kasba Peth, Near Surya Hospital,
Pune 411 0110

© P. Cullet, A. Gowlland-Gualtieri, R. Madhav and U. Ramanathan
First published 2010

ISBN 978-81-7596-634-5

All rights reserved. No reproduction of any part may take place without the written permission of Cambridge University Press India Pvt. Ltd., subject to statutory exception and to the provision of relevant collective licensing agreements.

Cambridge University Press India Pvt. Ltd. has no responsibility for the persistence or accuracy of URLs for external or third-party internet websites referred to in this book, and does not guarantee that any content on such websites is, or will remain, accurate or appropriate.

Published by Manas Saikia for Cambridge University Press India Pvt. Ltd. and printed at Sanat Printers, Kundli.

Contents

Acknowledgments	v
Contributor Biographies	vii
Introduction	1
Philippe Cullet	

I: Water Law, Policy and Institutional Reforms in India   9

1. Water and Questions of Law: An Overview   11
   Ramaswamy R. Iyer
2. Water Law – Evolving Regulatory Framework   26
   Philippe Cullet
3. Discourses in Water and Water Reform in Western India   53
   Priya Sangameswaran
4. The Slow Road to the Private: A Case Study of Neoliberal Water Reforms In Chennai   80
   Karen Coelho

II: Ongoing Irrigation and Ground Water Reforms in India   109

5. Canal Irrigation, Water User Associations and Law in India –Emerging Trends in Rights-Based Perspective   111
   Videh Upadhyay
6. Customary Rights and their Relevance in Modern Tank Management: Selected Cases in Tamil Nadu   129
   A. Gurunathan and C.R. Shanmugham
7. Ground Water – Legal Aspects of the Plachimada Dispute   159
   Sujith Koonan

III: Perspectives on Privatisation   199

8. Tirupur Water Supply and Sanitation Project: A Revolution in Water Resource Management?   201
   Roopa Madhav

9. The World Bank's Influence on Water Privatisation in
   Argentina: The Experience of the City of Buenos Aires   230
   Andrés Olleta

10. The Linkages between Access to Water and Water
    Scarcity with International Investment Law and
    The WTO Regime   269
    Francesco Costamagna and Francesco Sindico

11. More Drops for Hyderabad City, Less Crops for Farmers: Water
    Institutions and Reallocation in Andhra Pradesh   299
    Mattia Celio

IV: Environment and Human Rights 331

12. Balancing Development and Environmental Conservation
    and Protection of the Water Resource Base:
    The 'Greening' of Water Laws   333
    Stefano Burchi

13. The Right to Water as a Human Right or a Bird's
    Right: Does Cooperative Governance Offer a Way
    Out of a Conflict of Interests and Legal Complexity?   359
    Jonathan Verschuuren

14. South Africa's Water Law and Policy Framework:
    Implications for the Right to Water   388
    Alix Gowlland-Gualtieri

15. Respect, Protect, Fulfill: The Implementation of the Human
    Right to Water in South Africa   415
    Inga T. Winkler

V: Comparative Perspectives on Reforms 445

16. Learning from Water Law Reform in Australia   447
    Poh-Ling Tan

17. Law and 'Development' Discourses About Water:
    Understanding Agency in Regime Changes   477
    Radha D'Souza

18. Marginal Remarks Concerning Water Policy Regimes;
    Governance, Rights, Justice, and Development:
    An Epilogue   510
    Upendra Baxi

Bibliography   531

Acknowledgments

The editors would like to acknowledge the support of the Swiss National Science Fund (SNF), which has backed, since 2006, the Indo-Swiss research partnership on water law (www.ielrc.org/water) in whose context the two workshops that provided the basis for this book were organised.

A number of people have made significant contributions to the editing process. These include Arati Prasad who diligently edited most chapters as well as Lovleen Bhullar and Monica Kohli. Additionally, colleagues who worked on the Indo-Swiss research partnership on water law made a significant contribution to the overall editing process, in particular Adil Khan, Sujith Koonan, Andrés Olleta, Jessy Thomas and Irina Zodrow.

Contributor Biographies

Upendra Baxi is a Professor of Law in Development at the University of Warwick. Previously he served as a Professor of Law, University of Delhi (1973–1996) and as its Vice-Chancellor (1990–1994.) He also served as Vice-Chancellor, University of South Gujarat, Surat (1982–1985); Honorary Director (Research), Indian Law Institute (1985–1988). He was the President of the Indian Society of International Law (1992–1995). Professor Baxi holds LLM degrees from the University of Bombay and the University of California at Berkeley, which also awarded him with a Doctorate in Juristic Sciences. He has been awarded Honorary Doctorates in Law by the National Law School University of India, Bangalore, and the University of La Trobe, Melbourne. He has authored dozens of books, articles and book chapters on a variety of topics. His most recent books are *Human Rights in a Posthuman World: Critical Essays* (Oxford University Press, 2007) and *The Future of Human Rights* (Oxford University Press, 2nd ed. 2006).

Stefano Burchi was the chief of FAO Development Law Service in Rome until he retired at the end of 2008. Earlier, he had been a Legal Officer with FAO Development Law Service (specialising in water resources law) since joining FAO in 1983. He has authored several published articles and two books on water resources law and administration. He is a regular contributor of the item *Fresh Water* to the Year in Review feature of the *Yearbook of International Environmental Law* (since 1995) and a regular contributor of the item Year-end-review to the *Journal of Water Law* (since 2001). He has worked on the review and analysis of existing water resources legislation and institutional arrangements for its administration, and the drafting of principal and subsidiary water resources legislation. He holds an LLM from Harvard Law School and an MS from the University of Michigan.

Mattia Celio is Counsellor for Science and Technology at the Swiss Embassy in New Delhi. He worked in the ground water sector at the University for Applied Sciences of Southern Switzerland from 1998 to 2003 and was an Associate Expert at the International Water Management Institute in Hyderabad from 2004 to 2007. His main area of research is urban water supply, and intersectoral water allocation. He holds a Master's degree in hydrogeology from the University of Neuchâtel, and submitted in June 2008 his PhD thesis at the Water Engineering and Development Centre, Loughborough University.

Karen Coelho is an Assistant Professor at the Madras Institute of Development Studies. She has a PhD in cultural anthropology from the University of Arizona in Tucson. Her work focuses on critical explorations of urban governance, particularly in the infrastructure sectors, anthropologies of the urban state and reforms, and the constitution of urban 'civil societies' particularly in the city of Chennai and its peripheries.

Francesco Costamagna is a Lecturer in International Law at the Faculty of Foreign Languages and Literatures of the University of Turin. He holds an LLB in international law from the University of Turin, a Master's degree in the management of development jointly offered by the ILO Training Centre and the University of Turin and a PhD in international economic law from the L. Bocconi University, Milan. His thesis focused on the international protection of public infrastructure investments and state sovereignty. His research interests are international investment law, international trade law and human rights law. He works mainly on the relationship between state regulatory autonomy and international economic norms in the field of public utilities, such as water and energy services. He is the author of sections 2.1, 3 and 5.

Philippe Cullet is a Reader in Law at the School of Oriental and African Studies – University of London (SOAS) where he teaches law related to the environment, natural resources and intellectual property. He is also a Founding Programme Director of the International Environmental Law Research Centre (IELRC) and the Editor-in-Chief of the Journal of Law, Environment and Development (LEAD-journal.org). He studied law at the University of Geneva and King's College London (LLM). He received an MA in development studies from SOAS and his doctoral degree in international environmental law from Stanford Law School. He is the author of *Differential Treatment in International Environmental Law* (Ashgate, 2003), *Intellectual Property and Sustainable Development* (Butterworths, 2005) and *Water, Law and Development in the context of Water Sector Reforms in India* (Oxford University Press, 2009) and the editor of *The Sardar Sarovar Dam Project: Selected Documents* (Ashgate, 2007).

Radha D'Souza is a Reader in Law at the University of Westminster, London. Her research interests include water conflicts, law and development, colonialism and imperialism, socio-legal studies in the 'Third World' and global social justice. She is the author of *Interstate Conflicts Over Krishna Waters: Law, Science and Imperialism* (Orient Longman, 2006). She teaches law and development and public international law, and has previously taught sociology, development studies, human geography besides public law and legal theory. Earlier she practiced as a barrister at the High Court of Mumbai in India. Radha has worked as a social justice activist in India, New Zealand and internationally.

Alix Gowlland-Gualtieri is a Research Fellow at IELRC. She holds a law degree from the University of Geneva, Faculty of Law, and an LLM from the University of California, Berkeley School of Law. In 2003, she received her doctoral degree (DPhil) in international environmental law from the University of Oxford. She is currently working on water issues in the framework of IELRC's project on water sector reforms. Other research interests include trade and environment, accountability of international financial organisations, and human rights and environment. She has also worked at the Environment Department of the United Nations Economic Commission for Europe (UNECE) and the Secretariat to the Aarhus Convention on Access to Information, Public Participation in Decision-making and Access to Justice in Environmental Matters. She is co-editor of *Key Materials in International Environmental Law* (Aldershot: Ashgate, 2004) and *Water Law for the 21st Century: National and International Aspects of Water Law Reforms in India* (Routledge, forthcoming 2009).

A. Gurunathan, a rural management professional graduated from the Institute of Rural Managament Anand (IRMA) in 1993, currently leads a thematic programme on Water of DHAN Foundation having an outreach of 150,000 families in South India. He also teaches Project Management to the students of Tata DHAN Academy. He is a resource group member of water community of UN Solution Exchange. After graduation in agricultural engineering at Tamil Nadu Agricultural University in 1988-89, he worked as research assistant in few research projects at Water Technology Centre until May 1991. His area of interest is integrated water resources management, GIS Application for water resources development and development finance.

Ramaswamy R. Iyer is an Honorary Research Professor at the Centre for Policy Research, New Delhi. A former civil servant, he was Secretary, Minister of Water Resources in the Government of India from 1985 to 1987. After leaving the Government, he worked as a Research Professor in the Centre for Policy Research from 1990 to 1999. He has done consultancy assignments for a number of organisations, including the World Bank, the World Commission on Dams, the International Water Management Institute, UNDP and the European Commission. He has written extensively on water-related issues. His recent books include *Water: Perspectives, Issues, Concerns* (Sage: 2003) and *Towards Water Wisdom: Limits, Justice, Harmony* (Sage: 2007).

Sujith Koonan holds an LLM from Cochin University of Science and Technology in human rights and environmental law. Currently, he is pursuing an MPhil in International Law at Centre for International Legal Studies of Jawaharlal Nehru University, New Delhi. His research interest includes human rights, environmental law and international economic law in relation to human rights and natural resources.

Roopa Madhav worked on the Indo-Swiss water law research partnership from 2006 to 2008. Her primary research interests include labour law, human rights and environmental law. She holds an LLM from New York University and a BA/LLB from National Law School of India University, Bangalore. She has been a visiting faculty at the National Law School, has had extensive work experience with trade unions and was the President and Founder Member of the Alternative Law Forum, Bangalore.

Andrés Olleta is a Law Researcher at IELRC and a PhD candidate at the international law section of the Graduate Institute of International and Development Studies (GIIDS), Geneva. He holds a Diploma of lawyer from the Universidad Nacional del Sur, Bahía Blanca, Argentina and a Master's degree in International Relations from the GIIDS. His PhD thesis is on the permeability of the processes of international norm creation to power and the challenges that international law-making poses to developing countries for the incorporation of their interests to new norms. He is a member of research groups in international law in both Argentina and Switzerland and a consultant for the Competition Research for Economic Development project of the International Development Research Centre, Canada.

Usha Ramanathan is an internationally recognised expert on law and poverty. She studied law at Madras University, the University of Nagpur and received her PhD from Delhi University. She has taught in many universities around the world, including the Indian Law Institute in Delhi and the School of Oriental and African Studies (SOAS) in London where she is a regular guest professor. Her research interests include human rights, displacement, torts and environment. She has published extensively in India and abroad. In particular, she has devoted her attention to a number of specific issues such as the Bhopal gas disaster, the Narmada valley dams or slum eviction in Delhi. She is a frequent adviser to non-governmental organisations and international organisations, in particular concerning human rights issues. She is also the South Asia Editor of the Law, Environment and Development Journal (LEAD Journal). She has written numerous contributions, many of which can be found at http://www.ielrc.org/about_ramanathan.php. She is also the co-editor of *Water Law for the 21st Century: National and International Aspects of Water Law Reforms in India* (Routledge, forthcoming 2009).

Priya Sangameswaran holds a PhD in economics from the University of Massachusetts Amherst. She is currently a Fellow in environmental studies in the Centre for Studies in Social Sciences,

Calcutta. She works broadly at the intersection of developmental studies and environmental studies, drawing on heterodox economic theory, post-development theory and political ecology to understand how community-based efforts and questions of equity (particularly in the realm of water) have been conceptualised and practised. Currently, she is researching sector reforms in water in the state of Maharashtra in western India, and more specifically, how these changes link to larger debates about neoliberal development, rights, equity and commodification of resources.

C.R. Shanmugham is a civil engineer from Anna University, Chennai. He holds a Master's degree in Agricultural Engineering from the University of Saskatchewan, Canada. He has worked in Tamil Nadu Public Works and Agricultural Engineering Departments in various capacities and last as Superintending Engineer. Thereafter he served in Tamil Nadu Agricultural University as a Professor and then as the Dean, College of Agriculture Engineering. He is presently working in DHAN Foundation as Programme Advisor. He has been a consultant to the projects sponsored by various international agencies, including the World Bank, the Ford Foundation, the European Union, the International Water Management Institute and national organisations like the Central Institute of Agriculture Engineering, Bhopal, the Administrative Staff College, Hyderabad, and Water and Power Consultancy Services, New Delhi. He was presented with the best worker's award by the Soil Conservation Society of India for meritorious work. He is a Fellow of Indian Society of Agricultural Engineers.

Francesco Sindico is a Lecturer in Law at the University of Surrey where he teaches environmental law. He has an undergraduate degree in law from the University of Turin and an LLM in international law and international relations from the Autonoma University in Madrid. In his doctoral thesis at the Universitat Jaume I, Castellon de la Plana, Spain, he explores the relationship between the climate change and international trade regime. He has also worked on the linkage between climate and security and, more recently, he has started a project on the linkages between energy security, rural development and sustainable tourism. He is a former Marie Curie Fellow at the Institute for Environmental Studies of the Free University in Amsterdam and Visiting Researcher at the University College London and at the University of Geneva. He is the author of sections 1, 2.2 and 4.

Poh-Ling Tan is an Associate Professor of Law and a member of the Socio-Legal Research Centre, Griffith University. She practiced law in Kuala Lumpur, Malaysia before moving to Australia where she has taught in law schools for the last 16 years. Her research interests include water law reform particularly property rights, conflict management and the sustainable management of water. Currently she leads a multidisciplinary team researching collaborative water planning in Northern Australia. Her expertise also lies in the areas of property law and Asian legal systems.

Videh Upadhyay is an advocate and legal consultant and specialises in development and natural resources law. He practices in Delhi and has been a legal consultant to the Ministry of Environment and Forest, the UNDP, the World Meteorological Organization, DFID, the University of Cambridge, the University of Birmingham and NIRD. He has drafted legislation on specific aspects of water and irrigation laws for state governments and provides regular legal advice to civil society organisations. He has been a Visiting Fellow, Centre for Law and Governance, Jawaharlal Nehru University and an India Visiting Fellow at the Golden Gate University, San Francisco and at the University of California, Berkeley. His books include *Public Interest Litigation in India: Concepts, Cases, Concerns* (Butterworths, 2007), *Public Interest Litigation for the Halsbury's Laws of India* (2005) and has co-authored *Forest Laws, Wildlife Laws and Environment; Water Laws, Air Laws and Environment* ; and *Environment Protection, Land and the Energy Laws* (Butterworths, 3 volumes, 2002).

Jonathan Verschuuren holds a PhD degree 'cum laude' on the basis of a dissertation on the constitutional right to environmental protection. He is a full professor of European and international environmental law at Tilburg University. He is currently the Vice Dean for Research at the Faculty of Law, as well as the Director of the Tilburg Graduate Law School. His research mainly focuses on the role of international and EU environmental law in legal practice. Verschuuren has written more than 200 publications in the field of environmental law, including several books, many articles in outstanding refereed journals throughout the world. He is a member of the Commission on Environmental Law, as well as of the Academy of Environmental Law of IUCN, and a deputy judge at the District Court of Arnhem (on criminal environmental law cases). He is a visiting professor at the University of Connecticut (USA), North West/Potchefstroom (South Africa), Leuven (Belgium).

Inga Winkler is a doctoral candidate in public international law at the Faculty of Law of the Heinrich-Heine-University in Düsseldorf. Her doctoral thesis focuses on the human right to water and she has written several articles on different aspects of the topic. Her research is supported by a scholarship of the Heinrich Böll Foundation. In 2008, she was a visiting researcher at the Faculty of Law of the University of Stellenbosch in South Africa for several months. She has been involved with work for the UNDP Water Governance Programme in New York, the Right to Water Programme of the Centre on Housing Rights and Evictions and the German section of Amnesty International, where she works on economic, social and cultural rights and is responsible for the human right to water.

Introduction

PHILIPPE CULLET

Water law in India has experienced significant changes over the past few years and is still evolving. These changes have an internal and an external dimension. On the one hand, water law grew over time into a relatively amorphous set of laws, principles, rules and judicial decisions. The lack of clear direction given, for instance, by an overall water legislation, setting principles governing all water in all its uses, explains in part the relative lack of clarity that water law exhibited. The need to give more clarity to this regulatory framework together with the changing situation in the water sector in general called for reforms. On the other hand, there has been a growing push among international institutions over the past two decades for water sector reforms and water law reforms in many countries. International water policy making has been very influential and a great number of developing countries have adopted reforms that closely follow the model proposed at the international level. The level of compliance with the set of principles propounded by international agencies can be partly ascribed to the fact that lending institutions, such as the World Bank, have strongly endorsed these principles and included them as part of their lending instruments.[1] Since lending institutions such as the World Bank and the Asian Development Bank have been quite

---

[1] *See* World Bank, Water Resources Sector Strategy – Strategic Directions for World Bank Engagement (Washington, DC: World Bank, 2004). *Source:* http://www.ielrc.org/content/e0413.pdf.

active in India and have included a number of law reform conditionality in their loans, a number of ongoing reforms in India are directly linked to international principles.[2]

Water Sector Reforms

Changes to water law are being introduced under the general context known as water sector reforms. These reforms are promoted as a comprehensive set of changes that have the potential to address the current and forthcoming water crisis. Water sector reforms promote the integrated management of water resources and are thus conceived as having the potential to redress the shortcomings of earlier sectoral water policies.

While water sector reforms are broadly conceived, they are not as comprehensive as may appear at first sight. Rather, they are based on a relatively narrow set of principles that emphasise, in particular, the need to conceive water as an economic good or commodity. This seeks to ensure its 'efficient' use, which is seen as the only way to tackle water scarcity. While the economic dimension of water is emphasised, water sector reforms tend to avoid the language of human rights and prefer to talk about basic needs rather than basic rights. This is, for instance, reflected in the Dublin Statement that conceives of a human right 'within' the principle that water is economic good.[3] Another important dimension of water sector reforms is the drive for institutional reforms. Broadly, water sector reforms seek a reduction in the influence that the state has in the water sector and suggest that decentralisation and participation of water users should be introduced. This fits with a focus on 'management' issues within the water sector and also includes calls for private sector participation.

Over the past couple of decades, water sector reforms have spread to a number of countries. A number of different aspects have been addressed. In agriculture, the main focus has been on

---

[2] A list of water law conditionality in India. *Source:* http://www.ielrc.org/water/doc5.htm.

[3] Principle 4, Dublin Statement on Water and Sustainable Development, International Conference on Water and the Environment, Dublin, 31 January 1992. *Source:* http://www.ielrc.org/content/e9209.pdf.

the introduction of participatory irrigation management whose main rationale is to bring the management of irrigation infrastructure closer to water users. The main institutional innovation proposed is the setting up of Water User Associations. In the drinking water sector, for urban areas the privatisation of water services or the commercialisation of the operations of public sector service providers has often been proposed.[4] In rural areas, participation by water users in the management of their own infrastructure is suggested. In both cases, the primary principle that drives reforms is the desire to introduce full cost recovery of drinking water infrastructure, either immediately or as an eventual goal.

Water sector reforms have been widely implemented in a variety of countries and contexts. From the point of view of reform proponents, they have been successful to the extent that they have brought about significant changes, for instance, in the context of participatory irrigation management. Yet, at the same time, water sector reforms have been found lacking because they were not mainstreamed to the extent expected and were subject to significant opposition in certain cases, such as outright privatisation of water services.[5] These two combined factors have led to the development of new strategies to foster water sector reforms. On the one hand, the introduction of water sector reforms on a project-by-project basis was found insufficient to ensure the mainstreaming of reforms. On the other hand, the big bang approach to water privatisation was found to be politically unacceptable in many places.[6]

The new strategy put in place since the 1990s, but much more systematically this decade, has been to introduce water law reforms as a precursor to the reforms themselves. This second generation of reforms is based on the same principles that have driven water sector reforms for the past two decades but suggests new ways of

---

[4] *See* Naren Prasad, 'Privatisation of Water: A Historical Perspective', 3/2 *Law, Environment and Development Journal* (2007), p. 217. *Source:* http://www.lead-journal.org/content/07217.pdf.2

[5] *See* Olleta in this volume.

[6] This is, for instance, illustrated by the case of a privatisation project for Delhi that had to be suspended in 2005 because of public protests. *See* Amit Bhaduri and Arvind Kejriwal, 'Urban Water Supply: Reforming the Reformers', 40/53 *Economic & Political Weekly* 5543 (31 December 2005).

bringing them about. This is, for instance, the case of legislation making the setting up of Water User Associations compulsory throughout a state,[7] or legislation setting up independent water regulatory authorities whose mandate includes among other things to take the privatisation of water services agenda forward.[8]

The introduction of legislation as a way to introduce a set of reforms is not particularly noteworthy. However, in the water context, two issues stand out. Firstly, water-related laws that have been introduced in India over the past decade tend to be problem specific in the sense that they address either one specific issue such as participatory irrigation management, one specific type of water such as ground water or a specific part of the regulatory framework such as institutional reorganisation. There has thus been no effort to introduce water legislation that would provide an integrated, comprehensive framework setting out all the main principles that apply to water in general.[9] Assuming that the Union feels unable to persuade states to do this on their behalf as in the case of the Water Act, 1974, or to suggest concurrent listing, individual states could use their own constitutional prerogatives to do so. Secondly, the laws that have been introduced over the past decade display an uncanny similarity. This is not completely unexpected since a number of these laws are adopted in the context of water projects sponsored by the World Bank or Asian Development Bank. It nevertheless constitutes an important feature of ongoing reforms in India. Indeed, the marked diversity between states in terms of water situation is not reflected in the acts adopted. This is surprising, since legislation adopted in other parts of the world in the context of water sector reforms also look quite similar, as in the case of participatory irrigation management legislation.

---

[7] In the case of Andhra Pradesh, see Roopa Madhav, 'Irrigation Reforms in Andhra Pradesh: Whither the Trajectory of Legal Changes?' (IELRC Working Paper 2007-04, 2007). *Source:* www.ielrc.org/content/w0704.pdf.

[8] *See* Shripad Dharmadhikary, 'Maharashtra: Water Regulatory – A Flawed Model for Water Regulation', *India Together* (23 May 2007). *Source:* http://www.indiatogether.org/2007/may/env-mwrra.htm.

[9] Note that water policies adopted at the union level and in a number of states are not an alternative to comprehensive water legislation.

Water Law Reforms in India

Water law reforms are necessary in India as in many other countries. This is due to a combination of factors from changing conditions in water availability and capacity to access water to outdated principles and legislation in place. In other words, the necessity for changes is nearly universally accepted. For instance, it is widely acknowledged that rules that link access to ground water and control over land are neither socially equitable nor environmentally sustainable. Differences emerge, however, in the responses that need to be given to these problems. On the one hand, water sector reform proponents support delinking water and land to ensure that water rights can eventually become tradable entitlements. On the other hand, other proposals emphasise the need for ground water to be given – like surface water – the status of a public trust to ensure that no individual can appropriate the water and thus lead to more equitable and environment-friendly outcomes.

Ongoing reforms do not reflect the different reform options that exist. Most reforms follow a similar pattern modelled on the principles of water sector reforms. This is rather surprising in a democratic polity where each state faces remarkably different challenges in the water sector and where legislation is adopted by each state individually. The 'surprise' soon dissipates because most states adopt new legislation in the context of projects based on the principles of water sector reforms. The relative uniformity in water law reforms is a major characteristic of ongoing reforms.

Ongoing water law reforms can be divided into two main groups. Firstly, legislative assemblies have adopted a number of new acts over the past decade. These tend to focus on the same specific areas with the greatest number of new legislation concerning participatory irrigation management followed by ground water legislation and legislation for institutional reforms in the water sector. All these reforms are easily identifiable because they are adopted according to the formal procedures set in place for the adoption of legislation at the state level. Secondly, there are a host of reforms which do not qualify as legislation but are very significant. This is, for instance, the case with regard to drinking water where there is little by way of legislation. Changes to the policy framework for the provision of drinking water are, however, also a central part of ongoing reforms. This takes the form of government resolutions and other subsidiary instruments. Thus, a major reform initiative, the Swajaldhara Guidelines were adopted in the form of an administrative decision of the Ministry of Rural Development.[10] Despite the relatively innocuous instrument, the Swajaldhara Guidelines sought to introduce what is probably the single biggest change in drinking water policy at least since the early 1970s. In other words, water law reforms are not limited to legislation but also include a host of other instruments. Government Orders can, for instance, be used to take decisions that have widespread repercussions, whether or not the measures derive from an existing legislation. This is, for instance, the case of transfer of the responsibility of operation and maintenance responsibilities to Gram Panchayats under the guise of the principle of full cost recovery.[11]

This book provides an early attempt to map out and analyse ongoing water law reforms. This will need to be supplemented in years to come by much additional research that will examine the implementation of existing reforms, such as the impacts of the new independent water regulatory authorities as well as analyse the many new laws that are upcoming.[12]

Water Law Reforms in India – National and International Perspectives

This book attempts to analyse water law reforms in India in their multiple dimensions. It thus includes a series of chapters that focus specifically on India, chapters that focus on the international dimensions of national reforms and chapters analysing reforms in other countries whose experience is relevant for understanding changes in India.

This book finds its origins in a series of two workshops organised by the International Environmental Law Research Centre (IELRC)

---

[10] Ministry of Rural Development, Guidelines on Swajaldhara, 2002. *Source:* http://www.ielrc.org/content/e0212.pdf.

[11] For Maharashtra, *see* RWS-1002/ CR-639/ WS-07, 22 May 2003.

[12] First attempts at going further include the following two books: P. Cullet, A. Gowlland-Gualtieri, R. Madhav and U. Ramanathan eds., *Water Law for the 21st Century: National and International Aspects of Water Law Reforms in India* (Abingdon: Routledge, forthcoming 2009) and Ramaswamy Iyer ed., *Water and the Laws in India* (New Delhi: Sage, forthcoming, 2009).

in Delhi in December 2006 and Geneva in April 2007.[13] Both workshops took place in the context of the Swiss National Fund (SNF) sponsored Indo-Swiss research partnership on legal issues related to water sector reforms.[14] Each workshop shed light on a number of distinct but related aspects of water law reforms. In particular, the Geneva workshop provided an opportunity to explore the links between the principles of water sector reforms developed at the international level and water law reforms in different parts of the world. One of the most interesting lessons from the case studies that were presented was the significant similarities that can be found between reforms implemented in completely different parts of the world. Interestingly, while the differences between reforms implemented in different countries can be ascribed to local specificities, they can also be ascribed to chronological factors. Indeed, certain types of reforms widely proposed in the 1990s, such as city specific privatisation of water services, are still found in this decade, but have largely been overtaken by more systemic reforms such as the setting up of completely new water institutions to guide a broad process of reform in the water sector.

This book seeks to provide a cross-sectoral analysis of ongoing water law reforms. It highlights a number of the most significant reforms currently taking place and being implemented in India and in other countries to illustrate the breadth of changes that are proposed. The first part examines the framework within which water law reforms are implemented and the links between water sector reforms and water law reforms.[15] It examines some of the concepts that underlie water sector reforms, the process of change as well as the basic characteristics of water law reforms.

The second part focuses on two of the most sensitive areas of water law, irrigation and groundwater.[16] The former is of central importance because agriculture remains by far the primary user of water. Ground water regulation is increasingly sensitive because it has become the primary source of water for all water uses over the

---

[13] Information concerning these workshops can be accessed respectively at http://www.ielrc.org/activities/workshop_0612/index.htm and at http://www.ielrc.org/activities/workshop_0704/index.htm.
[14] For more information, visit http://www.ielrc.org/water.
[15] Chapters by Iyer, Cullet, Sangameswaran and Coelho.
[16] Chapters by Upadhyay, Gurunathan and Shanmugham, Koonan.

past few decades. Irrigation and ground water illustrate one peculiarity of water law, which is its fragmentation. Thus, irrigation legislation deals nearly exclusively with agricultural use of water and ground water principles and legislation addresses only ground water, potentially creating legal structures that are parallel but different from the rules governing surface water despite the close links between surface and groundwater.

The third part examines, more specifically, issues related to privatisation of water services.[17] This is the part of water sector reforms that has been the most visible in many cases. A number of legal issues arise in this context. Indeed, privatisation is often undertaken on a scheme-specific basis without necessarily amending the existing legal framework. Further, privatisation is a broad construct, which applies not only to water services at the local level but also has a number of international law ramifications of relevance to individual countries, as in the case of developments concerning water services or water as a good in the context of the World Trade Organisation.

The fourth part examines the broader context within which existing reforms are conceived and framed.[18] It focuses first on the environmental dimension of water law reforms, an all-important part of water sector reforms that are largely premised on addressing environment-related aspects of the water sector. Secondly, it addresses the question of the human right to water, a right which has been repeatedly confirmed as a fundamental one in India and is overwhelmingly recognised at the international level. One of the interesting aspects of water law reforms is that they tend to place themselves within the context of environmental sustainability and social equity. However, environmental considerations and the human right to water do not figure prominently in ongoing reforms. In fact, the human right to water is sometimes reduced to a discussion on basic needs.

The last part puts ongoing water law reforms in India in a broader perspective.[19] It analyses reforms undertaken in Australia, a country that has gone relatively far in its process of change and where the impacts of the reforms can already be identified. It also places water law reforms in the broader development discourse paradigm that informs the process of change.

---

[17] Chapters by Madhav, Olleta, Costamagna and Sindico, Celio.
[18] Chapters by Burchi, Verschuuren, Gowlland-Gualtieri and Winkler.
[19] Chapters by Tan, D'Souza.

UNIT I
Water Law, Policy and Institutional Reforms in India

1

Water and Questions of Law: An Overview

RAMASWAMY R. IYER

I

This paper aims at presenting a general overview of some of the important legal questions that arise in the course of the current international debates on water. It is not a detailed technical paper meant for a scholarly audience but a broad presentation, in simple terms, for the interested general reader. It is explanatory in nature, but it is also partly prescriptive, and in the latter aspect it is a personal statement. It draws upon Indian experience, but is not India-specific.

II

In the past, questions of law relating to water tended to arise mainly in three contexts:

(i) in the riparian context, regarding claims to the use of river waters arising from location in relation to a river, conflicts over such claims, complaints of injury caused by other (mainly upper) riparians, and so on (at the level of households, farms, villages, provinces or states, or countries);

(ii) in the context of wells (ground water, i.e., water lying under one's land), giving rise to the idea of ownership of or property rights over water; and

(iii) in the context of small ponds and lakes or springs near a village, perceived as 'commons' belonging to the community, giving rise to the idea of a common property resource.

All these were, of course, subject to the overarching sovereign powers of the state.

These three water sources were perceived as distinct, and governed by different legal principles. Sharp distinctions were drawn by the courts (in England) between flowing surface waters and underground water which was perceived as stationary. The management of village ponds and lakes (in India) was largely in the hands of the community, with rules often informal and sometimes written down, having the force of law through custom and consent, and backed by social sanctions of varying degrees of effectiveness.

That relatively simple picture has been greatly complicated over the years by many factors.

First, advancements in knowledge and technology have altered the assumptions on which the old laws and court decisions were based. Greater knowledge about ground water has rendered obsolete the old distinction between flowing surface waters and ground water; possibilities of storage, diversion and pumping of river waters have reduced the significance of riparianism to some extent, though it continues to be important in transboundary contexts; and power-driven tube wells and bore wells have enormously increased water-extracting capabilities of their owners, rendering land-related ownership of ground water highly inequitable as well as harmful to the resource.

Secondly, with the change in the ruling economic philosophy and the dominance of the capitalist ideology of free enterprise and market forces, water tends to be increasingly perceived in economic terms.

Thirdly, the growing sense of a looming water scarcity or crisis leads to (a) the engineering response of large supply-side water projects (this in turn leading to the postulation of private sector participation); and (b) the advocacy of water markets as the answer to future needs.

Fourthly, that dominant engineering-cum-economic approach is challenged by certain other factors and forces – environmental and ecological concerns and movements; the growing 'human rights' discourse; movements that champion the cause of tribes, communities and groups that have been the victims of historical injustice or otherwise disadvantaged; and NGOs that seek to empower civil society vis à vis the state.

Against that background, it is not surprising that the current debates about water are characterised by considerable divergences of perceptions and views. Those divergences are reflected in all the important international events such as the triennial World Water Forums, the annual Stockholm Water Week and Symposium, and so on. In all these gatherings, the participants can be broadly classified into three or four loose groups – government delegations (Ministers, Secretaries, technical experts); international organisations (World Water Council, Global Water Partnership and its regional and national affiliates, UN agencies, World Bank, ADB, etc); the big corporations with an interest in water; and an assemblage of NGOs, 'Greens' and social activists. There are many debates and controversies, but the major ones revolve around the following themes or clusters of themes – the projected water scarcity or crisis; big water projects and the related issues of environmental impacts and human displacement; conflicts relating to water; water markets, trade in water and the privatisation of water services; ecological concerns, critiques of 'development', and (in recent years) climate change.

III

A major divergence underlying these various debates is regarding the nature of water. One view is that water, as essential to life, and therefore a basic need and right, is something special, something *sui generis*. The other view is that water is a commodity like any other, and subject to the same economic laws and market forces. Many of us would be more sympathetic to the first view, but the second too has a degree of validity and cannot be dismissed lightly. Broadly speaking, one can think of a threefold categorisation:

(i) Water for life (i.e., for drinking, cooking, washing, personal hygiene);
(ii) Water as economic good (for agriculture, industry, hotels and other commercial uses, navigation);

(iii) Water as social good (for schools, hospitals, other public institutions, municipal uses such as fire fighting, sanitation).

(The above categorisation does not cover the historical, cultural and sacred aspects.)

It is hardly possible to rule out any of the above uses totally. However, choices may have to be made from time to time, and priorities are therefore necessary. The absolute priority must, of course, be for water for life; water as social good and water as economic good must take second and third places respectively. However, it is not enough to say that water for life must take precedence over water for other uses. It is necessary to go further and say that the economic or other uses of water by some must never be allowed to jeopardise water for life for others.

(A question that might arise about the threefold categorisation is the following – must we not distinguish agriculture or industry pursued as a commercial activity for profit from similar activities pursued for a bare subsistence, and if so, must we then introduce a separate category of 'water for livelihoods' between 'water for life' and 'water as economic good'? That sounds plausible, but it is very difficult to apply the principle in practical terms. Where does one draw the line between 'subsistence livelihood' and 'gainful economic activity'? A simple distinction between 'water for life' and 'water for economic use' is easy to understand and operate; the introduction of an intermediate category of 'water for livelihoods' will reduce that clarity and make things very difficult.)

Let us look at those propositions in terms of the 'rights' discourse. We need to distinguish 'the right to water' from 'water rights'. When we talk about 'the right to water' we have in mind water as basic need and therefore as a basic right. The right to water is a fundamental right in India by judicial interpretation of the 'right to life' laid down in the Constitution. It is an explicit constitutional right in some other countries, for instance, South Africa. The other term 'water rights' refers to certain use rights, say, agricultural use, industrial use, etc. Broadly speaking, these are economic rights. The World Bank, IMF, ADB, and neoliberal economists in general tend to talk about these rights in terms of property rights. The general slogan of economic 'reformers' is – 'Define property rights in water and make them tradable'. There are serious difficulties here, and we could go into them later, if necessary. For the present, we could translate the propositions put forward in the preceding paragraphs into the language of rights. We then have the following statements:

(i) If any question of choice or conflict arises, then fundamental rights must take precedence over economic or other use rights.
(ii) The economic rights of some must not be allowed to jeopardise the fundamental rights of others.

The broad approach tentatively adumbrated above might be of some use in dealing with the current controversies relating to water in the international arena.

IV

One set of controversies clusters around the theme of a looming water scarcity or crisis. There are three separate, though interlinked, controversies: (i) regarding the projection of a crisis or severe scarcity; (ii) regarding the right answer to that crisis or scarcity; and (iii) regarding the impacts and consequences of big water-resource projects for the storage/diversion/long-distance transfer of river waters, if these are to be undertaken. All these are water-policy debates and not legal ones. However, there may be some important legal aspects to such debates.

In the first place, any talk about a crisis begs the question of the way in which we use water for various purposes. The crisis, if one is coming, is partly of our own creation through the gross mismanagement of our water endowment. Strenuous efforts and campaigns are needed towards the promotion of demand-restraint, economy in water-use, resource conservation, and so on. These may not get rid of the crisis altogether but may minimise its severity. However, these are matters for persuasion, advocacy and inspirational leadership. It seems doubtful whether law by itself can do much towards those objectives. What it *can* do is perhaps impose limits on use and prescribe duties of conservation on the citizens.

Despite such measures some augmentation of supplies may be necessary. There are only three ways in which water available for use can be augmented, namely, big water-resource projects, drilling for ground water, and local rainwater-harvesting and watershed development. All three bring benefits and all three have problems associated with them. Some of these may need legal measures.

For instance, big water-resource projects generally require Environmental Impact Assessment (EIA) studies. It is this writer's view that EIAs must be made a statutory requirement where this is not the case at present; that the practice of EIA must be made a proper profession like medicine or chartered accountancy and provided with a statutory charter; and that it must be made truly independent of project-sponsors and approvers. Again, the Appraisal Committees that go through the EIAs and make a recommendation on the projects (positive, negative or conditional) need to be insulated from bureaucratic, political and corporate pressures and influences. In other words, we must devise legal means of ensuring that the processes of decision-making on such projects are open, objective and professional, and conform to the guidelines and criteria suggested by the World Commission on Dams to the extent possible.

The displacement of people by such projects must also require a statutory clearance. The consideration of options and alternatives and the choice of a non-displacing or least-displacing option must be a mandatory requirement. There should be no forced displacement. The law must incorporate the principle of 'free, informed prior consent' recommended by the World Commission on Dams. It should also include enlightened resettlement and rehabilitation policies and packages, and the project-affected families should be given the first claim on the benefits arising from the project.

Turning to ground water, its extraction through tube wells and bore wells needs to be regulated to ensure equity, social justice, resource-conservation and ecological sustainability. The fact that there are practical and political difficulties in dealing with large numbers of owners and operators of tube wells and bore wells cannot be an argument against regulation. Regulation is rendered difficult not merely by the numbers of tube wells and bore wells but also by the British common law tradition (also prevalent in India) that the ownership of land carries with it the ownership of the water lying under the land. This link needs to be broken and ground water treated as a community resource.

In regard to local community-based water-harvesting and micro-watershed development, the point is that such initiatives cannot be left entirely unrecognised by state and law. They need some support from formal law and from officialdom, without changing their non-official, informal and flexible nature. There must also be a constructive and harmonious relationship between the state and civil society, and between formal law and customary law. It must be added that even such small and local activities do need: (i) some overseeing to ensure that there are no adverse downstream impacts; (ii) informal institutional arrangements for obviating or resolving conflicts and disputes; and (iii) means of harmonising local initiatives and the overall basin hydrology.

V

What has been said above about big projects needs to be supplemented by a few special words about rivers. The damming and diverting of rivers are serious interventions in river regimes, and in some cases the rivers no longer reach the sea. Further, the processes of 'development' (urbanisation, industrialisation, commercial agriculture) involve an enormous generation of waste that in some instances has turned rivers into sewers. If dying rivers are to be rescued and if other rivers are to be saved from dying, certain principles need to be observed, and if necessary, given statutory backing.

The shocking state of many of the rivers in India led to the constitution of a National River Conservation Authority some years ago, and very recently, there has been some talk of legislation for the protection and conservation of rivers. As the point of departure for any such legislation (in India or elsewhere), this writer would propose four catchphrases or slogans: (i) a river is not a drain; (ii) a river must flow; (iii) a river must have space; and (iv) a river is an ecological system in itself, and part of a larger ecological system. An elaboration of these slogans follows.

> (i) 'A river is not a drain': In the language of engineers, a river is a drain in the sense that it drains a catchment. That is an accurate technical statement, but it does not give us an idea of the multiple dimensions of a river. A river is not just a conduit taking the water that falls on a

watershed to the sea. It performs many other functions in the ecological system and on Planet Earth. It is a sustainer of aquatic life and the ecological system, and it has a life and a personality of its own; it is part of a people's history and culture; it is also a sacred resource. It is only when we remember all this that we are likely to respect a river. If we think of the river in reductionist terms as a drain, we are unlikely to respect it, and we will not flinch from throwing waste into it and polluting it.

(ii) 'A river must flow': Doubtless a river has certain self-purifying capabilities, but only if there is water in it. That is why a river must flow. Indeed, if it does not flow, it is not a river. It is not a question of 'minimum flows'. This expression implicitly regards extraction from the river as the norm and leaving some water in the river as a necessary evil. We have to reverse this and regard flows as natural and extraction or diversion as a deviation from the norm, to be kept to the minimum. In other words, what we need is not minimum flow; but minimum interference with the flow.

(iii) 'A river must have space.' Floods are natural phenomena. They occur from time to time, and will continue to occur with varying severities. We must learn to live with them and minimise damage. When floods come, the river needs to spread and accommodate them. In other words, a river needs space. If we keep reducing the space available to a river the consequences will be serious. The natural flood-plains of a river must be respected. (Unfortunately, this principle has been seriously violated in India: the flood-plains of the Yamuna have been heavily built upon, first by extensive and large-scale housing developments and then by a massive temple complex; and now the forthcoming Commonwealth Games Complex is to be located on the flood-plains: a campaign against that location is proceeding but seems unlikely to succeed.)

(iv) 'A river is an ecological system in itself, and part of a larger ecological system'. This is obvious and needs no explanation. It follows that we cannot protect or conserve a river unless the ecological system as a whole is protected and conserved. This calls for a re-examination of lifestyles and our understanding of what constitutes 'development'.

Incidentally, much of the water supplied for any use returns to plague us as waste of one kind or another. The greater the supply of water, the greater the generation of waste. That is a very strong reason for minimising supply-side answers. This will not merely ease the pressure on a scarce resource but will also reduce the generation of waste and the consequent pollution of rivers.

VI

An important part of the debate on water-resource projects is that of the needed investments and 'private sector participation' in such investments. Here again positions are sharply polarised. Those who support this idea will advance the arguments of the greater efficiency of the private sector, the poor track record of the public sector, the need to find resources for investment, and of course the ideological position that the state has no business to be involved in such projects. The opponents of private sector participation in big water resource projects have many concerns, some of them doubtless ideological, but they are also worried about the corporate control of the natural resources of the community. They are further worried that control by domestic corporations could pave the way under the WTO regime to control by giant multinationals. Maude Barlow and Tony Clarke, as also Vandana Shiva and others, have written extensively about the 'corporate theft of the world's waters'. These criticisms need to be given serious consideration.

That brings us to the subject of privatisation of water, which is a major controversy in international forums. The term 'privatisation' is used here in a broad sense to cover both the entrustment of water *services* to private agencies and the acceptance and encouragement of private sector participation (i.e., investment) in major water-resource *projects*. The advocacy of privatisation is part of the dominant economic philosophy of capitalism, free enterprise, an implicit faith in market forces and a minimal government (though very recently, the financial crisis that is spreading from the USA to the rest of the world has led to some doubts about that philosophy). Privatisation has been advocated in relation to industry for a long time, and by analogy the advocacy has been extended to water in the last decade or two. There are serious difficulties with the analogy. However, without entering into an ideological debate, it seems to this writer that four important points of a legal nature must be kept in mind. *First*, the entrustment of the water supply service to a private entity must in no way compromise every citizen's basic right (fundamental or human or whatever we wish to call it) to water as life-support. (This has two implications – the private supplier must supply rich and poor areas alike, and must not be allowed to 'cherry-pick' the former; and whatever pricing principles are adopted, no one should be denied water because he or she cannot afford the tariff.) *Secondly*, the state's ultimate responsibility in this regard remains, and must remain, despite privatisation. *Thirdly*, the difficult distinction between the service and the resource must be maintained, and the privatisation of the former must not lead to the privatisation of the latter; even if private participation in a water-resource project is allowed, care must be taken to ensure that the community's or the state's control over natural resources is not compromised. *Fourthly*, given the nature of water, the privatisation of water services is generally likely to be accompanied by certain conditions (social obligations), and it needs to be ensured that the normal corporate primacy to profitability does not over-ride them.

VII

Related to but not identical with the advocacy of privatisation is the advocacy of water markets. If one believes that market forces will provide answers to all problems and that the outcome of their operation has an implicit validation, then one would certainly advocate water markets as the route to follow for the future. Hence the advocacy of the creation of well-defined tradable property rights in water.

No one can quarrel with a clear definition of rights, whether it be the fundamental right to water as life-support or the economic rights of farmers to irrigation water or industrialists to water for industrial use. The difficulty lies in the proposition that the rights must be converted into property rights and made tradable. Taking first the right to water as life-support (i.e., for drinking, cooking and washing), it is in India (by judicial interpretation) a part of the right to life, in South Africa a constitutional right, and in UN terms a human right. How can it be regarded as a property right and made tradable? This objection seems unanswerable. As for the water rights of a farmer for irrigation, or those of an industry for industrial uses, these are use rights and tied to certain uses; by strict implication the right must cease if the use ceases. If so, how can they be divorced from the use and made into property rights, and further, made tradable? Temporarily, a possibility of trading in water may arise. A farmer may, for certain reasons, decide not to cultivate his or her land for a year or two, or may temporarily change to crops demanding less water. He or she may then have surplus water for sale during that period. An industrial house may decide to suspend operations for a certain period for various reasons, and may have water to spare. However, these are temporary situations. If the industry closes down for good, or if the farmer makes a long-term change to crops that need much less water, or decides to move out of agriculture altogether, should the old water entitlements still hold and be allowed to be traded in? To this writer, that too seems an unanswerable objection.

As an illustration, consider the case of Chennai, India, where Metro Water (the governmental water supply organisation) has been buying water for supply to the city from farmers in adjoining rural areas. As a temporary expedient this may be in order, but can these farmers permanently trade in water? Can they become water-sellers instead of farmers on a regular basis? When they cease to be farmers, do they still own the water? Is it their water to sell? Besides, it is reported that they are willing to sell even more. Next, they will doubtless put in more bore wells and offer still more water to the market. These are some of the disturbing implications of the advocacy of water markets.

Water markets tend to emerge particularly in the context of ground water extraction through tube wells and bore wells, and they serve some useful purposes. They make possible the practice of irrigated agriculture by the poorer or less affluent farmers through the purchase of water from those who can afford to invest in tube wells or bore wells. However, there are dangers of unsustainable extraction as also of inequitable relationships between sellers and buyers.

It is not being argued here that water markets must not be allowed to exist; but there is indeed a case for arguing that they should not be allowed to proliferate and get out of hand; and they need to be carefully regulated in the interest of equity, resource-conservation and ecological sustainability.

VIII

That brings us to the subject of regulatory authorities. Why should we talk about a regulatory authority for water? In India, regulatory authorities came into being in the telecommunications and electricity distribution areas, in the context of the privatisation of these services. Historically, the origins of sectoral regulatory authorities can be traced back to the Margaret Thatcher era in Britain when a wave of ideologically driven privatisations began in that country. With privatisation came the need to regulate the private service-providing parties, for the purpose of ensuring adequate and fair competition, reasonable tariffs and the protection of the consumers' interests. Against this background, any talk of regulatory authorities in relation to water could be interpreted (or misinterpreted) as being based on the assumption of the privatisation of water services.

It may be argued that the case for a regulatory authority for water is independent of the case for the privatisation of water services and does not presuppose the latter. The question then arises – what is to be regulated? With electricity or telecommunications the regulation is of the private corporate entities providing the service. With water, given the projection of a water crisis, what we need to regulate is the use of water. This broad distinction needs to be kept in mind, though it undoubtedly needs some qualifications.

Further, underlying the talk about regulatory authorities is the view of water as an economic good or commodity subject to market forces; in that view, the principal instrument of regulation is market-led pricing aiming at 'full-cost recovery' limited only by competition. This is a seriously narrow and deficient, if not distorted, view. Incidentally, neoliberal economics is not particularly concerned about the poor. This philosophy places its faith in the market; regulation takes place within the market; and the very poor are outside the market. Breaking out of the narrow neoliberal economic view, we need to understand and grapple with water in all its complexities and multiple dimensions before we try to regulate water.

What then do we need to do about water? If we think that 'demand' is sacrosanct and that the essential thing to do is to make more water available for every kind of use, and if we think further that the responsibility for this can be shifted to private agencies or public-private partnerships, then it would, of course, be necessary to regulate the suppliers to ensure fair competition, good service and reasonable tariffs. On the other hand, if we think that the 'demand' for water cannot be allowed to grow unchecked but needs to be restrained, and that what is primarily called for is not supply-side augmentation but an acceptance of the finite quantity of the supply in nature and the limiting of our draft on this scarce and precious natural resource through sensible and equitable use and careful conservation, then 'regulation' takes on a different meaning.

In this writer's book *Towards Water Wisdom: Limits, Justice, Harmony* (Sage, 2007), a Declaration on water has been set forth, and a series of action points derived from it. Among other things we need to do the following – restrain the growth of demand; promote equity, efficiency and economy in water-use; foster a consciousness of a scarce and precious resource; promote rainwater-harvesting and micro-watershed development extensively; limit recourse to big projects to the minimum, treating them as projects of the last resort; arrest the present disastrous over-exploitation of ground water; and arrest and reverse the loss of good water to pollution and contamination. Some of these actions can be described as 'regulation' but this regulation is very different from what the regulatory authorities in the telecommunications and electricity sectors do.

IX

A legal question of some importance is the role of the state in relation to water resources. This is often loosely referred to as 'eminent domain' (a term current in American law), but it may be better to use the term 'sovereign power'. This often comes in the way of enabling the community or 'civil society' to undertake initiatives and perform its appropriate role in relation to water management. If civil society is to be empowered, then the state's sovereign power must be moderated. At the same time the state has its own role to play in relation to water and needs to be enabled to do so. How then can both the community and the state be empowered? The answer to that conundrum lies in the 'public trust' doctrine. Under this doctrine, the state is perceived, not as owning the water resources of the country, but as holding them in trust for the people (including future generations). As a trustee, the state will, of course, have to be empowered to legislate, regulate, allocate, manage, and so on, and all this must involve a degree of control. That control, however, is not inherently confrontational, and may permit a constructive relationship between the state and civil society.

We are not concerned here with the historical evolution of the public trust doctrine in different ways, under different circumstances, and reflecting different concerns, in England, Europe and the USA. What needs to be noted is that it is not part of the law in all countries. (For instance, it is by no means clear that it is part of the law in India, though it was so stated in a certain judgment.) The theory that the state holds water and other natural resources in trust for the community seems attractive and persuasive. It reconciles the position that the resources belong to the people or community with the evident need for the state to play certain roles. It seems desirable that it should be universally adopted.

X

Finally, turning to the international arena, we need to consider two aspects: (i) inter-country relations over rivers that straddle boundaries; and (ii) the global governance of water-resources.

In so far as inter-country water relations are concerned, these are matters for negotiations between the countries concerned, and there are many Treaties and Agreements. However, from the mid-1960s onwards there have been some international understandings. First, we had the Helsinki Rules of 1966 on the Uses of the Waters of International Rivers. Later, in 1997, the UN General Assembly adopted the Convention on the Law of the Non-Navigational Uses of International Watercourses (but it has not so far been ratified by the required number of countries). Putting the two together, we can say that what has commanded a fair degree of international acceptance is the principle of equitable sharing for beneficial uses (in the Helsinki language) or of utilisation in an equitable and reasonable manner (UN Convention). What is 'equitable' has, of course, to be determined with reference to many criteria, and there is enormous scope for differences here, but there is at least a consensus on the principles of equity and reasonableness. Similarly, there is agreement that the upper riparian must not cause harm to the lower riparian, though the wording has changed from 'substantial harm' in the Helsinki Rules to 'significant adverse effects' in the UN Convention. Beyond the generalities of these international documents, the countries concerned have to reach agreements through detailed negotiations.

It must be noted that both the old Helsinki Rules and the present UN Convention are concerned only with a limited aspect, namely inter-country relations over transboundary river waters, and not with the larger question of water as a scarce and precious natural resource. Concerns over the mounting pressure on the world's freshwater resources because of the growth of population, the kind of 'development' that the countries of the world are pursuing, and the growing commercial exploitation of water in the expectation of profitability arising from scarcity, are now accentuated by the reality of climate change and the impact that it might have on the availability of water. These concerns lead to increased interest in the global governance of water. Should there be a Global Freshwater Convention to protect this resource from wasteful use and unconscionable commercial exploitation and ensure resource–conservation and ecological sustainability? Ricardo Petrella, for instance, has proposed a Manifesto. There is, indeed, a strong case for a global convention on fresh water. However, great care is needed to ensure that any such exercise is not hijacked by the more powerful countries or the big international water corporations for establishing their own hegemony over the world's water.

2

Water Law – Evolving Regulatory Framework

PHILIPPE CULLET

Introduction

In the words of the United Nations Development Programme (UNDP), water is 'the stuff of life and a basic human right'.[1] Thus, water is an essential element for life – including human life – on earth and as a result is a core concern in law. From a legal perspective, the UNDP rightly emphasises the importance of the human right dimension of water. Yet, in practice, water law is made up of a number of elements comprising a human right dimension, as well as economic, environmental or agricultural aspects. In particular, historically, one of the central concerns of water law has been the development of principles concerning access to and control over water.

Drinking water is directly essential for human life. Water is also indirectly essential, for instance, as an indispensable input in agriculture. Yet, despite the central role that water has always played in sustaining life, human lives and human economies, the development of formal water law has been relatively slow and often patchy. At the domestic level, colonial legislation first focused on the regulation of water for economic reasons, for instance,

---

[1] United Nations Development Programme, *Human Development Report 2006 – Beyond Scarcity: Power, Poverty and the Global Water Crisis* 1 (New York: UNDP, 2006).

through the development of legislation concerning irrigation and navigation. Over the past few decades, increasing water pollution and decreasing per capita availability have led to the development of other measures such as water quality regulation and an emphasis on water delivery, particularly in cities, as well as environment-related measures. Yet, water law remains largely sectoral till date. At the international level, water regulation first focused mostly on navigation in international watercourses. It has progressively evolved to encompass issues concerning the sharing of international waters. International water law has, however, not yet reached the stage where it provides an overall regime for the regulation of water uses.

In India, water law is made of different components. It includes international treaties, central and state acts. It also includes a number of less formal arrangements, including water and water-related policies as well as customary rules and regulations. This chapter maps out the relevant legal framework concerning water in India. The first section delineates water law as it evolved until recently. The second section then examines proposed and ongoing water law reforms that are in the process of redrawing India's water legal framework.

I. Existing Water Law Framework

Existing water law is made up of a number of different instruments that do not necessarily make up a comprehensive framework. This is the case at the international level where only certain aspects of water law have been developed and where no international water law treaty exists. This is also the case in India where it remains difficult to identify a coherent body of comprehensive law concerning water. This is related to the fact that distinct concerns have been addressed in different enactments. This is also due to the division of powers between the centre and the states and the fact that water regulation is mostly in the hands of the states.

This section first highlights some of the salient international instruments that are relevant in India. It then moves on to examine existing water regulation in India and the different principles that govern different types of water.

A. International Framework

International water law includes a number of instruments. They may not all apply directly in India but contribute in various ways to the development of water law at the international as well as national levels.

For many years, international water law included mostly treaties concerning navigation in international rivers, which constituted one of the early areas of collaboration among states. This has been expanded to many non-navigational aspects over time but the focus on international watercourses remains an important part of water law, as exemplified in the Farakka treaty.[2] Indeed, the only multilateral treaty in the field of water is a convention concerning non-navigational uses of international watercourses.[3] The 1997 UN Water Convention provides a framework for cooperation among states on international watercourses concerning the use of their waters apart from navigational aspects.[4] The basic principle it proposes for using international watercourses is equitable and reasonable utilisation of water.[5] The basis for watercourse use is therefore agreement among concerned states concerning their respective needs. While there was substantial debate concerning the place of environmental aspects and sustainability, the principle of sustainable utilisation has not been adopted as a principle that would override equitable and reasonable utilisation.[6]

The adoption of the UN Water Convention was in itself a landmark development since it took member states many years

---

[2] Treaty on Sharing of the Ganges Waters at Farakka, New Delhi, 12 December 1996, 36 *International Legal Materials*, 519 (1997).

[3] Convention on the Law of the Non-navigational Uses of International Watercourses, New York, 21 May 1997, reprinted in P. Cullet and A. Gowlland-Gualtieri eds., *Key Materials in International Environmental Law* 481 (Aldershot: Ashgate, 2004).

[4] Ibid., Article 1.

[5] Ibid., Article 5.

[6] *See* Patricia Wouters, The Legal Response to International Water Scarcity and Water Conflicts: The UN Watercourses Convention and Beyond 20 (Dundee, 2003). *Source:* http://www.dundee.ac.uk/iwlri/Documents/Research/IWLRI%20Team/Wouters/GYIL.pdf.

to adopt this text.⁷ Nevertheless, the difficulties encountered in negotiating the Convention are reflected in the fact that its scope is relatively limited. Thus, it only applies to international watercourses and is therefore not a convention addressing freshwater in general. Further, its operative principles are relatively outdated as it fails to break clearly with the traditional principle of equitable and reasonable use in favour of a sustainability based approach. While the Convention does not break much new ground at the conceptual level, only 14 states have ratified it so far. Further, only 21 countries (including those that have ratified) have signed the Convention. India has not even signed it yet. Freshwater remains an issue over which states are fearful of losing control. As a result, even relatively weak coordination measures appear threatening to many.

Besides the UN Water Convention, there exist a number of international treaties that are directly or indirectly concerned with water. The UNECE Convention on impact assessment applies, for instance, in the case of dams and other water-related infrastructure projects.⁸ The Desertification Convention clearly links water and desertification. In fact, its objectives provision recognises that rehabilitation, conservation and sustainable management of water are key to combating desertification.⁹ The Convention on wetlands of international importance (Ramsar Convention) is intrinsically concerned with water.¹⁰ It is particularly noteworthy because it

---

[7] The mandate for the development and codification of the law of non-navigational use of international watercourses was first given to the International Law Commission in 1970. *See* General Assembly Resolution 2669 (XXV), Progressive Development and Codification of the Rules of International Law Relating to International Watercourses, 8 December 1970.

[8] Convention on Environmental Impact Assessment in a Transboundary Context, Espoo, 25 February 1991, reprinted in Cullet and Gowlland-Gualtieri, note 3 above at 29. This convention is open for global membership though India has not joined yet.

[9] Article 2, United Nations Convention to Combat Desertification in Those Countries Experiencing Serious Drought and/or Desertification, Particularly in Africa, reprinted in Cullet and Gowlland-Gualtieri, 2004, note 3 above at 267.

[10] Convention on Wetlands of International Importance Especially as Waterfowl Habitat, Ramsar, 2 February 1971, reprinted in Cullet and Gowlland-Gualtieri, note 3 above at 248.

goes beyond the main water treaties insofar as it considers water, which is entirely under national sovereignty. Indeed, the scope of the Ramsar Convention is not limited to transboundary wetlands but includes wetlands that are entirely within the territory of a member state.

Apart from treaties focusing on water or having a water dimension, there are a multitude of non-binding instruments concerning water. These include instruments focusing on water like the Dublin Statement that laid down principles for water sector reforms in the early 1990s.[11] These also include instruments not directly concerned with water, like the Declaration on the Rights of Indigenous Peoples specifically recognising that the prior informed consent of indigenous peoples is necessary for any project affecting their water resources.[12]

Overall, international water law is both an old and highly developed area of law as well as an area in need of significant development. International water law is well developed with regard to cooperation among states concerning issues and activities that are clearly transboundary in scope such as navigation on international watercourses. In recent decades, the importance of collaboration on non-navigational aspects of international watercourses has rapidly grown and is now recognised as a core objective of international water law. However, international water law is yet to be effectively developed with regard to cooperation on issues related to water found within national boundaries. While this still seems to be beyond what most states can agree on at present, water is no different from biodiversity, which is also nearly entirely found under national jurisdiction. Yet, it is now already fifteen years since UN member states recognised that biodiversity is a common concern of humankind, which is under state sovereignty but requires a degree of cooperation in conserving

---

[11] Dublin Statement on Water and Sustainable Development, International Conference on Water and the Environment, Dublin, 31 January 1992. *Source:* http://www.ielrc.org/content/e9209.pdf.

[12] Article 32(2), United Nations Declaration on the Rights of Indigenous Peoples. Report to the General Assembly on the First Session of the Human Rights Council, UN Doc. A/HRC/1/L.10 (2006).

and sustainably using it.[13] Further, while international water law has at least started integrating an environmental perspective, the social and human rights dimension of water remain largely absent. The absence of a human right perspective in water law has been addressed from the perspective of human rights law through the adoption of General Comment 15 of the first Covenant.[14]

B. Legal Framework in India

National water law is more developed than international water law. Nevertheless, India lacks an umbrella framework to regulate freshwater in all its dimensions. The existing water law framework in India is characterised by the co-existence of a number of different principles, rules and acts adopted over many decades. These include common law principles and irrigation acts from the colonial period, as well as more recent regulation of water quality and the judicial recognition of a human right to water. The lack of an umbrella legislation at the national level has ensured that the different state and central legal interventions and other principles do not necessarily coincide and may, in fact, be in opposition in certain cases. Thus, the claims that landowners have over ground water under common law principles may not be compatible with a legal framework based on the human right to water and the need to allocate water preferentially to domestic use and to provide water to all, whether landowners or not, on a equal basis.

In terms of statutory development, irrigation laws constitute, historically, the most developed part of water law. This is in large part due to the fact that the colonial government saw the promotion of large irrigation works as central to its mission. This also included the need to introduce a regulatory framework in this area. As a result, some of the basic principles of water law applicable today

---

[13] Convention on Biological Diversity, Rio de Janeiro, 5 June 1992, reprinted in Cullet and Gowlland-Gualtieri, note 3 above at 169.

[14] Committee on Economic, Social and Cultural Rights, General Comment 15: The Right to Water (Articles 11 and 12 of the International Covenant on Economic, Social and Cultural Rights), UN Doc. E/C.12/2002/11 (2002). *Source:* http://www.ielrc.org/content/e0221.pdf [hereafter General Comment 15].

in India derive from irrigation acts. The early Northern India Canal and Drainage Act, 1873 sought, for instance, to regulate irrigation, navigation and drainage in northern India. One of the long-term implications of this Act was the introduction of the right of the Government to 'use and control for public purposes the water of all rivers and streams flowing in natural channels, and of all lakes'.[15] The 1873 Act refrained from asserting state ownership over surface waters. Nevertheless, this act is a milestone since it asserted the right of the Government to control water use for the benefit of the broader public. This was progressively strengthened. Thus, the Madhya Pradesh Irrigation Act, 1931 went much further and asserted direct state control over water – 'All rights in the water of any river, natural stream or natural drainage channel, natural lake or other natural collection of water shall vest in the Government'.[16]

Colonial law in this area remains relevant to date because acts like the 1931 Madhya Pradesh Act are still in force. Further, in Madhya Pradesh again, the Regulation of Waters Act, 1949 reasserted that 'all rights in the water of any natural source of supply shall vest in the Government'.[17] The much more recent Bihar Irrigation Act, 1997 still provides that all rights in surface water vest in the Government.[18]

Statutory water law also includes a number of pre- and post-independence enactments in various areas. These include laws on embankments, drinking water supply, irrigation, floods, water conservation, river water pollution, rehabilitation of oustees and displaced persons, fisheries and ferries.

In general, water law is largely state based. This is due to the constitutional scheme, which since the Government of India Act, 1935 has, in principle, given power to the states to legislate in this area. Thus, states have the exclusive power to regulate water supplies, irrigation and canals, drainage and embankments, water storage, hydropower and fisheries.[19] There are, nevertheless,

---

[15] Preamble, Canal and Drainage Act, 1873 (Act VIII of 1873).
[16] Article 26, Madhya Pradesh Irrigation Act, 1931.
[17] Section 3, Madhya Pradesh Regulation of Waters Act, 1949.
[18] Section 3(a), Bihar Irrigation Act, 1997. *Source:* http://www.ielrc.org/content/e9703.pdf.
[19] Schedule 7, List 2, Entries 17 and 21, Constitution of India.

restrictions with regard to the use of inter-state rivers.[20] Further, the Union is entitled to legislate on certain issues. These include shipping and navigation on national waterways as well as powers to regulate the use of tidal and territorial waters.[21] The Constitution also provides that the Union can legislate with regard to the adjudication of inter-state water disputes.[22] While no substantive clauses could be adopted at the time of the adoption of the Constitution, a specific act, the Inter-State Water Disputes Act was adopted in 1956.[23] This introduces a procedure for addressing disputes among states concerning inter-state rivers that have not been solved through negotiations. It provides for the establishment of specific tribunals to adjudicate such conflicts and has been used in several cases.[24] Parliament also enacted the River Boards Act, which provides a framework for the setting up of river boards by the Central Government to advise state governments concerning the regulation or development of an inter-state river or river valley.[25] River boards can advise state governments on a number of issues including conservation, control and optimum utilisation of water resources, the promotion and operation of schemes for irrigation, water supply or drainage or the promotion and operation of schemes for flood control.[26] This Act has, however, never been used in practice.

While the intervention of the Central Government in water regulation is limited by the constitutional scheme, the importance of national regulation in water has already been recognised in certain areas. Thus, with regard to water pollution, Parliament did adopt the Water Act in 1974.[27] This Act seeks to prevent and

---

[20] Schedule 7, List 1, Entry 56, Constitution of India.
[21] Schedule 7, List 1, Entries 24, 25 and 57, Constitution of India.
[22] Article 262, Constitution of India.
[23] Inter-State Water Disputes Act, 1956. *Source:* http://www.ielrc.org/content/e5601.pdf.
[24] *See* Narmada Water Disputes Tribunal, Final Order and Decision of the Tribunal, 12 December 1979, reproduced in Philippe Cullet ed., *Sardar Sarovar Dam Project: Selected Documents* 47 (Aldershot: Ashgate, 2007).
[25] River Boards Act, 1956. *Source:* http://www.ielrc.org/content/e5602.pdf.
[26] Ibid., Section 13.
[27] Water (Prevention and Control of Pollution) Act, 1974. *Source:* http://www.ielrc.org/content/e7402.pdf.

control water pollution and maintain and restore the wholesomeness of water. It gives powers to water boards to set standards and regulations for prevention and control of pollution.

Besides statutory frameworks, a number of common law principles linking access to water and rights over land are still prevailing in India. These include separate rules for surface and ground water. With regard to surface water, existing rules still derive from the early common rule of riparian rights. Thus, the basic rule was that riparian owners had a right to use the water of a stream flowing past their land equally with other riparian owners, to have the water come to them undiminished in flow, quantity or quality.[28] In recent times, the riparian right theory has increasingly been rejected as the appropriate basis for adjudicating water claims.[29] Further, common law rights must today be read in the context of the recognition that water is a public trust.[30] If the latter principle is effectively applied in the future, it would have important impacts on the type of rights and privileges that can be claimed over surface water.

Common law standards concerning ground water have subsisted longer. The basic principle was that access to and use of groundwater is a right of the landowner. In other words, it is one of the rights that landowners enjoy over their possessions. The inappropriateness of this legal principle has been rapidly challenged during the second half of the twentieth century with new technological options permitting individual owners to appropriate not only water under their land but also the ground water found under neighbours' lands. Further, the rapid lowering of water table in most regions of the country has called in question legal principles giving unrestricted rights to landowners over ground water. Similarly, the growth of concerns over the availability of drinking water in more regions has led to the introduction of social concerns in ground water regulation. As a result of the rapid expansion of

---

[28] *Hanuman Prasad v. Mendwa*, AIR 1935 All 876.

[29] *See* Chapters 8 and 9, Report of the Narmada Water Disputes Tribunal with its Decision in the Matter of Water Disputes Regarding the Inter-State River Narmada and the River Valley Thereof Between the States of Gujarat, Madhya Pradesh, Maharashtra and Rajasthan (New Delhi: Government of India, Vol. 1, 1979).

[30] *M.C. Mehta v. Kamal Nath*, 1997 1 Supreme Court Cases (SCC) 388.

ground water use, the Central Government has tried since the 1970s to persuade states to adopt ground water legislation.[31] It is only over the past decade that some states have eventually adopted ground water acts.[32] The legal framework concerning ground water is still in rapid evolution. It is likely that common law principles will be increasingly challenged despite the fact that the Plachimada High Court decision seems to uphold landowners' rights to a large extent.[33] Further, ground water is increasingly likely to be linked to surface water in the context of the setting up of water regulatory authorities that are called upon to manage surface and ground water.[34]

The existing legal framework concerning water is complemented by a human right dimension. While the Constitution does not specifically recognise a fundamental right to water, court decisions deem such a right to be implied in Article 21 (right to life).[35] The right to water can be read as being implied in the recognition of the right to a clean environment. In *Subhash Kumar v. State of Bihar,* the Supreme Court recognised that the right to life 'includes the right of enjoyment of pollution-free water and air for full enjoyment of life'.[36] The Supreme Court went further and directly derived the right to water from Article 21: it stated that '[w]ater is the basic need for the survival of the human beings and

---

[31] *See* Model Bill to Regulate and Control the Development and Management of Ground Water, 2005. *Source:* http://www.ielrc.org/content/e0506.pdf.

[32] *See* Kerala Ground Water (Control and Regulation) Act, 2002. *Source:* http://www.ielrc.org/content/e0208.pdf; Andhra Pradesh, An Act to Promote Water Conservation, and Tree Cover and Regulate the Exploitation and Use of Ground and Surface Water for Protection and Conservation of Water Sources, Land and Environment and Matters, Connected Therewith or Incidental Thereto, 2002. *Source:* http://www.ielrc.org/content/e0202.pdf; and Goa Ground Water Regulation Act, 2002. *Source:* http://www.ielrc.org/content/e0201.pdf.

[33] *Hindustan Coca-Cola Beverages (P) Ltd. v. Perumatty Grama Panchayat*, M. Ramachandran and K.P. Balachandran (JJ), 7 April 2005. *See also* Koonan in this volume.

[34] *See* Maharashtra Water Resources Regulatory Authority Act, 2005. *Source:* http://www.lead-journal.org/content/05080.pdf.

[35] *See* S. Muralidhar, 'The Right to Water: An Overview of the Indian Legal Regime', *in* Eibe Riedel and Peter Rothen eds., *The Human Right to Water* 65 (Berlin: Berliner Wissenschafts-Verlag, 2006).

[36] Paragraph 7, *Subhash Kumar v. State of Bihar,* AIR 1991 SC 420.

is part of right of life and human rights as enshrined in Article 21 of the Constitution of India'.[37] While the recognition of a fundamental right to water by the courts is unequivocal, its implementation through policies and acts is not as advanced.

Water law includes a number of other laws and regulations that are directly or indirectly concerned with water. One example concerns dams. Two major aspects of dam building are regulated by laws and regulations, which are only partly concerned with water. With regard to environmental impact assessment, the Environmental Impact Assessment Notification provides a framework for assessing the environmental impacts of planned, big hydropower and irrigation projects.[38] Further, there are Guidelines for Environmental Impact Assessment of River Valley Projects, which provide a general framework since 1985 for assessing the impacts of planned big dam projects.[39] With regard to displacement, the main act that applies is still the Land Acquisition Act, 1894. This colonial Act, which was enacted with the interests of the colonial government rather than the interests of displaced people in mind, gives the government significant control over the process of eviction and oustees very few rights.

In addition to all the laws, rules and regulation that make up water law, there is a substantial body of additional rules and regulations at the local level. These include the multiplicity of written or unwritten arrangements that regulate access to and use of water for domestic purposes or irrigation. An array of different rules govern, for instance, access to existing sources of drinking water. They run in many cases along caste lines even though other rules of access also exist. With regard to irrigation water, all human structures such as tanks and check dams include a system of allocation.[40] Rules of access and control have often evolved over long periods of time but are often unwritten or not formally

---

[37] Paragraph 274, *Narmada Bachao Andolan v. Union of India*, Writ Petition (Civil) No. 319 of 1994, Supreme Court of India, Judgment of 18 October 2000, AIR 2000 SC 3751, reproduced in Philippe Cullet ed., *Sardar Sarovar Dam Project: Selected Documents* 138 (Aldershot: Ashgate, 2007).

[38] Notification on Environmental Impact Assessment of Development Projects, 2006.

[39] Guidelines for Environmental Impact Assessment of River Valley Projects, 1985. *Source:* http://www.ielrc.org/content/c8503.pdf.

[40] For Tamil Nadu, *see* Gurunathan and Shanmugham in this volume.

recognised in the legal system. As a result, they often run in parallel to 'formal' water rules and regulations. Another consequence of the lack of visibility of local level arrangements is that they can easily be displaced or extinguished by new laws that may fail to even acknowledge their existence.

The general picture which emerges is that of a multiplicity of principles and rules, a multiplicity of instruments and the lack of an overall framework. While certain principles have remained relatively constant until recently like the assertion of the state's right to use surface waters in the public interest, there have been a number of changes over time in the basic structure of water law, from the recognition of a human right to water to the introduction of the public trust doctrine. One general trend, which can be highlighted, is the gradual formalisation of water law. In most cases, this has had the effect of displacing or extinguishing existing local rules and arrangements. In other words, the introduction of water laws is often not done in a vacuum, as might be the case in certain other fields. This is due to the fact that water has always been of central importance in most communities and formal or informal rules, based on social, religious or castes have existed in most places for centuries.

II. Towards Water Law Reforms

Water law has been continuously evolving. Yet, the evolution witnessed over the first four decades after independence must be distinguished from recent and ongoing trends. While until the 1970s, water law can be seen as a field growing organically around issues and principles that were largely well settled, the past couple of decades have witnessed the beginning of a fundamental shake-up of water law. This is taking the form of reforms, which are changing and will change existing water law as well as expand the scope of regulation.

The requirements for water law reforms include physical as well as institutional reasons. Over the past decades, the water situation has become increasingly dire in many parts of the country. This is due to increased use of water by all categories of water users, to increased demand due to economic and population growth. This is also due to increased pollution of existing finite water resources, which not only restrict potential uses of available water but also threaten future use. One of the specific problems that have arisen is the dramatic increase in ground water use, which has led to depletion in many areas.

Increasing use of water has led to a number of suggestions to remedy the situation. This includes new strategies to cope with all the various water-related issues. Water pollution has been addressed through the introduction of environmental measures to control and reduce it. Access to domestic water has been the object of various governmental and other programmes. The provision of irrigation water and water to cities has, for instance, been taken up in the context of the construction of large dams.

There have also been progressive calls for changes of the law and policy framework concerning water. This is due to two broad factors. First, the water law and policy framework was for a long time the object of relatively little attention. While many water-related laws were adopted over several decades, comparatively little was done to provide a broader integrated framework for water. Secondly, the recognition that there is a water crisis in most countries of the world and that availability of and access to freshwater will be a challenge for nearly all countries in coming decades has led to a number of international initiatives to reform water governance, law and policy in most developing countries. In other words, domestic and international factors have contributed to ongoing water law and policy reforms.

Water sector reforms have been proposed as a way to address diminishing per capita availability, increasing problems in water quality and increasing competition for control, access and use of available freshwater. They seek to comprehensively reform governance in the water sector. Current reforms seek, in particular, to reduce the role played by the public sector and to emphasise the direct contributions of individuals to their water needs and the participation of the private sector.

These governance changes are underpinned by a number of principles, which guide the reform process. This section highlights some of the main principles guiding the reforms and the kinds of measures and instruments adopted to implement them.

A. Water as a Natural Resource and Economic Good

The first central principle that is guiding the reform process is that all uses of water should be seen from the perspective of its economic value because the absence of an economic perspective in the past explains existing unsustainable uses of water.[41] As a result, the emphasis is on water as a natural resource, which must be harnessed to foster the productive capacity of the economy, from irrigation water for agricultural production to water for hydropower. Thus, the National Water Policy laments the fact that an insufficient percentage of water is currently harnessed for economic development and even calls for 'non-conventional' methods of water utilisation such as inter-basin water transfers and seawater desalination as large-scale, high technology solutions to improve overall water availability.[42] This message is also found in a World Bank report stressing that India has not developed enough big water infrastructure.[43]

Beyond the relatively old characterisation of water as a natural resource, the underlying proposition for water sector reforms is that water is to be seen as an economic good. This implies an important shift in terms of the rights of control over and access to water. In fact, this leads to a complete policy reversal from the perspective that water is a public trust to the introduction of water rights and the possibility to trade water entitlements. As such, water-related rights are not new and there is already a vast corpus of law related to control over water. This includes, for instance, the absolute rights that the state may claim over water.[44] This also includes the rights and privileges that common law principles bestow over landowners. The novelty introduced by the reforms is that water rights are now created in favour of water users.[45] These rights are the necessary premise for participation in the management of water resources, for the setting up of water user associations and for the introduction of trading in entitlements.[46]

---

[41] *See* Dublin Statement, note 11 above.
[42] Section 3(1–2), National Water Policy, 2002. *Source:* http://www.ielrc.org/content/e0210.pdf.
[43] John Briscoe and R.P.S. Malik, *India's Water Economy: Bracing for a Turbulent Future*, New Delhi: The World Bank and Oxford University Press, 2006.
[44] *See* Section 26, Madhya Pradesh Irrigation Act, 1931 and Section 3, Madhya Pradesh Regulation of Waters Act, 1949.
[45] *See* Section 17(1)d, Uttar Pradesh Water Policy, 1999. *Source:* http://www.ielrc.org/content/e9904.pdf.
[46] Section 4(2), Maharashtra State Water Policy, 2003, http://www.ielrc.org/content/e0306.pdf.

Another important change brought about by the notion that water is an economic good is that all water services must be based on the principle of (full) cost-recovery.[47] In a situation where the provision of drinking and domestic water as well as irrigation water is substantially subsidised, this implies a significant policy reversal. At the national level, the policy is now to make water users pay at least for the operation and maintenance charges linked to the provision of water.[48] This strategy is already being implemented in the context of irrigation water where farmers are made to pay for operation and maintenance costs.[49] This has also been introduced under the Swajaldhara guidelines, which suggest that water users have to take partial responsibility for the capital cost of new drinking water infrastructure and full responsibility for operation and maintenance.[50]

The notion of cost recovery is directly linked to the environmental component of water sector reforms. Indeed, they are conceived as part of a single strategy.[51] Further, cost recovery is, for instance, seen by the Asian Development Bank as the first instrument for conserving water.[52]

B. Decentralisation and Participation

Water sector reforms are also based on the need to foster decentralisation and participation that involves water users.[53] This is meant to provide a framework for decentralising decision-making to the lowest level and to allow 'beneficiaries and other

---

[47] *See* World Bank, India – Water Resources Management Sector Review – Report on the Irrigation Sector (Report No. 18416 IN, 1998).

[48] *See* Section 11, National Water Policy, 2002.

[49] *See* World Bank, India – Water Resources Management Sector Review – Report on the Irrigation Sector (Report No. 18416 IN, 1998).

[50] Section 3(1), Ministry of Rural Development, Guidelines on Swajaldhara, 2002. *Source:* http://www.ielrc.org/content/e0212.pdf.

[51] Section 2(b), World Bank, Water Resources Management (OP 4.07, February 2000).

[52] *See* Section E, Asian Development Bank, Water for All – The Water Policy of the Asian Development Bank (2003) whose first sub-section – number 43 – is entitled cost recovery.

[53] Dublin Statement, note 11 above.

stakeholders' to be involved from the project planning stage.[54] The rationale for decentralisation is the perceived inability of the state to deliver appropriate benefits. The state is thus called upon to change its role from that of a service provider to that of a regulator.[55] In the case of irrigation, for instance, this implies transferring part or full control of irrigation systems to users by both allowing them and forcing them to take responsibility for the upkeep of irrigation systems as well as for the financial costs involved and for sharing the water allocated among themselves.[56]

In principle, participation is conceived as an umbrella term that covers participation from policy planning and project design to the management of water infrastructure. In practice, the focus is on participation at the tail-end of the process. In fact, the word 'participation' is some sort of a misnomer. On the one hand, what is envisaged is not so much the possibility for farmers and users to participate in taking decisions affecting them but the blanket imposition of a new system of local water use and control scheme based on commercial principles even where there may be successful systems of water governance already in place. On the other hand, the participation, which is envisaged at the local level, is not the participation of everyone using water. With regard to irrigation, the focus has been on land ownership and occupation as a basis for governing the use and control of water. With regard to drinking water, new measures put the ability to pay as the governing principle. Both measures are likely to reinforce existing inequalities in access to water.

Two different types of measures have been introduced to foster participation with regard to irrigation water and drinking water. The rest of this section examines Water User Associations set up to foster participation in irrigation and Swajaldhara, a scheme devised to foster participation of users in drinking water provision.

Water User Associations schemes (WUAs) have been introduced in different forms in different parts of the country and different areas of the world. However, a number of common characteristics can be identified in many schemes. This includes the fact that

---

[54] *See* Section 6(8), National Water Policy, 2002.
[55] Section 37, Asian Development Bank, note 52 above.
[56] *See* Section 17(1), Uttar Pradesh Water Policy, 1999.

WUAs are meant to be governed and controlled by people that both pay for the services the association offers and receive benefits. WUAs are not commercial entities but they have to be financially independent and therefore need to receive an income that is sufficient to allow them not to go bankrupt. Further, WUAs are in most cases subject to regulatory control by the state because they are deemed to provide a service of benefit to the public.[57]

The setting up of Water User Associations (WUAs) has been taken up with increasing intensity over the past decade and a number of states have introduced WUA legislation. These range from Andhra Pradesh and Madhya Pradesh to Orissa and Rajasthan.[58] These acts have been adopted at different points in time and the schemes proposed have evolved over time even though the basic principles are fairly similar in each situation. This section does not seek to provide a comparative analysis of these different acts and focuses on the latest act adopted in Maharashtra because it is unlikely that other states that are yet to adopt legislation in this field will go back to older schemes.

WUAs under the Maharashtra Management of Irrigation Systems by Farmers Act, 2005 are set up to foster secure equitable distribution of water amongst its members, to maintain irrigation systems, to ensure efficient, economical and equitable distribution and utilisation of water to optimise agricultural production as well as to protect the environment.[59] While the Act provides a decentralisation scheme towards farmer involvement in irrigation at the local level, it also gives significant powers to the Maharashtra Water Resources Regulatory Authority or other designated authorities. In particular, they have the power to determine the command area of an irrigation project for which a WUA must be constituted. Further, the same authority can also amalgamate or divide existing WUAs on a hydraulic basis and 'having regard to

---

[57] *See* Stephen Hodgson, Legislation on Water Users, Organizations – A Comparative Analysis (Rome: FAO, FAO Legislative Study 79, 2003).

[58] Andhra Pradesh Farmers Management of Irrigation Systems Act, 1997; Madhya Pradesh Sinchai Prabandhan Me Krishakon Ki Bhagidari Adhiniyam, 1999; Orissa Pani Panchayat Act, 2002 and Rajasthan Farmers' Participation in Management of Irrigation Systems Act, 2000.

[59] Section 4, Maharashtra Management of Irrigation Systems by Farmers Act, 2005. *Source*: http://www.ielrc.org/content/e0505.pdf.

the administrative convenience'.⁶⁰ In other words, the power granted at the local level is limited by the fact that authorities have the largely discretionary power to make and break WUAs.

The system set up under the act is constraining insofar as once a WUA has been set up, no water will be supplied to anyone individually outside the WUA framework and the scheme is binding on all landholders and occupiers. In this sense, WUAs are forced to take on the burden of administering the irrigation system and are largely left to sort out ways in which they want to achieve this. Further, the act provides a uniform model of WUAs regardless of existing arrangements at the local level and regardless of their success at equitably and sustainably using water.

The framework provided under the Act seeks to balance benefits and burdens. On the one hand, WUAs are meant to benefit from a more assured water supply and more control over water allocated to them. They also have the right to use ground water in their command area on top of the entitlement they receive from canals. On the other hand, the Act gives WUAs a number of powers, which are in fact responsibilities. This includes a number of functions such as the regulation and monitoring of water distribution among WUA members, the assessment of members' water shares, the responsibility to supply water equitably to members, the collection of service charges and water charges, the carrying out of maintenance and repairs to the canal system and the resolution of dispute among members.⁶¹ These are extensive and possibly burdensome powers. WUAs are not only given the task to manage the infrastructure but also to provide an institutional structure that equitably provides all the services that a public authority would provide. While such arrangements would be an appropriate choice if WUAs were linked to Panchayati Raj Institutions (PRIs), it is difficult to see how an association of landholders that has no democratic legitimacy can perform all these tasks in an equitable and sustainable manner for its members and for the broader society around it. To take one example, while there are now a number of rules attempting to ensure the participation of women and lower castes in PRIs, it is likely that WUAs will

---

⁶⁰ Ibid., Section 5(5).
⁶¹ Ibid., Section 52.

generally be dominated by male upper caste members. In other words, the existing legislation is both onerous on WUAs who seem to be saddled with more responsibilities than rights and is at the same time unlikely to provide a framework leading to a more socially equitable access to and sharing of water.

The section concerning the powers and responsibilities of WUAs is complemented by a section concerning financial arrangements. As specified under Section 54, the main sources of funding for WUAs will not come from the Government. WUAs are meant to meet their expenses from the proceeds of water charges, borrowings and donations. In other words, the Act seeks to ensure that WUAs are financially independent and financially viable, a fact which is confirmed by the encouragement given to WUAs to engage in additional remunerative activities, including the distribution of seeds, fertilisers and pesticides or marketing of agricultural produce which are only indirectly related to irrigation.[62]

In addition to the setting up of WUAs, the Union Government has proposed a scheme known as Swajaldhara, which proposes to foster new types of intervention to ensure better drinking water availability in villages. The guidelines on Swajaldhara are the direct outcome of a World Bank-sponsored pilot project called Swajal and adopt the same philosophy.[63]

The guidelines are meant to foster a change in the role of the Government from direct service delivery to that of facilitating activities largely undertaken by people themselves. In other words, the guidelines propose the progressive withdrawal of the state from the provision of the fundamental right to drinking water. The argument put forward by the Government is that people perceive water as a fundamental right in part because it has been provided free by the Government. The Government estimates that the public has, therefore, not understood that water is scarce and is a socio-economic 'good'. It is, therefore, proposed to shift from what is seen as a supply driven approach to one which focuses on the need of end users who will then get the service they want. The fundamental change of approach required by this demand-focused strategy is that people will get the service they 'are willing to pay

---

[62] Ibid., Section 4(2).

[63] On the Swajal project, *see* World Bank, Staff Appraisal Report – Uttar Pradesh Rural Water Supply and Environmental Sanitation Project (Report No. 15516-IN, 1996).

for'.[64] In fact, the basic economic rationale of Swajaldhara is that people should be made to pay for part of the capital costs of drinking water projects and for the whole cost of operation and maintenance.

Swajaldhara is premised on a number of principles. First, it proposes the introduction of a demand-focused approach, which involves some level of community participation. Secondly, it seeks to devolve ownership of drinking water assets to the appropriate panchayat, which are given the power to undertake all activities, related to water supply and sanitation from planning to maintenance. Thirdly, Swajaldhara imposes on communities a contribution of at least 10 per cent of the capital costs for a service level of 40 litres per person per day and imposes that they take complete responsibility for operation and maintenance. It also imposes that the contribution of the community to capital costs should be at least 50 per cent in cash. Further, under Swajaldhara, only individuals or households that make the first 10 per cent contribution will benefit from the schemes being implemented. Other people are simply not part of the scheme.

Swajaldhara was implemented throughout the tenth plan. In the eleventh plan, there was a proposal to change the pattern in a way that would have both mainstreamed the Swajaldhara principles to all projects under the Accelerated Rural Water Supply Programme and relaxed some of the conditions, in particular with regard to the capital cost contribution. Eventually, these changes were not carried out and the Government is again suggesting that 20 per cent of projects should be implemented under Swajaldhara principles.[65]

C. Redefinition of the Role of the Government

Water sector reforms include several proposals that affect the role that the government plays in the water sector. This includes both measures restricting the role that the government is playing as well as measures seeking to increase governmental control.

---

[64] Section 1(2), Ministry of Rural Development, Guidelines on Swajaldhara, 2002.
[65] Rajiv Gandhi National Drinking Water Mission, Allocation of funds under Accelerated Rural Water Supply Programme (ARWSP) during 2008–09, No. G-11011/5/2008-DWS.I (2008).

On the one hand, the main thrust of water sector reforms is to transform the role of the government by transferring part of existing governmental prerogatives to users and private actors. This includes, for instance, the transfer of operation, maintenance, management and collection of water charges to user groups.[66] This is meant to foster a sense of ownership at the user level that the overbearing presence of the government in the water sector has not been able to foster. A second thrust of the reforms is to set up new bodies at the local and state level to take over part of the functions of the government. This includes the setting up of Water User Associations to locally manage irrigation schemes instead of local bureaucrats and also includes the much more broad-ranging setting up of new water regulatory bodies.

The reduction of the role of the state in the water sector is also linked to the promotion of the use of incentives to ensure that water is used more efficiently and productively.[67] The main consequence, which is derived from this, is the call for private sector involvement in all aspects of water control and use from planning to development and administration of water resources projects.[68] An area, which is singled out for private sector participation, is urban water supply.[69]

On the other hand, some of the existing reforms seek to foster increased state involvement in the water sector. In a number of areas, the state seeks to either maintain its de facto prerogatives or extend them. In the national water policy, a clear statement is made to the effect that the government should be able to provide for the transfer of water from one river basin to another.[70] This is now being taken up in the context of the mammoth river inter-linking scheme.[71] At the state level, an increasing number of states are seeking to control and regulate ground water to foster its conservation and sustainability in its use.

---

[66] *See* Section 6(7), Karnataka State Water Policy, 2002.
[67] Section 1(3), Maharashtra State Water Policy, 2003.
[68] *See* Section 38, Asian Development Bank, note 52 above and Section 13, National Water Policy, 2002.
[69] *See* Section 9, Rajasthan State Water Policy, 1999.
[70] *See* Section 3(5), National Water Policy, 2002.
[71] *See* Government of India – Ministry of Water Resources, Resolution No. 2/21/2002-BM, New Delhi, 13 December 2002.

The redefinition of the role of the government in the water sector has, for instance, been taken up in the context of the setting up of water regulatory authorities meant to take over part of the functions of existing government departments. The first experiment undertaken in India in this regard took place in Andhra Pradesh where a Water Resources Development Corporation Act was adopted as early as 1997.[72] This Act largely sought to devolve existing governmental powers to a new institutional structure entrusted with the mandate of pushing water sector reforms forward.

Since 1997, there has been a lot of thinking in policy-making circles concerning water sector reforms and the type of measures that need to be taken to move the agenda forward. As a result, a more recent act setting up an independent water institution, the Maharashtra Water Resources Regulatory Authority Act, 2005 is quite different from the former and it is, in fact expected that the latter Act will be amended in view of the new scheme.

First, under the Maharashtra Act, it has been attempted to completely exclude political leaders from the power structure. However, while the Act takes a clear stand on paper to insulate the authority from political interference, the bureaucracy still has an important (in)direct role. The actual independence of the authority will thus have to be judged in practice rather than on the basis of the Act.

Secondly, the Maharashtra authority has broad prerogatives to establish a regulatory system for the water resources of the state, including surface and ground waters, to regulate their use and apportion entitlements to use water between different recognised categories of use.[73] Concurrently, the authority has to promote the efficient use of water, to minimise wastage and to fix reasonable use criteria. The authority also has the task of allocating specific amounts to specific users or groups of users according to the

---

[72] *See* Act to Create the Andhra Pradesh Water Resources Development Corporation for Promotion and Operation of Irrigation Projects, Command Area Development and Schemes for Drinking Water and Industrial Water Supply to Harness the Water of Rivers of the State of Andhra Pradesh and for Matters Connected Therewith or Incidental Thereto Including Flood Control, Act No. 12 of 1997. *Source*: http://www.ielrc.org/content/e9702.pdf.

[73] Ibid., Section 11.

availability of water. It is further required to establish a water tariff system as well to fix the criteria for water charges. This is to be done based on the principle of full cost recovery of management, administration, operation and maintenance of irrigation projects. The authority is also called upon to lay down criteria for the issuance of water entitlements. Further, it has to set up criteria for trading in water entitlements or quotas.[74]

One of the important consequences of the setting up of a water regulatory authority concerns the strengthened control over water resources, which is proposed. The Act provides, as a general principle, that any water from any source can only be used after obtaining an entitlement from the respective river basin agency.[75] This is qualified by a few exceptions such as wells (including bore and tube wells) used for domestic purposes or the grandfathering of existing uses of water for agriculture, at least in an initial phase. This illustrates the fact that while the role of the government is curtailed through the setting up of an independent authority, this does not necessarily translate into less regulatory intervention as far as water users are concerned. The overall impact is therefore as much to reduce the government's role as to transfer and possibly strengthen control over water resources.

D. Conservation

The increasing depletion of water resources, in particular ground water, has led to the realisation that existing rules concerning the use of ground water were unadapted to a situation of scarcity. As a result, the Central Government has put significant emphasis on the development of ground water laws by the states. Regulatory intervention is premised on the need to control the use of ground water to ensure that it is not unsustainably mined.

Legislative interventions concerning ground water are significant for two main reasons. First, from a legal perspective they constitute a major organised attempt at redrawing the rules concerning control and use of ground water, which is still otherwise largely based on common law principles that make it part of the

---

[74] Ibid., Section 11(i)i.
[75] Section 14, Maharashtra Water Resources Regulatory Authority Act, 2005. *Source*: http://www.lead-journal.org/content/05080.pdf.

resources a landowner can use largely without outside control. Secondly, they constitute a response to the fact that over time ground water has, in various areas, become the most important source of water and provides in particular 80 per cent of the domestic water supply in rural areas and supports around 70 per cent of agricultural production.[76] This strengthens the case for ensuring the sustainable use of ground water.

Until recently, ground water has been largely governed by old legal principles linked to a large extent to land ownership. Further, like in many other countries, from a legal perspective, ground water has until now been largely treated independently from surface water even though links have increasingly been acknowledged. As a result, until a few decades ago, there was little by way of statutory provisions concerning ground water use and control and the Central Government's intervention in this area was even less prominent than with regard to surface water. The increasing use of ground water has led to a spurt of legislative activity, which seems to be accelerating.

At the national level, even though the Central Government would find it difficult to justify ground water legislation under the constitutional scheme, several attempts have been made over the past few decades to provide a model law that individual states can adopt. The first attempt dating back to 1970 did not have much success since virtually all states ignored it. More recent versions of the Model Bill, including the latest version unveiled in early 2005,[77] are having more influence on legislative activity because ground water regulation has become a priority in many states. In fact, several states have proposed ground water related laws, which are related to the Model Bill. This is, for instance, the case of the Kerala Ground Water (Control and Regulation) Act, 2002. As a result, the following paragraphs focus on the model bill since it provides the framework that a number of states are likely to adopt.

The basic scheme of the Model Bill is to provide for the establishment of a ground water authority under the direct control of the Government. The authority is given the right to notify areas where it is deemed necessary to regulate the use of ground water.

---

[76] United Nations World Water Development Report – Water for People, Water for Life (United Nations, Doc. E.03.II.A.2, 2003).

[77] Model Bill, note 31 above.

The final decision is taken by the respective state government.[78] There is no specific provision for public participation in this scheme. In any notified area, every user of ground water must apply for a permit from the authority unless the user only proposes to use a hand pump or a well from which water is withdrawn manually.[79] Decisions of the authority in granting or denying permits are based on a number of factors, which include technical factors such as the availability of ground water, the quantity and quality of water to be drawn and the spacing between groundwater structures. The authority is also mandated to take into account the purpose for which ground water is to be drawn but the Model Bill, mirroring in this the Acts analysed above, does not prioritise domestic use of water over other uses.[80] It is noteworthy that even in non-notified areas, any wells sunk need to be registered.[81]

The Model Bill provides for the grandfathering of existing uses by only requiring the registration of such uses.[82] This implies that in situations where there is already existing water scarcity, an act modelled after these provisions will not provide an effective basis for controlling existing overuse of ground water and will, at most, provide a basis for ensuring that future use is more sustainable.

Overall, the model bill constitutes an instrument seeking to broaden the control that the state has over the use of ground water by imposing the registration of all ground water infrastructure and providing a basis for introducing permits for ground water extraction in regions where ground water is over-exploited. Besides providing a clear framework for asserting government control over the use of ground water, the Model Bill also shows limited concerns for the sustainability of use. From this perspective, the model bill and the acts based on it are a welcome development that should provide scope for better control over the use of ground water in general. However, further thinking needs to be put in making the

---

[78] Model Bill, note 31 above, Section 5.
[79] Ibid., Section 6.
[80] Ibid., Section 6(5)a. Only provides that the purpose has to be taken into account while Section 6(5)h which is the only sub-section referring to drinking water only considers it as an indirect factor.
[81] Ibid., Section 8.
[82] Ibid., Section 7.

model bill sensitive to social concerns. Some important provisions are currently missing from the model bill. These include the need to prioritise among uses and to put drinking and domestic water as the first priority. Further, the Model Bill does not differentiate between small and big users of ground water, commercial and non-commercial uses and does not take into account the fact that non-landowners/occupiers are by and large excluded from the existing and proposed system, which focuses on the rights of use of landowners.

Conclusion

Water law is made of a number of formal and informal laws, rules and principles. It has evolved over time in a relatively uncoordinated and ad hoc manner. This started to change with the progressive realisation that existing laws were inappropriate to ensure access to water to all for domestic purposes and inappropriate because of the fast increasing use of a finite resource. Over the past couple of decades, a more coordinated effort at changing water law has been put in place. This is based on a relatively specific set of principles that are meant to guide the overall development of water law. This is meant to make water law suitable to face the challenges of the water sector in the twenty-first century.

While water law reforms are more than welcome given existing shortcomings of water regulation and changing conditions in the water sector, it is unlikely that law reforms based on the principles put forward in the water sector reforms constitute an appropriate response. Ongoing water law reforms may contribute to fostering better water management but they are conceptually incapable of addressing the human right, social, environmental and health aspects of water. This is regrettable because any water law, which is not based on the constitutional right to water and the principle of public trust, is bound to fail as a legal tool and in its implementation as far as the overwhelming majority of people is concerned on top of being open to legal challenges.

Yet, avenues do exist to broaden reforms of water law. At the international level, some treaties are leading the way towards conceiving water law more broadly. Thus, the UN Economic

Commission of Europe has adopted a convention, which is broader than the 1997 UN Convention in scope insofar as it applies to transboundary waters in general. It is also based on a more progressive set of principles. This includes not only the fact that it strongly emphasises the need to prevent and reduce transboundary harm but also that it is based on the precautionary principle and inter-generational equity. The UNECE convention reflects much more than the UN convention developments in environmental law and related principles that have come to inform all treaties concerning environment and development issues. The convention is also opened to universal membership even though other states have not ratified it yet. Similarly, at the national level, countries such as Brazil and South Africa have adopted water laws that seek to provide a comprehensive regulatory answer to the problems identified. While the adoption of a comprehensive federal water legislation is not a precondition to ensure that water law achieves its social, human rights and environmental goals, this would constitute an appropriate starting point to realise the right to water and the principle of public trust throughout the country. Individual states could also adopt similar legislation at their level.

3

Discourses in Water and Water Reform in Western India

PRIYA SANGAMESWARAN*

1. Introduction

Water policies at all levels are shaped by a variety of actors – governments, interest groups within nations, social movements, international institutions such as the World Bank, water multinational companies, and so on. But one often finds common threads in the views and actions of actors at different levels (for instance, in the kind of water reforms that have been advocated), which indicates the presence of dominant discourses that shape opinions and provide legitimacy to particular kinds of policies. This chapter looks at how two discourses – the Global Environmental Management (GEM) discourse and the rights-based discourse – have shaped water reforms in Maharashtra, a state in western India. Since the relationship between knowledge and policy is complex, the aim is not to show a precise relationship between discourses and policies at different levels (international, national, and sub-national). Instead, this chapter emphasises the commonalities in the discussions around one aspect of water

---

* I would like to thank Philippe Cullet of IELRC and K.J. Joy of SOPPECOM for comments on previous drafts of the chapter; the usual disclaimers apply. I would also like to acknowledge the support of the Centre for Interdisciplinary Studies in Environment and Development, Bangalore, where a review of the concept of right to water in different discourses – that this chapter draws upon – was undertaken during my term as Visiting Fellow (2005–06).

(delivery of water services) at different levels. As Adger and others point out in their analysis of the environmental discourses associated with deforestation, desertification, biodiversity use, and climate change, such an exercise is useful to show how adopting particular languages and rhetoric constrains the solutions proposed for specific issues.[1]

The arena of delivery of water services[2] is particularly interesting to study from this point of view because it has seen changing trends in recent times, which are due, in no small measure, to the influence of different discourses in water. Traditionally, it has been the state (or state-owned enterprises) that have undertaken delivery of water services, both in the context of drinking water in urban areas and irrigation water from canals in rural areas. This is because of the peculiar characteristics of water such as high degree of natural monopoly, high capital intensity and the presence of sunk costs, the multipurpose and hydrologically interconnected nature of the water resource itself, as well as the perception that public provision is the best way to guarantee universal access.[3] But currently, there are two dominant trends in the realm of delivery of water services – sectoral decentralisation and privatisation, both of which stem from particular kinds of water discourses.

This chapter starts with a discussion of major water discourses. The central messages of these discourses have been dealt with in section two. Section three discusses how the Indian government has encouraged state governments to undertake particular kinds of policies (with respect to delivery of water services), and how this, in turn, reflects the hegemony of the GEM discourse. Sections four and five extend the discussion on the influence of particular discourses on water reform to the specific case of Maharashtra; in particular, the concepts of decentralisation and entitlements in the

---

[1] W.N. Adger *et al.*, 'Advancing a Political Ecology of Global Environmental Discourses', 32 *Development and Change* 681 (2001).
[2] Broadly, delivery of water refers to building the necessary infrastructure as well as operational and managerial capacities, including the institutional mechanisms that are actually involved in the working of water rights at different levels.
[3] Lyla Mehta, 'Problems of Publicness and Access Rights: Perspectives from the Water Domain', *in* Lyla Mehta ed., *Providing Global Public Goods: Managing Globalization* 556 (Oxford: Oxford University Press, 2003).

new legislation in the state are critically analysed. This is followed by some concluding comments in section six.

2. Water Discourses at the International Level

There are a number of different discourses in water, that is, different ways of speaking and thinking about it as well as of acting on water-related issues. Each discourse has its own central messages and policy prescriptions. Further, water practices of different governments, institutions and actors draw on different elements of these discourses (although they cannot be reduced to that).[4] In this section, I undertake a brief discussion of water discourses at the international level and indicate which discourse(s) or which elements are hegemonic in the sense that they dominate thinking and have most often been translated into institutional arrangements.

Broadly, one can distinguish between four formulations of water at the international level – the Dublin-Rio principles, the advocacy of water markets and privatisation of water services by the World Bank and the Asian Development Bank (ADB), the approach of Integrated Water Resources Management propagated by the Global Water Partnership and the World Water Council, and the rights discourse (of which the most important articulation is the idea of a right to water). The first three formulations together can be taken to constitute what has been called a Global Environmental Management (GEM) discourse of water,[5] that is a discourse which presents a technocratic worldview requiring science-based solutions and external policy and/or managerial interventions. Each of the three formulations also correspond approximately to a distinct phase of convergence of views on water.

Mehta distinguishes between three such phases.[6] The first phase (between 1977 and 1992) saw the consolidation of the water decade,[7] and the declaration of water as an economic good at the

---

[4] Bill Derman and Anne Ferguson, 'Value of Water: Political Ecology and Water Reform in Southern Africa', 62(3) *Human Organisation* 277 (2003).

[5] *See* Adger *et al.*, note 1 above.

[6] Lyla Mehta with Oriol Mirosa Canal, Financing Water for All: Behind the Border Policy Convergence in Water Management (Brighton: Institute of Development Studies, Working Paper No. 233, 2004).

[7] The 1981–90 period was the World Health Organisation's International Drinking Water Supply and Sanitation Decade.

International Conference on Water and the Environment held in Dublin (the Dublin Declaration), the run-up to the United Nations Conference on Environment and Development 1992 (the Rio Earth Summit). The second phase (between the Dublin Declaration and the Hague Conference in 2000) witnessed the spread of the neoliberal agenda both geographically and in newer arenas such as water management, and the rolling back of the state through conditionalities of the IMF and the World Bank, as well as regional development banks such as the Inter-American Development Bank and the ADB. The third phase refers to efforts, in the twenty-first century, on the part of supra-national bodies such as the World Water Council and the Global Water Partnership, which are viewed by many as providing a new impetus to private sector involvement.

Let me start with the Dublin-Rio principles. The Dublin Declaration highlighted four key principles – (i) the importance of freshwater as well as its finiteness and vulnerability (ii) increased participation of users, planners, and policy-makers at all levels of water development and management (iii) the central role of women in the provision, management, and safeguarding of water and (iv) the recognition of water as an economic good, with an economic value in all its competing uses.[8] These principles significantly contributed to the Agenda 21 recommendations adopted at the Rio Earth Summit in 1992. In line with the Dublin principles, Agenda 21 also emphasised the importance of protecting the supply and quality of freshwater resources and of delegating water resources management to the lowest appropriate level. However, unlike the Dublin principles, it emphasised that water is an economic *and* social good.[9]

The advocacy of water markets and the privatisation of water services by the World Bank and the Asian Development Bank is based partly on the Dublin-Rio characterisation of water as an economic good, but also relates to the increasing influence of neoliberalism and the consequent reduction sought in the role of

---

[8] Dublin Statement on Water and Sustainable Development, International Conference on Water and the Environment, Dublin, 31 January 1992. *Source:* http://www.ielrc.org/content/e9209.pdf.

[9] Agenda 21, *in* Report of the United Nations Conference on Environment and Development, Rio de Janeiro, UN Doc. A/CONF.151/26/Rev.1 (Vol. 1), Annex II (1992) [hereafter Agenda 21]. *Source:* http://www.ielrc.org/content/e9211.pdf.

the government in the provision of basic services.[10] The third formulation which is becoming important in recent times is the concept of integrated water resource management or IWRM. The concept has been introduced (to varying degrees) in the water policies of a number of countries such as South Africa, Uganda, and Brazil, and is considered to be an advance over earlier sectoral and fragmented approaches of water management, at least in some respects.

The GEM discourse, which is represented by the above three formulations, has a number of core messages such as the notion of water scarcity, the need to treat water as an economic good, water security, and the importance of sustainability. While the ideas represented by the messages are not entirely new, they have either become stronger in the last two and half decades or are being used in new ways (for instance, to justify particular kinds of policies). My focus here is on two of the messages. One is the notion of an existing or impending water scarcity. Agenda 21, for instance, refers to water as a 'scarce vulnerable resource',[11] and to the condition of widespread scarcity of water.[12] This, in turn, leads to a crisis rhetoric that is based, at least in part, on neo-Malthusian perspectives concerning environment and development. Thus, one of the justifications that the World Bank uses for its increasing engagement in the water sector and for the prescription of particular kinds of water reform is the growing scarcity of water (and the problems resulting from it).[13] However, it has been argued that scarcity is often manufactured by anthropogenic interventions or discursive constructions, and is not always real in the sense of having biophysical or social manifestations.[14] Similarly, many international, national, and regional conflicts over water are caused by other factors such as ethnic rivalries, nationalism, and power politics that extend to the cultural, political, and economic spheres.[15]

---

[10] *See* Mehta, note 6 above.
[11] *See* Agenda 21, note 9 above.
[12] Ibid.
[13] World Bank, Water Resources Sector Strategy: Strategic Directions for World Bank Engagement (Washington, DC: World Bank, Report No. 28114, 2004).
[14] Lyla Mehta, Water for the Twenty-first Century: Challenges and Misconceptions (Brighton: Institute of Development Studies, Working Paper No. 111, 2000).
[15] Riccardo Petrella, *The Water Manifesto: Arguments for a World Water Contract* (London: Zed, 2001).

The implication of the idea that scarcity of water is that a theoretical construct is a crisis rhetoric and recommendations of technocratic solutions to improve water availability (such as inter-basin water transfers and seawater desalinisation) may not be appropriate. Similarly, the argument that a universal right to water is not feasible because there is not enough water to go around, is not tenable if the notion of scarcity often found in the GEM discourse on water is problematised.

Apart from the notion of scarcity, another message that forms the core of the GEM discourse on water is the view that treating water as an economic good would result in improved efficiency, equity, and sustainability. This, in turn, calls for putting in place market-based delivery systems, the establishment and enforcement of an effective (individual) property rights regime, and pricing of water at its economic value.[16] Reforms that emphasise the principle of cost recovery, setting up of water rights, participation, decentralisation, privatisation of particular functions in water delivery, redefinition of the role of the government, and demand management (quantifying the amount of water available and then managing it within these limits using pricing options and other measures), all stem, at least in part, from this perspective, though, in each case, there are other influencing factors also. Further, these different aspects are also often mutually contradictory. For instance, as Cullet points out,[17] water sector reforms have included both measures that restrict the role of the government as well as measures that seek to increase government control.

The second major discourse at the international level is the rights discourse, and more particularly, the idea of right to water.[18]

---

[16] See, for instance, R. Maria Saleth, *Water Institutions in India: Economics, Law, and Policy* (New Delhi: Commonwealth Publishers, 1996).

[17] Philippe Cullet, 'Water Law Reforms: Analysis of Recent Developments', 48(2) *Journal of the Indian law Institute* 206 (2006). *Source:* http://www.ielrc.org/content/a0603.pdf.

[18] The concept of right to water is much broader than the concept of water rights. Right to water includes a variety of dimensions such as access to water, affordability, ownership, delivery, and participation in decision-making processes, while water rights refer specifically to the particular sub-set of these dimensions that are pertinent from the point of view of the right-holder.

The right to water is not fully defined by existing international law or practice; however, it is implicitly and explicitly supported by many human rights instruments.[19] For instance, implicit support for the right to water is provided by other human rights such as those relating to food, health, adequate housing, well being, and life, since water is necessary to secure these rights. Two human rights instruments also explicitly mention the right to water – the 1979 Convention on the Elimination of All Forms of Discrimination Against Women, where it is mentioned as a part of a right to adequate living, and the 1989 Convention on the Rights of the Child, where provision of clean drinking water is mentioned as a means to combat disease and malnutrition. However, the most explicit formal adoption of the right to water as an independent human right is in the General Comment No. 15 adopted in November 2002 by the United Nations Committee on Economic, Social and Cultural Rights.[20] The document provides guidelines for state parties on the interpretation of right to water under two articles of the International Covenant on Economic, Social and Cultural Rights 1966 (ICESCR): Article 11 (the right to an adequate standard of living) and Article 12 (the right to health). While the General Comment is not legally binding on the 146 States that have ratified the ICESCR, it aims to assist and promote the implementation of the Covenant and does carry the weight and influence of soft law.[21] The General Comment has also been supplemented more recently by the 2005 draft guidelines for the realisation of the right put forth in the Report of the Special Rapporteur of the United Nations Commission on Human Rights (the 2005 Draft Guidelines).[22] These guidelines emphasise the right to water for personal and domestic uses, in order to realise the right to adequate nutrition and the right to earn a living through work.

---

[19] Peter Gleick, 'The Human Right to Water', 1(5) *Water Policy* 487 (1999).
[20] Committee on Economic, Social and Cultural Rights, General Comment 15: The Right to Water (Articles 11 and 12 of the International Covenant on Economic, Social and Cultural Rights), UN Doc. E/C.12/2002/11 (2002).
[21] United Nations Department of Public Information, International Decade for Action: Water for Life 2005–2015 – The Right to Water, A Backgrounder (2004). *Source:* http://www.un.org/waterforlifedecade/pdf/righttowater.pdf.
[22] Report of the Special Rapporteur El Hadji Guissé, Realization of the Right to Drinking Water and Sanitation, UN Doc. E/CN.4/Sub.2/2005/25 (2005).

The core message of the rights discourse is that all human beings are entitled to a minimum amount of water for basic needs. Some strands in the rights-based approach also extend the right to all living beings (and to the ecosystem) and call for water to satisfy not just basic needs, but also economic needs.[23] This message, in turn, has led to calls for legal recognition of the right to water and corresponding changes in water/water-related policies and legislations of governments. However, while this message has been broadly accepted by many water conferences (such as the 1977 United Nations Water Conference held in Mar del Plata, Argentina and the Rio Earth Summit), consensus on an explicit right to water by governments has been difficult to come by. This is most evident in the ministerial statements at the World Water Forums, which recognise only the idea of water as a basic need and not the idea of water as a right, even when the latter has been debated in the Forums (for instance, at The Hague in the Second World Water Forum in 2000 and at Mexico in the Fourth World Water Forum in 2006). This, in turn, is a possible reflection of the lack of hegemony of rights-based discourses in water (and therefore of the widespread influence of the GEM discourse).

In general, the idea of a right to water has had limited official recognition at the international level (especially, in comparison to the principles advocated by the GEM discourse) and attempts to analyse the implications of different GEM policies from a rights perspective have been limited. As a result, although the idea of water as an economic good and of water markets has generated considerable controversy, particularly in its implications for pricing,[24] market remedies and privatisation solutions for water problems are still believed by some (especially donor countries) to be completely congruous with rights of the poor to water.[25]

---

[23] For a review of different conceptualisations of the right to water, *see* Priya Sangameswaran, Review of Right to Water: Human Rights, State Legislation, and Civil Society Initiatives in India (Bangalore: Centre for Interdisciplinary Studies in Environment and Development, Technical Report, 2007).

[24] *See* Mehta, note 3 above.

[25] Lyla Mehta and Birgit La Cour Madsen, Is the WTO after Your Water? The General Agreement on Trade in Services (GATS) and the Basic Right to Water (Brighton: Institute of Development Studies, 2003).

3. Water Discourses in India

Elements of the two discourses discussed at the international level as well as the hegemonic role of the elements of the GEM discourse are found in the water reforms undertaken in India too. This section considers the broad contours of the reforms that have been encouraged at the central level in the domain of delivery of water services. Although water is a State subject in India, the Centre does influence state policy with regard to water in two broad ways. First, the Centre plays an indicative role, that is, it indicates the direction in which states must move (for instance, putting in place groundwater legislation). In some cases, it may not apply pressure for the policy to be actually implemented or even discuss the direction in any great detail; in other cases, it does apply pressure (for instance, by making funding for projects conditional on adoption of particular measures). The second way in which the Centre influences state policies with regard to water is through binding legislation (for example, laws related to the environment).

In the specific context of delivery of water services, the first route is most relevant. The Centre has encouraged two kinds of policies, both of which have been taken up to varying extents by different states – sectoral decentralisation such as Participatory Irrigation Management (PIM) and privatisation. Sectoral decentralisation forms part of the policy prescriptions of both the GEM discourse and the rights discourses, although, as we will see in the ensuing discussion (particularly in the discussion of decentralisation and entitlements in the case of Maharashtra), the limited manner in which decentralisation has been undertaken means that it is not particularly commensurate with any notion of rights. Privatisation policies are also more a part of the GEM discourse and are related to the notion of water as an economic good. While some discussions of a right to water (such as in the General Comment and the 2005 Draft Guidelines) are relatively flexible about the system of water delivery and do not take an a priori stand for or against privatisation, many advocates of a right to water (particularly social movements in water) take a strong anti-privatisation position.

It is also important to note that international players such as the World Bank have played an important role in pushing for both kinds of policies. For instance, the World Bank's Country Strategy for India, which is applicable for lending from 2005–2008, lays down sector-specific guidelines for lending.[26] In the case of Urban Water Supply and Sanitation, one of the conditions is that the state/city in question agree 'to support actions to develop domestic private sector capacities for delivering urban water supply and sanitation services'.[27] In the case of Irrigation and Drainage, granting of loans is contingent on willingness to 'establish and operationalise decentralised service delivery mechanisms'.[28]

I turn now to the recommendations made at the central level with respect to the above two policies. In the case of irrigation, sectoral decentralisation has taken the form of PIM. Although this idea has been supported by the Government of India since the mid-1980s,[29] it is only recently that states have started taking measures to facilitate it. The precise nature and extent of powers and functions of Water Users' Associations (WUAs) varies from state to state, and is usually determined by a variety of factors internal to the state. For instance, in some states, the fixing of water charges has been kept outside the purview of WUAs, but in other states (like Gujarat), WUAs are free to decide the water rates to be charged from the beneficiary farmers.[30]

But one feature seems to be common to all WUAs, that is, the limited nature of the powers devolved to them. This, in turn, is very much in tune with the stand that central policies take with regard to water. For instance, while the 2002 National Water Policy (NWP) emphasises a participatory approach to water resources management, the aim of involving WUAs and local bodies is said to be 'to eventually transfer the *management of such facilities* to the

---

[26] World Bank, Country Strategy for India, Report No. 29374-IN, 15 September 2004. *Source:* http://www-wds.worldbank.org/external/default/WDSContentServer/WDSP/IB/2004/09/20/000160016_20040920102445/Rendered/PDF/293740REV.pdf.

[27] Ibid., p. 3.

[28] Ibid., p. 4.

[29] See, for instance, Government of India, National Water Policy of 1987. *Source:* http://www.ielrc.org/content/e8701.pdf.

[30] Videh Upadhyay, 'Water Management and Village Groups: Role of Law', 37(49) *Economic and Political Weekly* 4907 (2002).

user groups/local bodies';[31] there is no mention of *ownership* of the water facilities by local groups. Similarly, the NWP mentions that the involvement and participation of beneficiaries and other stakeholders should be encouraged right from the project planning stage itself, but the nature of this participation, as well as how and by whom beneficiaries and stakeholders are to be defined is unclear. Further, while participation at the level of the WUA might be encouraged, the question of participation in the process of irrigation policy-making at higher levels is not even mentioned.

In the case of drinking water too, the process of sector reforms, with decentralisation as one of its key features, was first started by the Centre in rural areas. Reforms were first introduced by way of the Sector Reform Program in 67 pilot districts covering 26 states, and were scaled up in 2002 in the form of Swajaldhara. Swajaldhara aims to provide direct access to central resources to communities and community institutions (*panchayats* and district water and sanitation committees), which want to develop and manage local water resources to meet their drinking water needs.[32] However, while the sector reform scheme of Swajaldhara was expected to replace the existing Accelerated Rural Water Supply Program (ARWSP)[33] by 2007, take-up of Swajaldhara has been slow and the role of different agents such as government technical support agencies and NGOs remains weakly defined.[34] Further, although the scheme purportedly rests on principles of social inclusion and governance, there are no mechanisms to actually ensure that the schemes are designed by including all sections of society.[35] In part, this could stem from eulogistic notions of 'community' (particularly of village communities) that do not take into account

---

[31] Government of India, National Water Policy of 2002, Section 12 [hereafter NWP]. *Source:* http://www.ielrc.org/content/e0210.pdf. Emphasis added.

[32] Government of India, Guidelines on Swajaldhara (New Delhi: Ministry of Rural Development, 2002). *Source:* http://www.ielrc.org/content/e0212.pdf.

[33] ARWSP is a supply-driven scheme introduced in 1972–73.

[34] WaterAid, Drinking Water and Sanitation Status in India: Coverage, Financing and Emerging Concerns (New Delhi: WaterAid, 2005).

[35] Sara Ahmed, 'Why is Gender Equity a Concern for Water Management?', *in* Sara Ahmed ed., *Flowing Upstream: Empowering Women through Water Management Initiatives in India* 1 (Ahmedabad: Centre for Environment Education, 2005).

power politics within the community. It could also be due to the fact that the goal of participation in these projects is itself very limited viz., to get local people to contribute (labour, for instance).

Another kind of change in delivery of water that has been encouraged by central policies is privatisation in the context of canal irrigation, minor surface irrigation, and drinking water systems (particularly in urban areas). For instance, the NWP points out that corporate sector participation in canal irrigation will help in 'introducing innovative ideas, generating financial resources and improving service efficiency and accountability to users'[36]. Further, it could include one or all of various aspects such as building, owning, operating, leasing, and transferring of water resource facilities.

In the arena of drinking water, the Chennai Metropolitan Water Supply and Sanitation Board, popularly known as Metrowater, was an early reformer in India, and negotiated its first big loan from the World Bank in the early 1980s, that is even before the central-level policy changes. But since the late 1990s, reform of the water sector has become an important part of the policy discourse in several cities such as Bangalore and Delhi. At the present juncture, however, there is little analysis of the precise forms that privatisation is taking and its implications, although concerns about equity (particularly as a result of the increase in prices that privatisation is likely to result in) as also the negative experiences of privatisation in other parts of the world have led to protests by civil society groups in many parts of the country.

The emphasis of central-level policies on both sectoral decentralisation and privatisation is in line with global trends discussed earlier – focus on cost recovery, limited role for the state, emphasis on water as an economic good, and so on. But the rights discourse is not reflected in policies, even though there is a constitutional basis for the right to water (in that it has been derived from the right to life by various judicial judgments). For instance, the NWP continues to call water a 'basic human need' as against a 'basic human right', in spite of many attempts by civil society agents (at the time that the draft was being circulated in the public domain) to change the nomenclature from need to right.[37] In a

---

[36] *See* NWP, note 31 above, p. 6.

[37] Anonymous, 'Water Commonwealth', *Times of India*, 16 March 2002.

sense, this (the NWP's stand) reflects tensions at the international level (discussed in the previous section) about whether water should be called a need or a right.

In fact, while the Centre does concede that water is an economic and social good, it also holds that some of the problems in the drinking water sector (such as lack of sustainability) are due to the perception of people that 'water is a social right to be provided by the government, free of cost'.[38] While the idea of water as a right need not necessarily imply free water in all cases, and conversely, the agenda of cost-recovery could potentially be undertaken in conjunction with the idea of water as a right, the lack of explicit engagement with the idea of a right to water means that the particular manner in which the Centre ends up shaping reforms is limited from the point of view of equity. Thus, as Cullet argues,[39] decentralisation of only limited number of functions has taken place and WUAs or drinking water committees have little say about surface water sources, whose control continues to be largely dependent on decisions taken at higher levels.

4. Delivery of Water: The Case of Maharashtra

4.1 Introduction

Maharashtra is a good example of the different kinds of changes that are occurring in the water sector, not just in India, but the world-over. These include a greater emphasis on WUAs for management of water resources at various levels, revision of water rates, corporate involvement in medium and major irrigation projects, demand-driven rural drinking water projects, and a focus on watershed projects as well as on river basin management in water policy. One realm in which change is evident is legislation; since 1990, a number of laws – the Groundwater (Restrictions for Drinking Water Purpose) Act 1993, the Maharashtra State Water Policy 2002 (MSWP), the Maharashtra Management of Irrigation Systems by Farmers Act 2005 (MMISFA) and the Maharashtra Water Resources Regulatory Authority Act 2005 (MWRRA) – have

---

[38] Government of India, Annual Report of the Ministry of Rural Development 136 (New Delhi: Government of India, 2003–04).

[39] *See* Cullet, note 17 above.

been passed. But before turning to the current changes in the water sector, it is useful to briefly consider the water situation in Maharashtra.

According to the 2001 census, 79.8 per cent of the households in the state have access to safe drinking water. This includes 68.4 per cent of households in rural areas and 95.4 per cent in urban areas. In terms of irrigation, although the percentage of gross irrigated area to gross cropped area has increased steadily since the time of formation of the state (from 6.5 per cent in 1960–61 to 16.6 per cent in 2000–01), it is still low as compared to the ultimate potential as well as to the all-India average of 38.7 per cent.[40] As in the rest of the country, there are problems with respect to efficiency, equity, and sustainability in the case of both drinking water and irrigation. The lack of efficiency is evident, for instance, in the fact that actual utilisation of the irrigation capacity created up to June 1999 was only 38 per cent for major and medium irrigation projects.[41]

There is also inequity in the distribution of water, both between districts and within the same district. For instance, sugarcane-growing areas get water even during droughts, while other areas lack water for subsistence crops or even drinking water. Sugarcane cultivation is problematic not only in terms of equity, but also in terms of environmental sustainability. Increased cultivation of sugarcane usually has gone hand-in-hand with lavish use of water for irrigation and use of fertilisers in excessive amounts (which further increases the need for water), and has resulted in waterlogging and salinity in many areas.[42] It is also important to note that the problems of efficiency, equity, and sustainability of water are inter-related. For instance, the growing problem of groundwater depletion means that the newer technology needed for pumping water is increasingly less accessible to poor farmers, resulting in inequity in the way different classes of people can cope with groundwater shortage.

---

[40] Government of Maharashtra, Economic Survey of Maharashtra, Directorate of Economics and Statistics, Planning Department, Mumbai (2000–01) [hereafter ESM].

[41] Ibid.

[42] Donald Attwood, Small is Deadly: Coping with Uncertainty at Different Scales (Paper presented at the Conference on the Culture and Politics of Water, University of Delhi, March 2001).

While at least some of the problems in the water situation are to do with topography (hard rock and undulating surface) and rainfall (wide variation across different parts of the state), many of the problems can be attributed to deficiencies in state policy with regard to water. In the case of irrigation, this is primarily reflected in the undue focus on large surface irrigation projects, and in the case of drinking water, in the piecemeal and target-oriented approach followed. For instance, successive state governments in Maharashtra have emphasised major and medium surface irrigation projects, so that the state now has the 'distinction' of having the largest number of on-going major and medium irrigation projects and extension/renovation/modernisation schemes in India (108 out of a total of 476 in the country).[43] The emphasis on large-scale dams stems in part from the goal of increasing agricultural production in India and in part from what Datar and Kumar call 'the psychological power of planning to reduce "scarcity" conditions';[44] in the specific case of Maharashtra, there is also a particular historical context which gave rise to this.

Since the 1970s, ground water development has also been emphasised, and tubewells have received considerable institutional credit. But on the whole, the attention directed towards minor irrigation has not been adequate, especially when one considers the fact that minor irrigation accounts for a large portion of the state's ultimate irrigation potential and much of this has still not been attained.[45] The bias of state policy in favour of major and medium surface works has exacerbated in the late 1990s because the Government of Maharashtra started trying to impound as much as possible of the water awarded to it by the Bacchawat Interstate Water Dispute Tribunal.[46] This resulted in a rapid process of dam

---

[43] *See* ESM, note 40 above.

[44] Chhaya Datar and Ajith Kumar, Rural Drinking Water in Maharashtra 45 (Tuljapur: Tata Institute of Social Sciences, 2001).

[45] R.S. Deshpande and A. Narayanamoorthy, 'Issues before Second Irrigation Commission of Maharashtra', 36(12) *Economic and Political Weekly* 1034 (2000).

[46] This Tribunal was set up to resolve the dispute on the sharing of the water of the Krishna river between the states of Andhra Pradesh, Karnataka, and Maharashtra. The state of Maharashtra was given an award of 560 Thousand Million Cubic feet of water in May 1976, which was to be used by May 2000. *See* Deshpande and Narayanamoorthy, note 45 above.

construction with considerable social costs (in that rehabilitation concerns in these dams were not met at all). Ironically, much of the water impounded in the dams remains unutilised to date because of incomplete canal work.[47]

In the case of drinking water in rural areas, as in the rest of the country, provision of water supply has been supply-driven, with an emphasis on norms and targets and on construction and creation of assets, rather than on management and maintenance of the facilities built or of the sustainability of the source itself; this, in turn, has led to a large gap between officially claimed coverage and actual coverage on the ground.[48] For instance, the most common form that drinking water schemes have taken is digging of borewells, neglecting other sources of drinking water like tanks. Further, during times of severe water shortages such as droughts, ad hoc measures (such as supply of water via tankers) are offered instead of seeking long-term solutions. Until recently, there has also been no systematic, comprehensive policy on recharging strategies such as water harvesting and watershed development, although soil and water conservation measures have been undertaken on a sporadic basis. Even in the limited cases where such practices have been adopted, emphasis is often more on irrigation water rather than on drinking water.

With this brief discussion of the water situation in Maharashtra, the paper now turns to the recent changes in the state, particularly with respect to delivery of water services.

4.2 Recent Changes in Water

The 2002 MSWP is the first water policy of Maharashtra, and as such, an important landmark. Even though state water policies do not have legal status, and there are usually gaps between policies, passage of enabling laws and rules, and implementation by the bureaucracy, they are still important because they provide overall guidelines. The MMISFA was passed in 2005 in order to provide

---

[47] Anant Phadke, "Thiyya Andolan" in Krishna Valley', 39(8) *Economic and Political Weekly* 775 (2004).

[48] Water and Sanitation Program, Alternate Management Approaches for Village Water Supply Systems (New Delhi: Water and Sanitation Program – South Asia, Field Note, 2004).

a statutory basis for the management of irrigation systems by farmers, which in turn is in tune with the recommendations made at the central and state levels. The Act aims to increase efficiency in the utilisation of irrigation capacity, as well as in the distribution, delivery, application, and drainage of irrigation systems.[49] The MWRRA, also passed in the same year, aims to establish a Maharashtra Water Resources Regulatory Authority (the Regulatory Authority) to regulate water resources within the state, as well as to facilitate judicious, equitable, and sustainable management of water resources.[50]

The aforementioned policy and legislations have been put in place to facilitate particular kinds of reforms in the water sector in Maharashtra; in the realm of delivery of water, these reforms primarily include (although they are not limited to) sectoral decentralisation and privatisation. Policy changes are the result of a complex inter-play of factors and it would be simplistic to claim that they are a direct result of particular discourses at the international and national levels. Yet, there is a fair amount of evidence in support of the claim that international water discourses, and particularly the GEM discourse, has provided an important impetus to the recent policy changes in Maharashtra. Firstly, the core messages of the GEM discourse – notions of scarcity and of treating water as an economic good – are also found in the MSWP, the MMISFA and the MWRRA. For instance, the need for the MSWP is justified, among other things, by the increasing scarcity of water.[51] Secondly, the World Bank, a key player in the formulation, propagation, and dissemination of the GEM discourse, has played an important role in the reform process in Maharashtra. More particularly, in June 2005, the World Bank approved a loan of US$ 325 million to assist the Government of India with the implementation of the Maharashtra Water Sector Improvement

---

[49] Government of Maharashtra, Maharashtra Management of Irrigation Systems by Farmers Act, Act No. XXIII of 2005 [hereafter MMISFA]. *Source:* http://www.ielrc.org/content/e0505.pdf.

[50] Government of Maharashtra, Maharashtra Water Resources Regulatory Act, Act No. XVIII of 2005 [hereafter MWRRA]. *Source:* http://www.lead-journal.org/content/05080.pdf.

[51] Government of Maharashtra, Maharashtra State Water Policy, 2002 [hereafter MSWP]. *Source:* http://www.ielrc.org/content/e0306.pdf.

Project, whose key components include institutional reforms such as the establishment of a Regulatory Authority and of water entitlements, as well as the promotion of effective participation by way of WUAs in the management of irrigation schemes.[52]

This brings one to a discussion of the working of sectoral decentralisation and privatisation in Maharashtra. In the case of drinking water, sectoral decentralisation has taken place in both urban and rural areas; however, the focus of this chapter will be on rural areas. Traditionally, government-owned agencies have been responsible for the construction and management of rural water supply systems. Although this approach has led to the creation of assets on a massive scale, the assets have often been of poor quality and service delivery has been inadequate. The Sector Reform Program pioneered by the Government of India and state-level projects directly funded by donors such as the World Bank have increasingly encouraged demand-driven projects in lieu of the older supply-driven projects. Their key feature is that management (and in some cases construction also) is undertaken via a representative committee called the Village Water and Sanitation Committee, which is supposed to be a sub-committee of the *gram panchayat*. The main sponsors are the World Bank, the Government of Germany, and the Government of India (via its Swajaldhara program); the Government of Maharashtra also funds some demand-driven projects, though it also continues to fund some older, supply-driven schemes.

In the case of irrigation, sectoral reform has taken the form of PIM in canal irrigation, and a move towards greater community participation in watershed development programs. The focus of the discussion here will be on PIM. While associations for managing water systems have existed for a long time in Maharashtra, the recent genesis of the PIM program can be traced to the formation of cooperatives in the late 1980s by NGOs such as the Centre for Applied Systems Analysis in Development (CASAD). Partly in reaction to the pressure exerted by NGOs, and partly in response to the widespread trend of decentralisation (including the central

---

[52] World Bank, World Bank Supports Water Sector Management (Press Release, 23 June 2005). *Source:* http://web.worldbank.org/WBSITE/EXTERNAL/ NEWS0,contentMDK:20554604~menuPK:34463~pagePK:34370~piPK:34424~ theSitePK:4607,00.html.

government's own encouragement of PIM), the Government of Maharashtra took a decision to encourage formation of cooperative WUAs for irrigation management in 1988. The rationale was to improve water use efficiency, increase agricultural productivity, and reduce work for the Irrigation Department. The policy of participatory management was also expressed in the State Government's *Cooperative Water Users' Association Guidelines* 1994. But bureaucratic hurdles to the setting up of WUAs continued to exist. A 2001 government notification made WUAs compulsory, and the MMISFA was finally passed in 2005. However, the process of formation of WUAs and actual handing over of control of irrigation facilities is expected to take a long time, partly because all the relevant administrative rules have still not been changed, and partly because at many levels of the state bureaucratic apparatus, devolution of powers to farmers continues to be met with resistance (either because it means a loss of 'under-the-table' income for bureaucrats, or because of continuing scepticism about the ability of farmers to manage irrigation systems on their own).

Under the new farmer managed systems in surface irrigation, water for irrigation is supposed to be supplied to farmers only through WUAs. Even Lift Irrigation Schemes are to be undertaken only by WUAs, and eventually sanctions to individual schemes of lift irrigation are to be cancelled.[53] In terms of the nature of rights given to WUAs, the most important change now is that WUAs have the freedom to decide the cropping pattern. Bulk entitlement of water to the WUA would then be decided by the Regulatory Authority on the basis of the cropping pattern designed and the designated command area. However, the right to distribute water to individual farmers would rest with the WUA. Further, the WUAs would pay for the water received on a volumetric basis, although individual farmers may continue to pay the WUA on an area basis.[54] Charges for surface water (primarily canal water) have also been revised a number of times in the last few years.

Apart from sectoral decentralisation, the other form that changes in delivery of water have taken is privatisation. This trend first began in the irrigation sector when the Government of Maharashtra

---

[53] *See* MMISFA, note 49 above.

[54] The discussion in this paragraph draws on a personal communication with K.J. Joy (12 December 2005).

established five Irrigation Development Corporations in order to accelerate the completion of irrigation projects. These corporations are allowed to raise funds through the open market for funding their construction activities. Although the irrigation corporations were set up with considerable fanfare, their working has not borne out the initial expectations. They also constitute an added financial burden for the state, since these corporations sometimes receive budgetary support from the state (such as in the case of the Maharashtra Krishna Valley Development Corporation). Further, if the promised rate of return on the corporation's fixed investment (17.5 per cent, a rate that is very high for irrigation projects) is not met, the state government has undertaken to meet the difference out of its own resources.[55]

Various forms of private sector participation are also increasingly being undertaken in the management of minor irrigation tanks and for water distribution in urban areas. For instance, in June 2001, the state issued guidelines encouraging private sector participation in urban water supply and sewerage, especially in areas such as metering, billing, collection, O&M, and repairs of the distribution system. This process has already begun to be undertaken in different parts of the state (for instance, in Mumbai and surrounding suburbs). But lack of transparency about these efforts as well as the absence of adequate regulatory mechanisms – both essential conditions for privatisation to work effectively – are already emerging as critical issues.

5. Analysis of the Policy Changes in Maharashtra

The hegemony of the GEM discourse is evident not only in the specific kinds of policies adopted in Maharashtra (such as PIM, demand-driven drinking water projects, and privatisation), but also in the details of their working – which aspects are privileged, which ones are ignored, and so on. In order to show this, this section will focus on the concepts of decentralisation and entitlements in the ongoing water reforms. But before turning to this task, it is useful to briefly consider the role of legislation (and

---

[55] *See* Deshpande and Narayanamoorthy, note 45 above. This in turn brings into question the extent to which the irrigation corporations even represent a trend towards privatisation.

more particularly, changes in the form and content of legislation) in the reform process.

At least some of the changes that have been introduced as part of the reform process in the water sector have already been in place for a while; one example of this is WUAs in the case of canal irrigation. But the current reforms are distinct from the earlier policies in a number of ways, such as the scale at which they have been undertaken (across different realms and in different states), the importance accorded to formalisation (especially via legal reforms), and the presence of certain all-pervasive themes (such as scarcity).[56] The significance of the process of formalisation, in particular, is evident from the fact that in recent years, international donors as well as the Government of India have been encouraging state governments to put in place a legislative framework that is conducive to reform in both water and other arenas. For instance, the guidelines for World Bank lending for 2005–2008 point out that the Bank would consider full scale investment lending in the urban water supply and sanitation sector only if states have an adequate *legislative* and regulatory framework.[57]

It has been argued that changes in legislation (that lead to a broad change in the legal regime) are a necessary part of neoliberalism, since market regulation requires a different kind of legal regime than state regulation.[58] The new legal regime would involve, among other things, a restructuring of relations between corporations, states, and social groups, as well as the setting up of regulatory authorities which operate under a distinct set of institutional rules, which are different from the 'conventional rules that govern state institutions comprising the civil service, the executive and rules of parliamentary procedures'.[59] The ensuing

---

[56] This point draws partly on a discussion comment by M. Roopa at the Workshop on Water, Law and the Commons organised by the International Environmental Law Research Centre in New Delhi, 8–10 December 2006. More information on this workshop can be found at http://www.ielrc.org/activities/workshop_0612.

[57] *See* World Bank, note 26 above, Annex 5, p. 3. Emphasis added.

[58] Radha D'Souza, Dams, 'Development' and International Law (Paper presented at the Workshop on Water, Law and the Commons organised by the IELRC at PRIA, New Delhi, 8–10 Dec. 2006).

[59] Ibid., p. 11.

discussion of decentralisation and entitlements, based on their conceptualisation in the recent legislation in Maharashtra, offers one example of the limitations that such a change in legal regime could entail.

The conceptualisation of decentralisation in the specific case of Participatory Irrigation Management (PIM) can be taken as the first point of discussion. On the one hand, PIM seems like a good example of user groups being given the power to undertake functions that are best performed at the local level. On the other hand, as indicated in the discussion of PIM at the central level, the limited extent of the powers granted to WUAs calls into question the very intent of the process of decentralisation. For instance, while the role of the government is sought to be reduced by PIM, this does not necessarily translate into less regulatory intervention as far as water users are concerned because the Regulatory Authority becomes the new body exercising control over water resources.[60] Although the Regulatory Authority is delinked from the government, and in that sense is supposed to be 'free of politics', the powers given to it are extensive and include, among other things, distribution of water entitlements for different categories of use, determination of priorities in distribution of water at different levels (basin, sub-basin and project), and establishment of water tariffs. In fact, the Regulatory Authority not only has the power to make regulations for matters that come under the MWRRA but also for 'all other matters for which provision is ... necessary for the exercise of its powers and the discharge of its functions under this Act'.[61] Further, confirming D'Souza's fears that such bodies may operate under different rules, the Regulatory Authority has powers equivalent to those vested in a civil court with respect to certain matters (such as summoning of witnesses, reception of evidence on affidavits, and so on) for the purposes of making any inquiry or initiating any proceedings under the MWRRA.[62]

There is also another important lacuna in the current conceptualisation of decentralisation. In order for decentralisation

---

[60] *See* Cullet, note 17 above.
[61] *See* MWRRA, note 50 above, Section 31.
[62] *See* MWRRA, note 50 above, Section 13.

to be meaningful, it should include provisions for participation in both policy-making and actual implementation on the ground. The presence of strong civil society groups in the state (both historically and in current times) has meant that there has been greater participation in Maharashtra than in many other states. But mechanisms to facilitate participation in state policy and legislation continue to be limited. For instance, although the idea of farmers' participation has influenced (at least in part) the formation of WUAs, specific provisions to ensure equity in participation do not exist in the government guidelines; only procedural aspects of internal functioning are mentioned.[63]

Similarly, in the case of the MSWP, there is precisely one reference to gender, and that too a nominal one: 'The women's participation in the irrigation management should also be considered'.[64] But if participation at the micro-level (such as in WUAs) is merely mentioned and not facilitated, the question of participation in the process of irrigation policy-making at higher levels is not even mentioned in any of the state policies or legislation. As a result, even though policy-making continues to be subject to pressures and lobbying from different groups, there are no formal mechanisms to ensure that all sections of society have a chance to participate in the process of policy-making, or that these inputs are actually taken into account. On the contrary, the space available for any kind of negotiation is increasingly being limited by conditionalities such as the World Bank's requirement that all rural water supply and sanitation projects, *irrespective of the source of funding*, need to incorporate certain reforms (such as decentralised service-delivery and recovery of O&M costs) for receipt of investment lending by the Bank in that sector.[65]

The experience of the recent water legislation is also interesting in this regard. For instance, in the case of the MSWP, not only was the adoption of the policy itself a result of considerable lobbying and pressure applied by individuals and organisations working in the field of water, but also three drafts of the policy were open to

---

[63] Government of Maharashtra, Irrigation Department, Cooperative Water Users' Association Guidelines, 1994.
[64] *See* MSWP, note 51 above, Section 2.2.2.
[65] *See* World Bank, note 26 above, Annex 5, p. 5.

public suggestion before the finalisation of the document, a practice that is highly unusual. The process was, of course, subject to a number of limitations: for instance, the state was not duty-bound to actually take into account these suggestions. As a result, the final version of the MSWP was retrogressive compared to the earlier drafts.[66] The two laws that were passed three years later to actually operationalise some aspects of the MSWP – that is, the MMISFA and the MWRRA – had different kinds of experiences in this regard. In the case of the MMISFA, at least some process of public consultation was undertaken. A draft version of the Act was circulated to obtain the views of various NGOs, even though, as in the case of the MSWP, these were not necessarily accepted.

However, the MWRRA was not discussed with anyone initially, although some NGOs tried to push for changes before it was tabled in the legislature. Sainath points out that the process of passage of the bill offers an interesting lesson on the working of parliamentary democracy.[67] When the bill was first introduced in the Nagpur session of the State Legislative Assembly in 2004, it was subject to criticism by a CPI-M legislator. It was then referred to a joint committee of both houses, though not all party members (including the one that originally critiqued it) were included on the committee. Following some modifications made by the committee, the revised bill was re-introduced in the Mumbai session in 2005 on the last day and passed by voice vote at the last minute, so that there was not enough time to read, let alone discuss, the bill.[68]

Sectoral decentralisation policies, whether in the context of PIM or demand-driven drinking water programs, also do not sufficiently engage with the multiplicity of bodies at the local level that deal with different kinds of functions (related to water and otherwise) and the related question of which is the most suitable body from the point of view of different objectives. For instance, different kinds of water programs deal with different kinds of 'water communities' and corresponding user groups; the village and the water and sanitation committee in drinking water programs, the command area and the WUA in the case of canal irrigation, and

---

[66] Interview with Seema Kulkarni on 11 June 2004.
[67] P. Sainath, 'Water: How the Deal was Done', *The Hindu*, 28 April 2005. *Source:* http://www.hindu.com/2005/04/28/stories/2005042804831100.htm.
[68] Ibid.

the watershed or the river basin and the corresponding watershed committee or river basin group in other contexts such as the integrated planning, development, and management of water resources. There has been no attempt to link these different kinds of 'communities' or deal with problems of division of labour and coordination between them and *panchayat raj* institutions.

In fact, sectoral decentralisation policies may potentially create new power centres at the local level. For instance, Cullet notes how WUAs are encouraged to become financially independent and viable by engaging in additional remunerative activities such as distribution of seeds, fertilizers, and so on;[69] these are only indirectly related to irrigation, but at the same time they are also likely to result in WUAs becoming new centres of power.

One can now move on to the concept of entitlements to water, which is mentioned in the state policy and the two subsequent laws. The MSWP mentions entitlements to water for the first time and grants water users' organisations and entities 'stable and predictable entitlements to water so that they can decide on the best use of water without bureaucratic interference'.[70] Further, it claims that a well-defined transparent system for water entitlements will be established, so that these cannot be changed unilaterally by any state agency or authority.[71] Both the MWRRA and the MMISFA, which were put in place three years after the MSWP, discuss entitlements in greater detail. While the term entitlement seems to evoke some notion of rights, an actual consideration of the concept shows that it is far from any concept of right to water for all and more in line with a tradable permits concept of water rights that reinforces the claims of current users of water.

For instance, entitlements in legislation refer to authorisation granted to use water, that is a usufructuary right. But this is not linked to any notion of *inherent* rights of farmers over water.[72] Even in the case of surface irrigation, where there is some degree of commitment by the irrigation authority of the state, the extent

---

[69] *See* Cullet, note 17 above.
[70] *See* MSWP, note 51 above, Section 1.3.
[71] *See* MSWP, note 51 above, Section 4.1.
[72] Videh Upadhyay, 'Confusing Water Rights with Quotas', *India Together*, October 2005. *Source:* http://www.indiatogether.org/2005/oct/vup-rights.htm.

to which this commitment is enforceable is limited. Prior to the reforms, the Memorandum of Understanding signed between the Irrigation Department and the Water Users' Association would usually specify how much water the WUA would be allocated, along with details of proportionate reduction in case of reduced storage or reservation of part of the water. This, in turn, created at least some basis for negotiation. With the change in regulations, it is not clear what space there will be for the kind of negotiations that used to take place in the past.[73] While there is some option for redress (via a dispute resolution mechanism), its adequacy in the face of the powers of the Regulatory Authority remains to be seen.

Further, the entitlements are granted only to landowners/occupiers, and there is no provision for transfer of entitlements to non-entitlement holders (such as the landless). At the same time, the MSWP permits transfer of all or a portion of water entitlement between entitlement holders in any category of water use. This has led to fears that water use claims are being delinked from land occupancy not from the point of view of equity but in order to result in progressive commercialisation of the water sector.[74] In fact, even in the context of landowners and occupiers of land, the question of access to water is complicated by the proposed hikes in charges for surface water (primarily canal water) under the MWRRA. These hikes have come in for a lot of criticism as they are likely to result in agriculture becoming unviable for a large number of small farmers. Although there is the claim that cross-subsidies could be allowed to alleviate the impact of such charges on the poor, the exact mechanisms for this have not been stated.

In the context of drinking water also, there is no mention (explicit or implicit) of a right or a guarantee of access by the state. In theory, drinking and domestic needs of water are prioritised (for instance, in the MSWP). At the same time, reforms do not deal adequately with water for drinking or domestic needs. On the contrary, the emphasis on demand-driven drinking water projects implies that water would be accessible only to people who can afford the charges being levied. In the case of Swajaldhara, for

---

[73] Thanks to Suhas Paranjape (personal communication) for drawing my attention to this point.
[74] See, for instance, Cullet, note 17 above.

instance, people are not only expected to pay for the water, but also to bear ten per cent of the capital cost and all operation and management expenses; those who cannot afford to pay this price would be unable to have access to funds in these projects, and may have to turn to private sources that entail greater expenses and burden in the long-run. Further, the emphasis on water rates (the revenue side) has not been accompanied by an equal emphasis on attempts to cut down unwarranted expenses (the expenditure side).[75]

6. Conclusion

This chapter discusses how water discourses play out at different levels in the realm of delivery of water services. The messages of the GEM discourse at the international level – the notion of scarcity and the importance of treating water as an economic good – have led to particular kinds of water reform in India. Even though the policy of sectoral decentralisation is, in theory, also commensurate with the rights-based discourse in water (with its central message of everyone being entitled to water), the manner in which it has been undertaken indicates the hegemony of the GEM discourse. The analysis of the notion of entitlements in particular reflects the tensions between the two discourses, especially because the language of entitlements evokes the idea of rights, which is present in both the GEM discourse and the rights discourse, albeit in very different forms (water rights in the first case and the right to water in the second case). The confluence of different trends in the water reform process in Maharashtra – privatisation (in the form of the irrigation development corporations), decentralisation (via the formation of WUAs), centralisation (via the provision to set up a Regulatory Authority which has no room for PRIs), cost recovery (by way of volumetric pricing and increased tariffs for surface water) and water rights (by the provision of entitlements) – is another reflection of the tension between these discourses. This chapter also briefly touches upon the importance accorded to the process of formalisation in the reform process as evident in the emphasis on the enactment of new laws. How these legislative changes work themselves out in actual micro contexts now remains to be seen.

---

[75] *See* Deshpande and Narayanamoorthy, note 45 above.

4

The Slow Road to the Private: A Case Study of Neoliberal Water Reforms in Chennai

KAREN COELHO

1. Introduction: Anomalies and Contradictions of Reform

A tragic conundrum of the turn of this century is that the changes so long awaited and demanded in our governing systems have appeared in the form of 'reform' – a term that serves as a euphemism for the unleashing of neoliberal orthodoxies across the spectrum of sectors and services. The word reform has come to index a politics of complicity between global commercial interests, international aid agencies, and national governments, aimed at transforming public resources into profitable enterprises. In the institutional arena, a reforming agency means one that has come under the financial and managerial disciplines of the commercial sector. The consensus on 'best practice' across sectors is one in which state agencies operate like profit-making businesses, firmly turning their backs on transfer-based or relief-oriented welfarism. This new order, then, spells an even greater alienation of public institutions from the public than was witnessed under bureaucratic regimes.

But reform of municipal water systems is tricky business in more ways than one. From a business perspective, these systems have scales of operation, capital intensity and tariff structures which are not easily amenable to commercialisation and/or privatisation. At a deeper level, the social, political and cultural contexts in which water as an element and water provision as a service are embedded pose a number of challenges to the project of commodification.

Reform of the water sector inevitably, then, becomes mired in a set of contradictions and tensions in practice. The case of Chennai's water utility, Metrowater is revealing. Its career of over twenty-five years on the path of reform has been marked by dilemmas and distortions that reveal the discrepancies between the stated goals of municipal water service reform (to ensure equitable, sustainable and efficient water distribution and management) and its methods (corporatisation, commercialisation and privatisation).[1]

On the surface, Chennai's Metrowater emerged by the early 2000s as among the most successful water utilities in India, and one of the most dynamic public infrastructure organisations in the city. In 2001 it was praised by the World Bank for achieving the principles of best practice widely held for water utilities around the world. In contrast to the huge deficits and government subsidy common in public sector utilities, Metrowater is a financially strong and viable organisation. By 2002, it reported a surplus on its revenue account for the tenth continuous year, and had been operating without State government grants for over seven years. Its capable performance allowed it to take over the running of water and sewerage systems for housing projects run by the Tamil Nadu Housing Board, the Chennai Metropolitan Development Authority and the Slum Clearance Board. Suburban townships and neighbouring municipalities relied on Metrowater for technical guidance and/or contract with it to run their water and drainage systems.

Since the late 1990s, Metrowater has streamlined operations, frozen hires, instituted audits in a wide range of operational sectors, expanded its network and coverage, modernised its systems, contracted out several components, and stayed on track with its Master Plan. It has made steady improvements in revenue collections and has become creditworthy in its own right (i.e., independent of government guarantees).

---

[1] This chapter is based on dissertation research on Metrowater, conducted from 2001 to 2003. *See* Karen Coelho, Of Engineers, Rationalities And Rule: An Ethnography of Neoliberal Reform in an Urban Water Utility in South India (Ph.D. Dissertation, University of Arizona, Tucson: November 2004). I gratefully acknowledge the assistance of the American Institute of Indian Studies, the Foundation for Urban and Regional Studies, and the Richard Carley Hunt Fellowship of the Wenner Gren Foundation for Anthropological Research in making this work possible.

Outside this circle of funders and investors, however, the image of the organisation and the service is far from rosy. By 2002–3, it was clear that water governance in the city was in serious crisis. As droughts accumulated, the approximately ten billion rupees (1000 crore) spent on infrastructure improvements and source augmentation efforts proved to have yielded meager results, and large sections of the middle classes stopped depending on Metrowater for their drinking water.[2] The city's waterways remain perennially choked with untreated sewage, and the abundant rains of late 2005 brought disastrous floods, turning plenty into a problem. Meanwhile, the agency's search for supply-side solutions to meet the city's growing water needs grows increasingly more desperate. Following the failure of the expensive inter-basin transfer schemes during the drought of 2002–3, Metrowater began relying increasingly on ground water extraction. Entrusted with protecting the region's ground water resources, the agency emerged as the greatest culprit in depleting the aquifers of nearby river basins through highly unsustainable extraction over a decade.[3] In 2001–2, when yields fell in its own deep bore wells in the Araniyar-

---

[2] Studies estimate that Metrowater provides only a small fraction of the city's water demand. While Metrowater claims that 98 per cent of the city is covered with piped water supply, a survey conducted in 2003–2004 of over 1500 households in Chennai found that only a third of the city's water demand was met by Metrowater, while almost two thirds of the demand was supplied through private means such as consumer-owned borewells, tankers and packaged water. A.Vaidyanathan and J. Saravanan, Household Water Consumption in Chennai City: A Sample Survey (Centre for Science and Environment, 2005).

[3] The Chennai Metropolitan Water Supply and Sewerage Act of 1978 vests the Board with all powers to 'control extraction, conservation and use of underground water in the Chennai Metropolitan Area'. In addition, the Chennai Metropolitan Groundwater Regulation Act of 1997, amended in 2002, identifies Metrowater as the Competent Authority for regulating ground water extraction in the metropolitan area. Nevertheless, extraction remained entirely unregulated. A consultancy study commissioned by Metrowater in 2002 found that few, if any, licenses or permits had been issued since the Act was passed. It also found that the annual average extraction from the AK Basin Aquifer over the past 30 years was about four times the sustainable yield, resulting in progressive depletion of aquifer storage and saline intrusion extending to 15 km inland from the coast. Scott Wilson Piesold, The Reassessment of Ground Water Potential and Transferable Water Rights in the A-K Basin (Inception Report on Phase II: January 2005).

Kosastaliyar (AK) basin (northwest of the city), the agency began purchasing water from farmers in the area. By late 2004, when Metrowater's extraction from private agricultural wells in the AK Basin reached about 100 million liters a day, crises erupted in the peri-urban areas.[4]

How does reform play into these crises of water management? The reform paradigm guiding Metrowater transforms the water service into an industry responsive to consumer demand. Such a framework propels the agency towards exploitative and short-sighted handling of water resources, in direct contrast to its legal mandate to protect and promote the long-term sustainability of the resource. Revenue-enhancing supply-side investments are privileged over conservationist strategies.[5] The reform slogans of transparency and accountability are conceived on a model of public relations, limited to complaint-response and consumer grievance-redressal measures. Not only does this serve as a substitute for genuine citizen consultation and information-sharing, it also absolves the agency from accountability to sections of citizens that do not fit the bill of revenue-paying consumers.

This paper outlines the context, provenance and character of reforms in Metrowater, demonstrating how they fit a hegemonic model of global 'best practice' in running infrastructure utilities. It outlines some of the effects produced by the reforms, including the ways that meanings are conscribed ('institutional strength' now

---

[4] Farmers, residents and women's organisations went on protest, claiming that the sale of water to Metrowater had depleted wells and damaged agriculture in the area. In 2004, farmers of Velliyur gave an ultimatum to Metrowater and to the water sellers of the village to stop pumping ground water. When Metrowater failed to heed the request, 400 farmers took action in February 2005, breaking Metrowater's pumping structures. Forty-four farmers were arrested, kept under judicial custody for 15 days, and later released on bail.

[5] The latest supply-augmentation measure is a 100 million dollar seawater desalination plant being installed in Minjur, north of the city on a DBOOT (Design-Build-Own-Operate-Transfer) basis at a cost of Rs 500 crore. The proposal was severely criticised by citizens' fora in Chennai on the grounds of its high installation and operating costs (water produced would cost about Rs 50 per 1000 litres), environmental impacts (primarily marine desertification) and the lack of transparency about how the water would be distributed. More crucially, they argued, Metrowater failed to first explore other, more sustainable long-term options such as recycling waste water, regeneration tanks and lakes, and promoting reuse and conservation.

refers to success in commercialising operations), the organisational culture is transformed (audit and finance dominate over engineering), and most importantly, subsidised or common-access components of the service are progressively marginalised.

2. Corporatisation: cleansing the water service of 'politics'

The package of reforms that would prepare the utility for eventual, if not immediate, privatisation, had its roots in the formation of the 'progressive' corporate-style parastatal in 1978. The Metrowater Board was established as an autonomous statutory body, removed from the jurisdiction of the Municipal Corporation and placed directly under a department of the State government. Thus did municipal drinking water, traditionally the domain of local self-government, become shaped and steered by a global regime of discipline. The Madras Metropolitan Water Supply and Sewerage Board (MMWSSB) was created as the result of a major pre-investment study commissioned by the Government of Tamil Nadu (GoTN) and sponsored by the World Health Organisation (WHO) and the United Nations Development Program (UNDP) in 1976, in response to the growing population and infrastructure needs of the city. The study comprised an engineering component which resulted in a 20-year water and sanitation Master Plan for the metropolitan area, and a component on Organisation, Management and Finance (OMF), carried out by the multinational accountancy firm A.F. Ferguson in collaboration with the British firm Peat, Marwick, Mitchell and Co. The key recommendation of the OMF study was the establishment of an autonomous board that would integrate under its jurisdiction the various scattered components of the city's water service. The Bill that would give statutory basis to the proposed Board was drafted by the consultants and enacted in June 1978.

The primary motive of this institutional innovation was autonomy, in particular financial autonomy – the power to manage, independently obtain, and invest funds, to set tariffs, and to contract with private parties. Also envisaged was substantial 'managerial autonomy, giving independence from short-term influences on its policy and finances'.[6] The OMF report envisaged the creation of

---

[6] A.F. Ferguson and Co., in association with Peat, Marwick, Mitchell and Co., Final Report on Organisation, Management and Finance (1978), p. 5.50 [hereafter OMF Final Report].

an 'efficient and virile service' with a staff that had 'the capacities and aptitudes needed to carry out the various duties of an expanding and modern water and sewerage undertaking'.[7] The documents outlining the structural details of this new Board carried a celebratory tone – its creation was portrayed as a highly progressive step in the movement towards excellence in the water and drainage sector.

Fiscal imperatives were stressed from the start, and provided legitimation for all change. The study recommended the formation of a Public Relations Committee and the launching of a full-scale public relations campaign aimed at 'attuning people to accepting that water is becoming a scarce and expensive commodity, thus providing a receptive climate for the acceptance of inevitably higher tariffs...'.[8] The campaign was also supposed to prepare people to accept disciplinary cut-off action for non-payment.

The immediate upshot of the formation of the Board was the interest of the World Bank. In fact, the World Bank was already active in the wings before the organisation was set up – the terms of reference for the pre-investment study required that its financial analysis follow World Bank requirements for project appraisal, and its draft report as well as the draft legislation for the formation of the Board were submitted to the Bank for review. This institutional transformation initiated the flow of funds for the agency, a flow that has continued ever since. As a retired senior engineer who had been active in the agency at the time recalled:

> Even at the time of finalisation of the Master Plan some aspects were picked up by the World Bank for funding – the first time an international agency funds a project before the document is even complete! ... So what was a pre-investment study turned into a set of proposals for funding – a milestone for the Board, wherein it crossed two stages in one step: the move from city corporation to Board, and the move of attracting the interest of funding agencies.[9]

---

[7] OMF Final Report, note 6 above, at 11.14.
[8] A.F. Ferguson and Co. *et al.*, Interim Report on Organisation, Management and Finance (September 1977), p. 8.39 [hereafter OMF Interim Report].
[9] Interview with retired senior Metrowater engineer, 6 February 2003. All interviews in this study were conducted on the basis of a guarantee of anonymity to the interviewee, and will therefore be referenced with an indication of the interviewee's position and the interview date.

This moment of formation, then, brought Metrowater, directly and indirectly, into line with the orthodoxy of infrastructure sector reforms that was emerging in the projects funded by World Bank since the mid-1970s and culminated in the Bank's influential report entitled *Infrastructure for Development* (World Development Report 1994). This report presented a general picture of failure in the predominantly state-run infrastructure utilities of the Third World, and attributed these failures to institutional, rather than economic, technological or even financial factors. It argued that infrastructure agencies in Third World countries had focused on investment at the expense of maintenance, resulting in massive under-utilised capacity, overstaffing, and inefficiency. The crux of the problem, it concluded, was that decisions were made on the basis of political expediency rather than sound utility-management principles.

The report identified three core instruments for reversing the failures of government utilities, short of privatising them. These were – *corporatisation*, which 'establishes the quasi-independence of public entities and insulates [them] from noncommercial pressures and constraints', a *pricing strategy* designed to ensure cost recovery, and *contracts* between governments and private entities.[10] Thus, managerial and financial autonomy – in effect autonomy from the political ('non-commercial') sphere – and an unwavering focus on commercial viability were held as the touchstones of a good service. The role of government in this scheme was seen as regulatory, limited to setting policies and goals.

While Metrowater was an early reformer in India, by the late 1990s these basic principles of corporatisation, commercialisation and privatisation had become part of the national discourse of reform in the water sector, as the World Bank shifted its lending strategies in India from projects to policy reform.[11] The Bank's

---

[10] World Bank, *World Development Report: Infrastructure for Development* (New York: Oxford University Press, 1994).

[11] The Eighth Five Year Plan (1992–97) outlined a key principle for the sector: water being managed as a commodity and not a free service. This thrust was carried over into the Ninth Plan (1997–2002). The Irrigation Sector Review produced in 1991 by the World Bank in collaboration with the Government of India outlined a reform agenda which included steep hikes in water charges, with Water Users Associations given responsibility for ensuring high collection rates to finance their activities. This document inspired water reform legislations in Andhra Pradesh (1997) – which raised irrigation charges by 300 per cent – and Maharashtra (2005).

India Water Resources Management Sector Review published in 1998 'aimed at initiating and sustaining water sector reforms undertaken in partnership with GoI'[12] and its reform proposals were discussed at five national workshops convened by GoI.[13] The National Water Policy of 2002 favours widespread private sector participation in the country's water management. The privatisation agenda is promoted by a consistent conflation of the private sector with 'community' and 'civil society', all of these shown in opposition to 'the state'. Prime Minister Vajpayee's speech at the Fifth Meeting of the National Water Resources Council in 2002, said:

> The (revised National Water Policy) should ... recognise that the community is the rightful custodian of water. Exclusive control by the government machinery, and the resultant mindset among the people that water management is the exclusive responsibility of the government, cannot help us to make the paradigm shift to that participative, essentially local management of water resources. ... Wherever feasible, public-private partnerships should be encouraged in such a manner that we can attract private investment in the development and management of water resources.[14]

The World Bank's 2005 Report contains a reiteration of the fundamental elements of its reform agenda for the Indian water

---

[12] World Bank, India Water Resources Management Sector Review – Initiating and Sustaining Water Sector Reforms (Report No. 18356–IN in 6 volumes, 1998).

[13] A key event was the State Water Ministers' Workshop held in Cochin in December 1999, organised by the Water and Sanitation Program, South Asia and the GoI's Department of Drinking Water Supply. The workshop resulted in the Cochin Declaration, which emphasised the principles of financial viability of services (including recovery of all O&M costs and at least ten per cent of capital costs from users) and a shift in the role of government from provider to facilitator.

[14] This conflation is evident also in the World Bank's 2005 report on India's water sector: 'India faces this challenge (of transforming public water services) with many assets and some liabilities. The assets include citizens, communities and a private sector who have shown immense ingenuity and creativity... critical for the new era of water management. The major liability is a public water sector which ... is not equipped to deal with the central tasks which only the government can do...(such as developing a regulatory framework and facilitating the entry of private sector and cooperative providers)'. World Bank, India's Water Economy: Bracing for a Turbulent Future (Report No. 34750–IN, 22 December 2005), p. xiii.

sector – the creation of a 'new water state' which will focus on developing instruments (including water entitlements, contracts between providers and users, and pricing) and incentives to control the use of water, stimulating market competition, and establishing a sound financial footing. It pushes for a 'dramatic transformation in the way in which public services are provided... in which the watchwords are water entitlements, financial sustainability, accountability, competition, regulation and entry of alternatives to government provision, including cooperatives and the private sector'.[15]

By 2001, the rationales of reform had been so successfully internalised within Metrowater that the majority of officials saw them as independently arising imperatives, arising both from the need for funds and the need for change. Many engineers ascribed the reforms to the initiatives of dynamic leaders within the organisation, or to state policies such as the freeze on recruitment. A few, however, who were openly critical of the direction of the reforms, attributed them squarely to donor conditionalities.

These varied perspectives within the organisation on the role of conditionality in bringing about reforms was partly due to a process of negotiation between the Board and the Bank that produced the 'consensus' on Metrowater as a commercial entity. This process of negotiation was at least partly textual – a study of the documentary history of the organisation reveals how local 'ownership' of the reforms was slowly organised, through a subtle shift in the World Bank Aide Memoires, from a language of conditionality ('items that are critical to satisfying the conditions of appraisal and negotiations') to that of shared agreement ('Discussions were held with the Government of Tamil Nadu and the Metroboard and an understanding was achieved of the importance of these measures and the reason for them') and back to one of mentorship ('Metrowater's proposals for reform should be completed by the time of appraisal so that the Bank may review it at that time').[16]

Metrowater's own documents are a study in apprenticeship, revealing the process through which the organisation was steadily shaped by World Bank orthodoxies over the years since its

---

[15] Ibid. xiii.
[16] World Bank, Preparation Mission Aide Memoire, 5 December 1985, at 2 and 5.

inception. Its Annual Reports and project proposals increasingly reflect or echo the World Bank's Aide Memoires, Staff Appraisal Reports and other official commentaries, which in turn reflect the World Bank's more foundational documents such as the *Infrastructure for Development Report (1994)*, the *Water Resources Management Policy* (1993) and the *India Water Resources Management Sector Review (1998)*.

While the basic thrusts of reform in Metrowater were set by the conditions of its formation in 1978, the consumer relations reforms introduced in the 1990s were simply extensions and elaborations of this move. As the retired senior engineer (mentioned above) puts it:

> The first set were macro-improvements, the new changes are micro, in-depth operational reforms. For example, five to six years back the bill collector went to people's houses, now the Board feels the employment of these guys costs a lot, so they define it as the duty of citizens to go and pay their bills... Another example: if you have a sewer block and approach your local Metrowater office, they will first check if your taxes and charges have been paid in full.[17]

The next section elaborates on how the ongoing transformation of the service culture, from a welfarist to a commercial paradigm, is achieved.

3. Projects of Commodification and Commercialisation

Like the commercialisation of a government service, the commodification of water is not instantly accomplished. Not only does it pose complex economic and legal challenges,[18] it also calls

---

[17] Interview with retired senior engineer, 6 February 2003.

[18] In legal and economic terms, water remains notoriously hard to commoditise. Developing water markets is challenging due to: (i) the non-exclusive character of water in piped systems (i.e., it is difficult to exclude individuals once they have entered the system); (ii) difficulties in defining tradable property rights in water; and (iii) especially in the case of ground water, difficulties in pricing. *See* Marcus Moench and S. Janakarajan, Water Markets, Commodity Chains and the Value of Water (Chennai: Madras Institute of Development Studies, MIDS Working Paper No.172, June 2002).

for extraordinarily detailed work in the domains of discourse, language and daily practice. Water in India and in Tamil Nadu, as in many places around the world, is a highly symbolic material, surrounded by thick systems of social and religious meaning, and located in rituals of gifting, exchange and rule.[19] As Shiva puts it, one aspect of the 'water wars' raging around the globe is the 'paradigm war' – a conflict over how water is perceived, valued and treated: 'The culture of commodification is at war with diverse cultures of sharing, of receiving and giving water as a free gift'.[20]

The banner of 'scarcity', once a favoured prop of bureaucratic patronage systems[21] is now a pennant of the movement for marketising water. In Chennai, with its heavy dependence on surface water sources and its unreliable monsoon patterns, water crises are both acute and chronic. However, given the political context within which the service is embedded, the push for market pricing of water as a solution to the problem of scarcity can only win limited acceptance. Solutions have thus tended to focus more on massive source-augmentation schemes, which raise the imperatives of attracting investment finance and hence of improving the financial viability of utilities. In Metrowater, then, the process of commodifying water proceeded concomitantly with – and through – the process of commercialising the service. Two major projects were implicated – first, 'institutional strengthening' of the utility, using financial and management disciplines modelled on commercial organisations, and second, turning clients into consumers through attempts at tariff reform and full cost recovery from users. This section traces the organisation's efforts along these lines over the 25 years of its existence, as evidenced in policy documents and project proposals as well as in the discourses of senior agency officials.

---

[19] Vandana Shiva, *Water Wars: Privatisation, Pollution and Profit* (Cambridge, Mass: South End Press, 2002), David Mosse, *The Rule of Water: Statecraft, Ecology and Collective Action in South India* (New Delhi: Oxford University Press, 2003) and Wendy Espeland, *The Struggle for Water: Politics, Rationality, and Identity in the American Southwest* (Chicago: University of Chicago Press, 1998).

[20] *See* Shiva, note 19 above at x.

[21] P. Sainath, *Everybody Loves a Drought: Stories from India's Poorest Districts*, (New Delhi: Penguin, 1996).

The central thrust of reforms in Metrowater, since its inception, was on 'institutional strengthening', a term that was systematically conflated with financial strengthening and commercialisation of operations, since the first preparation missions of the World Bank. The Bank's vision of Metrowater 'exercising leadership in the water sector' was grounded on its achievement of commercial viability. The (Second Chennai) project's 'strong emphasis on strengthening Metrowater' would be achieved through 'tariff increases and improvements in financial performance'[22] and also through 'increased worker and management productivity through the application of incentives...'.[23] Also listed under the overall goal of institutional strengthening are goals to 'ensure full cost recovery of Metrowater's investment and operational costs', and to 'improve the performance of Metrowater in key areas such as revenue mobilisation and utilisation, ... commercial accounting, consumer education, and sector management'.[24]

A key aspect of the commercialisation of the service was the effort to build 'management capability' in the organisation. In the late 1990s, three major consultancies were initiated to review the integrated functioning of the organisation – one, an 'Organisation Re-engineering Study' carried out by Osmania University; two, a 'Twinning Consultancy' with the Compagnie Generale des Eaux (GdE), a subsidiary of the French giant multinational water utility Vivendi, and three, a 'Strategic Review of Institutional Options' carried out by the multinational accounting firm, KPMG. The first was a diagnostic study of Metrowater's corporate performance, recommending measures for capacity-building towards 'a customer-oriented, demand-driven, financially sound and self reliant organisation'. The second consultancy aimed 'to guide CMWSSB towards providing a commercially minded customer orientated service that will operate in an efficient and cost-effective manner'. All three studies repeatedly reiterated the vision of an organisation on its way to becoming a commercially viable utility through the application of sound management principles. An official at the organisation's training and resource center explained how he facilitated shifts in the mindset of engineers towards what he called

---

[22] World Bank, Preparation Mission Aide Memoire, 2 March 1986, at 4.
[23] Ibid. at 3.
[24] Ibid. at 3.

'leaner and meaner government'. 'We introduce them to modern management techniques, sort of sugar coating the pill! We get Human Resources experts from private firms – these people have analysed systems thoroughly for working on a profit basis'.[25]

The Twinning Consultancy also sought to improve operational efficiency by bringing to bear on the public utility the experience, disciplines and best practices of a private sector water utility. A senior government official explained how the original idea was to have a public sector to public sector twinning, with the water utilities of Singapore or Malaysia, which were carrying out a number of internal reforms. 'Then the World Bank got into the picture and they always want the private sector. But it's not bad having Vivendi – your stock goes up, having Vivendi as consultants!'[26]

The Bank, apparently, also intervened in shaping the policy-making capabilities of the Metrowater Board in its early years. A World Bank Preparation Mission in 1985 expressed concern about:

> the position of the board of directors of Metroboard (sic) related to transforming Metroboard into a commercially viable public utility. ...It is not uncommon in public utilities ... for it to be prevented from carrying out its objective by well-intentioned but poorly informed board members.[27]

According to the Bank, 'Further work with the Board will need to be carried out to assist them to further identify policy issues...'. The same mission then acknowledged that 'the project is bringing into sharper focus a policy agenda for the Board of Directors...'.[28] The mission recommended a set of consultancy studies that would be presented to the Board members as policy briefs 'for their edification'.

An earlier mission in 1986 also commented on the absence of a Finance Director, and disapproved of the fact that the Chief Engineer of the Project Preparation Unit set up to manage World

---

[25] Interview with Metrowater training official, 6 November 2001.
[26] Interview with senior government official, 22 May 2002.
[27] World Bank, Madras Metropolitan Water Supply and Sanitation Project, Brief on Policy Making, 16 July 1986, at 5–6.
[28] World Bank, Preparation Mission Aide Memoire, March 2 1986.

Bank projects was shown reporting only to an Engineering Director and not to a Financial Director – 'A balance must be struck between engineering objectives and the financial and institutional objectives of the project. Agreement was reached that this balance should be sought in this project and indeed within Metroboard itself…'.[29]

By 2001, it appeared that the Bank's efforts at institutional capacity building had been successful at least in so far as many senior personnel in Metrowater had internalised the disciplines of thinking and acting in a commercial way. As one senior official described it, 'Metrowater has been functioning not like a government department but like a company for a while now!'[30] Internal reformers, for instance, had begun to recognise the potential of organisational re-engineering to address the problems of waste in the system. A former head of the organisation said: 'I introduced costing in every activity – even an ad for a tender had to be costed'.[31] By 2001, all spending was closely scrutinised for its potential returns; all engineers submitting budget requests were routinely asked to show what the benefit of the expenditure would be and the savings to be realised.

As part of the ongoing strengthening of financial and internal auditing functions, the Management Audit wing was set up in 2000. According to the Annual Report of 2000–1:

> The Board made this an integral part of overall financial systems… Various concepts such as transaction audit, …systems audit, management audit, energy audit, stores audit, etc., have been clearly defined and used as tools to enhance productivity.[32]

The deployment of internal audit resulted in a significant streamlining of expenditures and in cost-cutting. The energy audit, for example, resulted in negotiations with the Electricity Board for lower rates on high tension connections for pumping stations, based, ironically enough, on the claim that Metrowater was a non-commercial organisation! Budget control was carried out on a monthly basis, as compared to annually or biannually before.

---

[29] Ibid. at 7.
[30] Interview with senior government official, 22 May 2002.
[31] Interview with ex-Metrowater official, 5 December 2001.
[32] CMWSSB Annual Report 2000–1, at 83.

Accounting practices were changed from location-based manual accounting, to activity-based accounting, with each activity coded as a 'cost center' or 'profit center' and analysed for its profitability. This 'unbundling', the breaking down of integrated functions into units that lent themselves to easier commodification – such as sewerage, revenue collection, water distribution – is a classic strategy of commercialisation. Each of these 'strategic business units' could then be turned into limited companies or concessional contracts or privatised. This strategy was not only advocated by the World Bank, but was a key recommendation of the KPMG consultancy study.

By 2001, as a result of vigorous audits, cost-saving drives and the accelerated trend of contracting out as many 'cost centers' as possible, expenditure on Operations and Maintenance, a category that yielded the most budget flexibility, had declined both in absolute terms (Rs 320 million in 2000–2001 compared to Rs 394 million in 1990–91) and as a proportion of total expenditure (*see* Figure 1) because of the huge increases in debt servicing and depreciation caused by the Second Chennai Project's large capital investments.

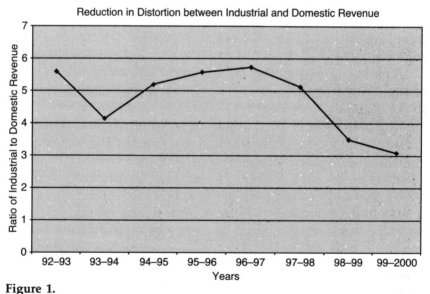

Figure 1.

Source: Chennai Metropolitan Water Supply and Sewerage Board, Annual Report 1999–2000, p. 55.

3.1 Organisational Relations

Inevitably, the transformation of the bureaucracy to a commercial entity called for changes in organisational relations. The organisational chart drawn up on the recommendations of the pre-investment study in 1977 proposed designating the heads of engineering divisions as 'managers'. The proposal failed, as, according to one informant, engineers consistently resisted this change in titling. However, while engineers strove to hold fast to their special identities and status in Metrowater, the winds of change were pushing for a dissolution of this engineering ethos in favour of a stronger role for finance and auditing wings. Against the backdrop of a general freeze in recruitment in the agency, the finance wing hired several new personnel, many from the private sector. Its strength increased from one Controller of Finance (CoF) and one Deputy CoF (DCoF) in 1991, to five DCoFs, two internal auditors, and one CCoF (Chief Controller of Finance) in 2001. A senior finance official corroborated that the finance department had substantially more say in the opening of tenders and the settlement of contracts. More significant for the culture of service delivery, however, was the expanded role of the finance department in the public sphere of the service, i.e., at the interface with clients. Internal auditors regularly made field visits to investigate complaints about water services, a shift highly resented by field engineers.

A former head of the organisation confessed that the centrality of engineering knowledge in the running of the service was being re-examined in recent years.

> To be honest, I feel that this kind of work does not require a great knowledge of engineering. There is this feeling that but for the engineers, the service cannot be run. But when we went into details of the different operations, we found that these were myths. There was quite a lot of resentment when non-engineers were brought into decisions that were earlier the prerogative of the senior engineering staff. I opened up a lot of technical decisions to be reviewed by a mixed team, with financial people and managers also given a say. Many of the senior engineering staff began to feel a bit redundant.
>
> For example, when it comes to the type of pipes – say we have about four options that may be suitable for our purpose.

You have to think about what the goal is – do you want something that will last a hundred years, or an option that will be good for 30 years, and then the next generation can replace it if needed. So, sometimes cost accountants' recommendations were selected over those of engineers, and later the engineers also came to feel that the decision was a right one.[33]

The de-centering of engineering knowledge in the organisation, then, was of a piece with the shortening of investment planning horizons with a view to cost-cutting. Engineers at all levels remained highly critical of the enhanced role of financial personnel in technical decision-making. Field engineers believed that investment on preventive maintenance was compromised, leading to problems in O&M. A senior engineer admitted that major schemes had not worked according to plan because engineering perspectives were not taken seriously. 'Cost-cutting in some cases is fine, but it often has wrong consequences! There have been many instances where projects are unsuccessful because the engineers are not listened to!'[34]

The atmosphere of horizontal distrust across departments appeared to have spilled over to cause tensions in the vertical relationship between senior and field engineers. Junior engineers complained that the buck was invariably passed down the line, and that senior engineers were afraid to speak out before the Managing Director and other administrators, even on technical issues. The climate of constraint and suspicion appeared to have lowered the collective morale of engineers in Metrowater, rendering them reluctant to take responsibility for decisions.

3.2 Citizens to Consumers: Towards Tariff Reform and Elimination of the Commons

The meanings of a 'good service' were increasingly associated in Metrowater with the creation of consumers. As Metrowater's 2000–2001 Annual Report put it, 'A sound tariff policy remains the backbone of any viable financial management system and also for (sic) improving the relationship of 'consumers as the user' (sic)

---

[33] Interview with ex-Metrowater official, 5 December 2001.
[34] Interview with senior Metrowater engineer, 7 March 2002.

and the Board as 'service provider'.³⁵ This section reviews the strategies deployed to produce consumers, particularly to separate the urban poor, who receive water free of charge, from the ambit of Metrowater's services.

Key to the creation of the efficient and virile service was the removal of all subsidies. This was a consistent theme in World Bank Aide Memoires from the start. An initial step towards 'full cost recovery' was a shift in the funding relationship between Metrowater and the Government of Tamil Nadu. World Bank loans were channelled via the State government to Metrowater and were initially received by the latter as part grant and part loan. Bank missions since the early 1980s opposed this pattern and insistently pushed Metrowater towards eliminating the grant component and funding its projects through a combination of loan and internally raised revenues. By 1996, the goal was achieved. Metrowater stopped receiving grants from the government and was financially self-sufficient, with debt service forming almost 25 per cent of its expenditure.

A second critical component of the project of full cost recovery was tariff reform. The pre-investment studies recommended that 'Eventually the total capital and operating costs of the water and sewerage system have to be borne by the consumer through the tariff'.³⁶ This principle was subjected to a stinging critique by a senior executive of an infrastructure financing institution in Chennai, as posing an unfair burden on the current generation of water users: 'Since the benefits are not accruing only to the current users of the system, it is unfair to bill them in the way the [World] Bank and others are doing. It is now fashionable to say that users have to pay. But this is nonsense! It's an orthodoxy, and a nonsense orthodoxy! Theoretically, there is no case in economics – even a first year economics student will tell you that when there are externalities, you cannot price the entire thing on to the consumer'.³⁷

This 'nonsense orthodoxy' of 'moving towards full-cost pricing of water services', however, is one of the five key actions that the World Water Council (a body dominated by private water corporations and international financial institutions including the

---

³⁵ CMWSSB Annual Report 2000–2001, at 85.
³⁶ OMF Interim Report.
³⁷ Interview with senior finance executive, 12 July 2002.

World Bank) identifies as necessary to achieve sustainable access of all people to safe and sufficient water. There is undoubtedly a need to re-examine the highly subsidised provision of water that has hitherto been the norm, especially since such subsidies tend to benefit wealthier people with access to piped water and storage facilities rather than the poor who rely on mobile sources often involving private providers. However, in the vast majority of cases, tariff reform occurs in preparation for, or as a concomitant of, privatisation. The notion of costs also differs radically between private companies and the public sector. Government costs provide protected employment with living wages and benefits to large numbers of public sector staff, while private company costs include the salaries of multinational corporate bosses and shareholders' profits. In 1986, a World Bank mission quoted findings from a consultancy study to suggest that Metrowater would need to raise average tariffs by 400 per cent 'to move towards commercial viability while at the same time maintaining affordability'.[38]

Meanwhile, the term 'equitable' derives a new meaning in World Bank usage, referring to the removal of the cross-subsidy built into Metrowater's tariff structure. The cross-subsidy kept domestic water rates low by charging high rates from industrial consumers. A 1986 Bank mission wrote: '[The tariff reform study should] result in a structure which is administratively cost-effective, equitable among all of the consumers, and efficient from an economic point of view'.[39] As a result of these pressures, cross-subsidy from industry to domestic consumers was substantially reduced.

This specific meaning of equitability is strengthened by some marked silences in the Aide-Memoires. The urban poor, largely absent from the documents, appear suddenly in a 1999 Aide Memoire under a section entitled 'Tariff Discrimination due to Cross Subsidies'. The Bank mission acknowledges that cross subsidisation is 'often used for the purpose of helping the poor have access to the service', but contends that the 'outcome, almost without exception, is that the poor seldom benefit from these subsidies'.[40] They argue that 'many countries are fast abandoning

---

[38] World Bank. Preparation Mission Aide Memoire, 2 March 1986, at 6.
[39] Ibid. 2. Emphasis added.
[40] World Bank, Proposed Third Chennai Metropolitan Water Supply and Sanitation Project, Preparation Mission Aide Memoire (14 June to 1 July 1999), at 7.

this practice as they realise that there are better instruments to subsidise the poor'. None of these instruments are described; the document instead goes on to detail how price distortions can be gradually removed.

The creation of differential categories is critical to projects of commercialisation and market formation. 'Unbundling' the service in Metrowater involved an attempt to create, apart from cost and profit centers, a clear distinction between consumers of the reformed service and the government's protégés, sections of the population that were supplied water free of charge through public fountains. This attempt dates from the pre-investment study, which, while recommending the integration of all aspects of the municipal water service, set up a separation between public fountains and the mainstream (piped) service, with the ultimate goal of removing the former from the responsibility of the Board. The study maintained that:

> the construction and maintenance of public fountains and public conveniences is essentially a civic function and should be discharged by local bodies. The MMWSSB as a commercial body should not be involved except to the extent of supplying water to the public fountains at a charge. Thus these assets should not be taken over by the Board. ... [If the Board continues to supply water to the public fountains, the charges] should be paid for in full by the appropriate authority. Any subsidy required to enable poor people to receive an adequate supply of water should be provided through these bodies and not by the Board.[41]

Thus, service to the poor was to be excluded from the major institutional innovation expected to enhance the quality of the service. The draft bill excluded public fountains, public conveniences and stormwater drains from the ambit of the Board's operations. However, the Government of Tamil Nadu, in reviewing the Bill, reinstated the care of public fountains under the Board.

The issue of service to the slums has remained a contentious theme in World Bank's relationship with Metrowater from the start. A 1986 mission pushed the organisation to re-examine its responsibility for supplying water to the slums. The mission also

---

[41] Ibid., at 7.10 and 14.6.

insisted that 'The principle of cost recovery, even if indirectly recovered from MMC, should be sought from slum dwellers especially those occupying illegal land since they pay no taxes nor water charges'.[42]

In 1989, the World Bank spelled out its opposition to the utility being directly involved in government schemes to provide water to the poor through unlevied public standpipes, especially as such involvement kept Metrowater reliant on grants from the government.

> The mission discussed the use of grants from the Government with MMWSSB, which pointed out that grants were used in part to cover the costs to MMWSSB of providing ... subsidised water ...through standpipes. The mission explained that the Bank's policy is not to reject the use of subsidies per se, but to require that such subsidies be transparent and explicit. Thus it would be better for MMWSSB to charge the public agencies a fair price (for example the actual cost) for standpipe supply, and for the GoTN [Government of Tamil Nadu] to pay for this service outright.... This would assist the authorities in understanding and recognising the cost of subsidising.[43]

Thus, while the Bank was 'not opposed to subsidies per se', it objected to the form in which they were given, a form that integrated all citizens into the domain of state service. Using the language of transparency and the discourse of 'recognising the true costs', the Bank sought to redefine the state's accountability for water provision as a commercial accountability to consumers, and to separate it from the government's accountability to the poor. This move to bring subsidies out into the open is also part of the World Bank's larger goal of freeing the service of 'politics'.

By the late 1990s, Metrowater had adopted a policy of gradually eliminating public standpipes. The policy was never publicly announced, and circumstances made it difficult to implement. A senior engineer told me:

---

[42] World Bank, Preparation Mission Aide Memoire (2 March 1986), p. 15.
[43] World Bank, Madras Metropolitan Water Supply and Sanitation Project, Review Mission, 13–21 April 1989.

There has been a decision to not provide public standpipes in new areas that are being served ... This was a decision taken internally by Metrowater in 1996 or so, because of the problems in maintaining these standpipes ... And also because the Board has turned towards revenue generation as the focus.[44]

A high-ranking official of the State government outlined the policy of eliminating public fountains as part of a more ambitious vision of promoting private water connections in slums.

I feel we have not marketed this idea enough. ...The political economy of water in slums is amazing. They already pay for water. Water is by no means coming free to them now! ... All are willing to pay for water, in some way or other. But Metrowater has not effectively marketed the concept of a private service. ... I really believe, if the quantities of water are sufficient, and it can be, then there is no reason not to give everybody a connection.[45]

In the reformist visions of the state, then, faith in the potential for endless supply augmentation is combined with a discourse of 'willingness to pay' and an assessment of poor people's capacity to pay, to portray universal private connections as a pro-poor solution. Yet, each of the assumptions underlying this vision is specious – supply augmentation efforts are fast reaching the limits of sustainability, the vast majority of the urban poor in slums do not live in conditions that encourage them to invest in private connections, and the shifting constitution of the urban poor predicts a continued reliance on public facilities. Meanwhile, these visions of reform feed into a discourse that consigns public taps, the commons[46], to the margins of order and citizenship.

In Metrowater, pressures to achieve full cost recovery and tariff reform have translated into punitive effects for clients as well as

---

[44] Interview with senior Metrowater official, 14 May 2002.
[45] Interview with senior government official, 22 May 2002.
[46] In municipal water systems, public fountains are arguably a type of 'commons', despite the fact that they are not naturally occurring resources. They constitute a public-access option for urban dwellers who lack private sources. In Chennai, pubiic fountains are heavily relied on even by people with private connections during drought periods when domestic pipes run dry.

frontline service-providers.[47] Field officials face sanctions if they fail to achieve ambitious revenue-collection targets; annual performance awards are based on success in meeting these targets; and clients are denied service until they meet all arrears, even if they have not received water for several months.

4. Privatisation

Although outright privatisation in the form of long-term concession contracts or disinvestment are not yet publicly on the cards in Chennai, the process of contracting out components of the service for maintenance and service on short- to medium-term leases has become standard practice. All new installations, from sewage pumping stations to water treatment plants are now constructed on BOOT or DBOOT arrangements. Reforms in contracting systems in Metrowater since 1997 have moved the organisation towards turnkey type contracts which favour single large contractors over the many small firms that the agency traditionally partnered with. Middle-level engineers in Metrowater confessed that holding contractors accountable on the ground was often harder with large corporations like L&T than with small local firms.

Privatising O&M depots, the nodes of direct services to the public, is regarded as particularly challenging, as these depots handle far more complex and sensitive functions (including micro-level allocation and distribution of water, policing irregularities in the grid, and public relations) than technical facilities such as treatment plants or pumping stations. In 2000, Metrowater initiated a pilot project of contracting an O&M depot to a small local private firm to manage. The cost savings to the agency were significant, but the depot was back in the hands of the Board by 2007, as companies failed to evince interest in bidding for this contract.

But privatisation of water in Chennai has been long underway in one way or another. Apart from the high and increasing reliance on private groundwater sources by individual households, more than a fourth of the city's households purchase packaged water

---

[47] For a fuller account of this dynamic, *see* Karen Coelho, 'Unstating 'the Public': An Ethnography of Reform in an Urban Water Utility in South India', *in* David Mosse and David Lewis eds., *The Aid Effect: Giving and Governing in International Development* 171 (London: Pluto Press, 2005).

for drinking, and about a fifth of the city's water supply comes from private suppliers that form a powerful lobby. Most commentators ascribe this situation to Metrowater's failure to manage water supply for the city. Many also claim that Metrowater's failure to enforce the law prohibiting commercial exploitation of groundwater is due to its own inability to supply adequate water to the city.

The reform orthodoxy of full-cost recovery is linked to the agenda of privatising water in ways that are not obvious. While cost-recovery is widely understood as the recuperation of the financial costs of treating and supplying water, the more radical long-term goal of reformers is to reach the full 'economic costs' of water. In this system, water is valued according to its opportunity costs, which in turn reflects its highest value across the spectrum of water use. In other words, the cost of drinking water to the average consumer would reflect the price that industrialists would be willing to pay for it. Economic pricing is promoted as a means of reducing water consumption. The vision of global water policy, as articulated by the World Water Council, is of the development of markets of transferable water rights and the reallocation of the limited resource to 'high value users of water' through 'treating water as a tradeable commodity'. A World Bank Strategy paper foresees that '... in case after case reformed utilities... (will) push for market-based rules for facilitating the voluntary temporary or permanent transfer of water rights from low-value to high-value users'.[48]

This brings us back to the links between municipal water reforms and the over-exploitation of the AK basin aquifers, outlined at the start of this paper. The practice of sucking resources out of

---

[48] Water Resources Sector Strategy: Strategic Directions for World Bank Engagement, *in* World Bank Report 28114, 2004. The increasing stress on ground water is among the foremost concerns of recent World Bank reports on India. The Bank's 2005 report pushes for demand management rather than the demand-responsiveness earlier advocated. However, this is proposed to be achieved through market instruments (such as pricing) and incentives for voluntary transfers of individual entitlements to achieve economically efficient allocations. '(A)n approach which begins with acknowledgement of and respect for the private interests of individual farmers will be far more successful than approaches which resort to command and control, or ones that are based on a communitarian ideal'. *See* World Bank, note 14 above at xiii.

rural hinterlands to cater to the ever-expanding urban appetite is now a globally recommended policy breakthrough, facilitated by the institution of 'modern water rights', which create markets in groundwater and permit individual farmers to profit from selling water commercially. This strategy fits into a larger 'development' vision of re-allocating water from low-end uses (like small-scale agriculture) to high-end uses (like urban growth). In 2002, Metrowater hired consultants Scott Wilson Piesold to study the introduction of a system of tradeable water rights in the AK Basin, which would allow the organisation to continue extraction of groundwater from these areas under a legal, ostensibly more controlled regime. The consultants' report met with mass opposition at the public meeting called to present the draft. The revised report, submitted to the government in 2006, has not yet seen the light.

The draft was one attempt to address what was an effective legal vacuum in the governance of municipal water services. The metropolitan groundwater legislations,[49] and the recent state-wide law, the Tamil Nadu Ground water (Development and Management) Act, 2003[50] are widely regarded as unworkable given the heavy administrative and policing infrastructure required to adequately implement them. More importantly, they fail to establish any clear principles of entitlement for different users, to identify and address possible conflicts and complementarities among different uses and to specify different kinds of rights. The Scott-Wilson Piesold consultancy went furthest in laying out draft legislation which specified a system of volumetric entitlement for individual users in the AK Basin.[51]

---

[49] *See* note 3 above.

[50] This Act envisages the establishment of a State Ground water Authority and the declaration of critical areas in which no drilling or deepening of wells would be permitted.

[51] Copies of the report, obtained informally, reveal that the consultants met their World Bank-drafted terms of reference in Appendix 9, by laying out a system of transferable water rights. In the light of public opposition to the tradeability clause, however, they also outline, in Appendix 9A, a system of volumetric entitlements with very limited scope for transfers. Farmers are permitted to sell to Metrowater in cases of emergencies. There are stringent conditions imposed on industries wishing to purchase groundwater, including demonstration of maximum efforts to recycle wastewater. *See* Piesold, note 3 above.

In terms of access to clean water for drinking and household use, no legal framework exists apart from the very general mandate given to the Board by the CMWSSB Act in 1978, to operate and maintain water and sanitation services in the Chennai Metropolitan Area 'to the best advantage of the inhabitants of that area'. The Act directs the Board to be 'guided by such instructions on questions of policy involving public interest as may be given to it by the Government'.[52] Sangameswaran notes that while the right to water is recognised by the Supreme Court (as part of the right to life), the implications of this for institutional choices and the role of the state remain debatable.[53] The trend of recent Supreme Court judgements to treat the urban poor as encroachers or illegal occupants of the city also contributes to denying them the rights of citizens to water, rendering them subject to executive decisions to provide – or, under pressure of reform, to roll back – public water fountains or tanks.

5. Stickiness along the Slippery Slope

In the early 2000s, the trajectory of reforms seemed to be clear to some senior officials in Metrowater, one of whom predicted:

> Slowly [the agency] will be privatised. Mainly in the form of small contracts. They are not yet talking about it, this is just my guess. Already so much has been privatised... Our lower levels [of staff] are not aware, have not understood the transformations that are coming within Metrowater.[54]

That he was right about the slow road to privatisation was revealed by a senior government official in the water sector, who claimed that the process of setting up a regulatory authority was already secretly underway.

---

[52] India, Chennai Metropolitan Water Supply and Sewerage Act 1978, Section 5.1 (d) and 5.2 (c).

[53] Priya Sangameswaran, Review of Right to Water: Human Rights, State Legislation, and Civil Society Initiatives in India (Bangalore: Centre for Interdisciplinary Studies in Environment and Development, Technical Report, 2007).

[54] Interview with senior Metrowater official, 14 August 2002.

[But] it is happening quietly, because once you start talking about a regulatory authority, people know there is privatisation in the offing, and all the shouting starts. There will be a huge debate... So we are working at this, setting up the regulatory authority, simultaneously preparing ourselves by privatising components that will benefit from private sector efficiencies.[55]

Several senior officials in the water sector had serious misgivings about privatisation, based on their experiences with private contractors in the past. Many of them believed that water provision was a public service that could not be turned into a free market operation. Yet, the global orthodoxy of privatisation as the route to better quality of service, and the pragmatics of the bottom line, overrode their misgivings about its potential threats to state sovereignty and accountability. The following comments from a senior government official reveal the continuity between the commercialising reforms in Metrowater and the larger agenda of privatisation:

Ultimately the public don't care who supplies them as long as they get water... This system anyway is headed for extinction – these huge government-run utilities will die like dinosaurs, they are on their way out. ... The bottom line is that we need to generate the resources to take the service to more people...The public sector is burdened with a long-established way of doing things, with a culture of all kinds of interference and claims... We nationalised everything a few years ago, and now we are disinvesting. There was good reason then, and there is good reason now. Metrowater will die like a dinosaur! Government organisations cannot be lean and mean, so they will die. We need to evolve new ways of responding to the needs. That's why I say we need to hold hands and create a private sector, evolve a new creature, part public, part private, something that combines the strengths of the two, that will meet our needs.[56]

Reforms, thus, appear to constitute, even for state officials, an inexorable and pre-determined evolutionary trajectory within which some limited creative options are possible. Yet, reforms in

---

[55] Interview with senior government official, 22 May 2002.
[56] Interview with senior government official, 22 May 2002.

Metrowater have not always moved as smoothly as the agency claims. Many observers who were involved with the agency in some form or other over the years commented that the basic goal of autonomy from political decisions had never been achieved – that key decisions, like those on hires and tariffs remained under the control of the State government. In 2006 and 2007, the agency reported its first deficits (of Rs 12 crore and Rs 35 crore respectively) in close to 15 years, partly due to reduced offtake from industries which had developed their own private sources (including desalination) and partly due to tariffs remaining unrevised since 1998. The World Bank was no longer in the picture, as the agency no longer received any loans from it. By 2008, measures such as providing subsidised water connections to low-income households were still under discussion, metering domestic supply was still being piloted, demand management was barely on the agenda and expensive supply-side measures such as seawater desalination were being prioritised. Meanwhile, increasing numbers of households, commercial establishments, industries and institutions sport the ubiquitous bubble-top water dispenser, sourced by private water suppliers and in 2007 the total consumption of private packaged water in Chennai touched 4 million litres a day.[57]

6. Conclusion: Alternative Models of Reform?

This chapter explored the translation by a public utility of a World Bank-inspired reform package, with all the anomalies and distortions that translation inevitably brings. This translation, however, exposes the pitfalls that a commercially biased reform package can create. In Metrowater, commendable achievements in financial management and cost-savings are offset by an institutional ethos dominated by the finance function to the detriment of long-term preventative maintenance. The pressures of demand threaten the availability of water for low-value uses such as subsistence needs of poor households and producers, and the long-term sustainability of the resource. Frontline engineers, seen as the corrupt and inefficient node in the service, and subject to the

---

[57] K. Lakshmi and Meera Srinivasan, 'Packaged Water Sells like Hotcakes', *The Hindu*, 17 April 2007. *Source:* http://www.thehindu.com/2007/04/17/stories/2007041715420500.htm.

stringent disciplines of audit and performance appraisal, tend to displace their victimisation onto their most vulnerable clients, the urban poor.

The long-standing need for reform in water management, on the other hand, has also been addressed through alternative models, notably the worldwide movements for 'reclaiming public water'. Innovative public delivery models, public-public partnerships, cooperatives and community-managed utilities have, in numerous cases, demonstrated their potential for improving water delivery, not least to the poorest, through strategies that reject privatisation and economic pricing of water.[58] These efforts, based on the paradigm of water as a human right, emphasise distributional equity and resource sustainability alongside the imperatives of financial viability and improved services. The reform journey undertaken by the Tamil Nadu Water Supply and Drainage Board in Chennai's own backyard is one such instance of 'democratisation of water management'[59] in which water engineers have become change agents, working in partnership with citizens to ensure equitable supply and coverage to marginalised sections, conserve natural resources and create sustainable water management as well as significant cost-savings and financial health for the organisation.

---

[58] About twenty such cases are described in Belén Balanyá *et al.* eds, *Reclaiming Public Water – Achievements, Struggles and Visions from Around the World* (Amsterdam: Transnational Institute and Corporate Europe Observatory, Second Edition 2005).

[59] V. Suresh and Pradip Prabhu, Democratisation of Water Management as a Way to Reclaiming Public Water: The Tamil Nadu Experience (Amsterdam: Transnational Institute, 2007). *Source:* http://www.tni.org/books/watertamilnadu suresh.pdf?

UNIT II

Ongoing Irrigation and Ground Water Reforms in India

5

Canal Irrigation, Water User Associations and Law in India –Emerging Trends in Rights-Based Perspective

VIDEH UPADHYAY

The present chapter seeks to examine the state of the formal Water User Associations that has been created in India through a series of state legislations in recent years from an essentially rights-based perspective. Given the bias in the legal system that recognises statutory rights to the exclusion of almost any other form of rights, the passing of these laws presents an opportunity to these Associations to see what effective rights have come their way through these laws. This has provided the immediate provocation for this chapter. The discussion below is premised in a larger context where first some conceptual links between decentralised water governance and rights are explored in the Indian context. This is followed by a brief comment on how the rights are a function of law making and whether the water users and the farmer identifies and owns the rights *created for them and vested in them*. The chapter then proceeds to examine the formal rights available to the Water User Associations in different states today and more particularly in Andhra Pradesh and Chhatisgarh while focusing on both their 'internal and external rights'. It also suggests whether and how far they are precipitating a 'group rights regime' while tracing its inevitable linkage to the critical question of water entitlements and the state of the irrigation systems.

I. Decentralised Water Governance and Rights: Exploring Some Conceptual Links

Harold Laski in his classic *A Grammar of Politics* points out that 'Every State is known by the Right it maintains' and clarifies further that 'If it (the State) is legitimately to exercise its functions, it must be upon the basis that it moves consistently to the realisation of rights.'[1] He adds that there are three conditions under which the authority of the State is to be exercised for realising the rights. The first and foremost of this is – 'The State must be a decentralised State'. The virtues of decentralisation to him are that it prevents the application of uniform solutions to different things and also by multiplying the centers of administrative act it ensures a fuller participation in the responsible business of government'.[2] The central point that Laski made that *decentralised State is important for realisation of rights* remains true today. Indeed in the Indian context the reverse can also be argued – that *Rights are important for realisation of a decentralised State*.[3] As we will see later the Water User Associations –formal farmer's bodies empowered to operate and manage irrigation systems – also need a perspective based on rights and their realisation in India today. The points made by Laski linking rights to a decentralised state is of great value as the coming of the state laws creating Water User Association for management and operation of the network of canals across India are at an interesting intersection of decentralisation and rights in India. The nature of powers and rights vested with the Water User Associations by various states in India shall make these aspects further clear and these are discussed in detail in the sections below.

The other critical aspect of the rights, especially human rights regime in India, is that they tend to be grounded mostly on fundamental rights. However, fundamental rights under the

---

[1] Harold J. Laski, *A Grammar of Politics* 88 (London: George Allen & Unwin, Fifth Edition, 1967).

[2] Ibid., 131–134.

[3] It has been argued that democracy itself is on the way of becoming a global entitlement and that this 'democratic entitlement' is gradually being transformed from a moral prescription to international legal obligation.' S*ee* Thomas Franck, 'The Emerging Right of Democratic Governance', 86 *Am. J. Int'l L.* 46, 53 (1992).

Constitution are rights to individuals and not to a group. In the context of decentralised governance and rights being required to be vested with Group entities, it is only fair to raise the question – Can an essentially individual rights regime be gainfully utilised to overcome the deprived status of the Groups? It is worth bearing in mind that one must not overestimate 'the significance of the language and the law that emphasises the vindication of individual rights'.[4] In the context of the fact that the recent initiatives of the government in the 'water sector' – whether under Rural Water Supply Scheme, Participatory Irrigation Management or Watershed Development – have all sought to vest powers to village groups and associations, this becomes an important point. Thus it is imperative that the water rights regime needs to evolve conditions under which a group and not an individual can become a right holder so that an entity like a legally constituted Water User Association can exercise such a right to its advantage. In strict rights parlance, these village level entities need to develop both *Internal Rights* that the members of the village group can exercise against each other as well as the *External Rights* that the group can enforce against anyone and everyone outside the group. It is clear that the Group Rights' regime needs to grow – and under a fast ageing process – because it could be central to the resolution of conflicts over water resources in future.

The other important conceptual point to note is that some of the specific rights discussed under the State Rules in the sections below tend to create a strange legal fiction – a system where water 'entitlements' exist without corresponding obligations to ensure one receives them! This phenomenon points to a larger concern – one can see that the word 'entitlements' is increasingly being used, mostly by economists who are part of a select and selective community of 'policy experts', in a manner that violates the basic meaning of the word. An entitlement is something that one 'has a title to', and more importantly, has this title as a matter of right, i.e., a *right to demand and receive*. The new legislations in the water and irrigation sectors never, as a rule, create any enforceable right

---

[4] *See* Nathan Glazer, 'Individual Rights Against Group Rights', *in* G. Mahajan ed., *Democracy, Difference and Social Justice* 416 (Oxford: Oxford University Press, 1998).

to water for farmers or other water users.[5] These aspects shall be further clear when specific discussion on some of the rights and entitlements of the Water User Associations and the water user are carried out in the subsequent sections of the chapter.

It is also useful to note at this stage that unlike most user association/beneficiary groups in other sectors the Water User Associations giving effect to 'Participatory Irrigation Management' are now founded under specially enacted State laws for them.[6] This makes it possible to speak in terms of definite rights and duties of the water user groups in a formal sense. Notwithstanding this, in the entire body of legal support to the water user groups in India, the mention of rights range from very little to non-existent. One feels that the concerns on water rights generally, and rights of the water user and the Water User Associations can be mainstreamed in the water discourse only if issues in water resources are seen as more than mere managerial concerns. So long as we keep on using phrases and principles like 'Water Resource Management' or 'Participatory Irrigation Management', we seem to be restricting the issues around water and irrigation to simply management issues. This limitation gets more obvious when we begin appreciating some larger issues on the question of preservation and sustainable use of water resources that need urgent redress. These include:

- *The need for right based approaches to access, use and conservation of water resources:* This is a critical corrective, badly needed in the entire discourse around the use and sustainability of water resources. In the absence of this corrective what the village groups at the local level are likely to have by way of water resources projects and participatory irrigation management, etc., is mere transfer of 'burden' of managing resources and responsibility for implementing decisions taken elsewhere.

---

[5] One is thus forced to wonder what kind of 'entitlements' are these, and from which dictionary does the word derive. For more comments on these aspects, *see* V. Upadhyay, Confusing Water Rights with Quotas, *India Together* (27 October 2005). *Source:* http://indiatogether.org/2005/oct/vup-rights.htm.

[6] What are Water User Associations and what constitutes Participatory Irrigation Management are explained in the subsequent section.

- *The need to examine who owns and who controls the water resources:* It is dangerous to dismiss this as an abstract enquiry as there is now a growing feeling that without a sense of ownership villagers, including farmers, will not participate in the maintenance of the water structures. This sense of ownership should not be an illusion but grounded on people's right to water and their ownership over local water harvesting structures.
- *The need for linking water rights with larger governance issues at the local level:* Rights are located in a system and the efficacy of the rights is often a direct function of the state of the system. Thus, well defined irrigation rights can be of little value in a dilapidated canal network. At another level the way water resources, policies and laws are conceived in the country can have direct implications for amicable centre-state relations. For example, there are schemes like Swajaldhara which, while being a flagship drinking water supply scheme of the centre and one that involves *panchayat* institutions, have little role whatsoever for the states. This is the case even when 'Water Supplies' is a State Subject under the Constitution of India. The reasons behind the schemes and management initiatives that have floundered in the past –whether due to systemic deficiencies or lack of motivation/accountability of the governments – and their implications for rights also needs to be explicated and discourse on 'Water Resource Governance' should provide space for it.
- *The need for mainstreaming natural resource issues with overall development concerns at the local level:* It is important to recognise that the way water resources are used, sustained and controlled locally can have direct implications for livelihoods, poverty alleviation and employment concerns at the local level. The emphasis on water augmenting projects /structures under the National Employment Guarantee Act, 2005 is a straightforward pointer. It is, therefore, imperative that issues around natural resources are not addressed compartmentally but holistically as there are linkages as much between the forest, land and water resources as with the other socio economic central concerns

at the local level. The legal and policy regime needs to mature to recognise this reality.

II

The discussion above on decentralised water governance and rights and efforts at exploring some conceptual links in that larger context provides a useful backdrop for appreciating some specific concerns on rights of the Water User Associations and the programme of Participatory Irrigation Management in India. This section and the subsequent one aims at exploring these concerns in some detail.

Participatory Irrigation Management and Formal Water User Associations in India: An Introduction

One economic historian described harnessing of India's rivers for irrigation purposes through the networks of canals as one of the 'greatest monuments to the British rule'[7]. Another historian gives a reason why canals could be seen as one of the 'greatest monuments' given that 'one acre in six was irrigated from government schemes by the late 1930s'[8]. However, it is perhaps Elizabeth Whitcombe who puts the above point in modern irrigation history of India in its right perspective, as below:

> These irrigation systems, the backbone of modern Indian agriculture, had in their time been major innovations. Through prolonged efforts to reconcile the essentially conflicting interests of equity and mechanical efficiency – to serve both government and community in a manner at once philanthropic and unabashedly commercial –they had become an institution, hedged

---

[7] W.J. MacPherson, 'Economic Development in India under the British Crown, 1858–1947', *in* A.J. Youngson ed., *Economic Development in the Long Run* 144–145 (London: G. Allen & Unwin, 1972), as quoted in Ian Stone, *Canal Irrigation in British India: Perspectives on Technological Change in a Peasant Economy* (Cambridge: Cambridge University Press, 1984).

[8] *See* Stone, note 7 above.

about with prescription and restriction, more conservative, perhaps, than the peasants...[9]

It is this conservative nature of the institution of irrigation systems that explains why there was very little done in terms of reforming the ways these irrigation systems were managed even up to four decades after Indian independence. However, over the last two decades there have been significant attempts made to involve the farmers – the beneficiaries of the irrigation canals – in operation and management of the irrigation systems in India, as in many parts of the developing world. It is instructive to note how this thinking took root in a largely conservative history of management of irrigation systems.

Beginning in the 1980s, there have been large-scale programs to turn over irrigation management from Government Agencies to organised Water User Associations (essentially, farmer associations) in a number of countries such as Philippines, Indonesia, Senegal, Madagascar, Columbia and Mexico[10]. This trend has been seen as the convergence of a number of policy trends including decentralisation, privatisation, participation and democratisation[11]. A result of this has been 'rolling back of the boundaries of the state within the irrigation sector. Participatory irrigation management refers to the programs that seek to increase farmers' direct involvement in system management, either as a compliment or as a substitute for the state role.

The acceptance of Participatory Irrigation Management (PIM) was powered by the dismal state of irrigation systems itself. Non-irrigated fields because of undependable water flows, indiscriminate use of water by head-enders depriving the tail-enders, inequitable distribution and resulting conflicts created a situation where farmers' participation was beginning to be seen as an answer. The Water User Associations (WUAs) was seen as a lasting response to

---

[9] E. Whitcombe, 'Irrigation', *in* D. Kumar and M. Desai eds., *The Cambridge Economic History of India* 737 (New Delhi: Orient Longman, Volume II, 2004).

[10] Vermillion Douglas ed., The Privatisation and Self Management of Irrigation (Final Report submitted to the GTZ Germany by the International Irrigation Management Institute, Colombo, Sri Lanka, 1996).

[11] *See* L.K. Joshi and Rakesh Hooja, *Participatory Irrigation Management: Paradigms for 21st Century* (New Delhi: Rawat Publications, 2000).

such systemic inadequacies. It was thought that where the state had failed the farmers will not, and that operation and management of irrigation system by the farmers themselves can change things around. The result was that state after state in India, much like other parts of the world, came up with policies, resolutions and then laws supporting PIM.

The Basic Legal Regime: Three-Tier Water User Bodies under Formal Laws

Before closely examining the rights of the WUAs and the individual water users it makes sense to first understand the overall structure of Water User Association as created by the various state laws in India. As pointed out above, several states in the recent past have come up with major policy and legal initiatives that have transferred some responsibilities of Irrigation Management from government agencies to the Water User Associations (WUAs).[12] While some of these WUAs have been founded under government resolutions, most states today have done so through enabling laws. In states like Andhra Pradesh, Rajasthan, Orissa, Madhya Pradesh, Tamilnadu, Maharashtra and Chattisgarh the law enabling farmers' participation in irrigation management has come by the enactment of specific 'Farmers' Participation in Management of Irrigation Systems' laws.[13] Surveys of these laws and rules made under them were taken for writing the present piece. Specifically, these laws included – The Andhra Pradesh Farmers' Management of Irrigation Systems Act, 1997, Madhya Pradesh Sinchai Prabandhan Me Krishkon Ki Bhagidari Adhinyam, 1999; The Tamil Nadu Farmers' Management of Irrigation System Act, 2000; Kerala Irrigation and Water Conservation Act, 2003; Orissa Pani Panchayat Act, 2002; Maharashtra Management of Irrigation System by Farmers Act 2005 and The Chhattisgarh Sinchai Prabandhan Me Krishkon Ki Bhagidari Adhinyam, 2006.

---

[12] The formation of these associations is now generally seen as the most effective strategy for ensuring farmer/users' participation in management of water for irrigated agriculture.

[13] Some states like Goa have provided for farmers' association by amending their Command Area Development Acts. Other states have adopted the principle of Participatory Irrigation Management through government resolutions and orders.

Typically all these laws empower the 'Project Authority' to delineate every command area under each of the irrigation systems 'on a hydraulic basis which may be administratively viable' and declare it as a Water User area. Every Water User area is to be divided into territorial constituencies. The laws then provide for establishing a Water User Association (WUA) for every Water User area. Every WUA is to consist of all water users who are landowners in such Water User area as members.[14] All the members constitute the general body of the WUA. There has to be a managing committee for every WUA and the project authority is responsible for election of President and members of the Managing Committee of the WUA by direct election from among its members by the method of secret ballots. Further, the project authority may also delineate every command area comprising two or more water user area as a Distributory Area. All the presidents of WUAs constitute the general body of the Distributory Committee. The general body of the distributory committee also elects the president and the members of the managing committee of the Distributory Committee. Likewise, the government may delineate any command area to be a project area while requiring it to form a project committee for every project area. All the presidents of the Distributory Committee constitute the general body of the Project Committee.

The WUAs at the primary level, the distributory committee at the secondary level and project committee at the project level is together referred to as *Farmers Organisation* under the laws. In some cases there is also a liberty for the managing committee of a Farmers' Organisation to constitute subcommittees to carry out their functions. Finally, Water User Associations have typical functions like (a) to prepare and implement a warabandi schedule for each irrigation season, (b) to prepare a plan for the maintenance, extension, improvement, renovation and modernisation of irrigation systems, (c) to regulate the use of water among the various outlets under its area of operation, (d) to maintain a register of landowners as published by the revenue department, (e) to monitor flow of water for irrigation, (f) to resolve the disputes, if any, between its members and water users in its area of operation.

---

[14] Though there are useful variations from these positions in some states and most notably by the Chhattisgarh Sinchai Prabandhan me Krishkon ki Bhagidari Adhinyam, Act No. 20 of 2006. *Source:* http://www.ielrc.org/content/e0605.pdf.

III

Before examining some specific 'rights' vested in the WUAs under state laws it is useful to note how and why the rights regime created by these laws are not owned by the water user/farmers themselves. It is important to appreciate this aspect at the outset because it has direct implications for pointing out the limitations of rights based approach to WUAs in India.

Regime without Farmers' Ownership of Rights

Over ten years on since the first State law on 'Farmers Participation in Management of Irrigation Systems' came into being, the euphoria and romance associated with WUAs have given way to hard realities. Ambitions have taken a beating and expectations from WUAs have been scaled down. A post mortem of ten years with PIM can explain why this happened – a post mortem that is due and is yet to be carried out. This paper is not in the nature of a post mortem – it only focuses on one aspect of WUAs that holds the key to their sustainability, their rights.

From an essentially legal standpoint two points from the legislative history of Participatory Irrigation Management in India stands out. One, the legal and management regimes for farmers were never owned by them. The laws were made for them not by or even through them and this despite the fact that they are at the centre of giving the law its' operative effect. Secondly, and more relevant to the present paper, the legislations did not establish clear water rights.

While the focus of the paper is on rights and we shall return to it, a few words on the first aspect can put the points to follow in better perspective. Why is the legal regime for Participatory Irrigation Management not owned by the farmers? The question becomes more troublesome when we know that India has had a long history of farmer-managed irrigation systems with a number of examples from the *Kuhls* of Himachal to the tanks of South India.[15] However, the traditional community managed systems

---

[15] *See* A. Agarwal and S. Narain eds., Dying Wisdom: Rise, Fall and Potential of India's Traditional Water Harvesting System (New Delhi: Centre for Science and Environment, Fourth Citizens Report, 1997).

has not been the motivation for shaping new policies involving local people for managing irrigation. In fact the laws creating Water Users Association are all part of a dominant trend of policy and law making as part of donor-driven projects across the country.[16] The adoption of this mode of law making has meant that a true demand-oriented and participatory mode for establishment of the WUAs has simply not been possible. The fact that this can be done has been shown by other countries where intensive preparatory steps are taken before enactment of a legislation beginning with the identification of potential participants as well as the area of operation of the WUA. For example, the Romanian legislation calls for the establishment of an 'initiation committee', composed of several potential members of the WUA. The committee must call a preliminary meeting to which all potential members are invited. At that meeting, decisions are taken on the proposed delimitation of the territory of the WUA, on the individuals to be responsible for drafting the WUA governing document and for taking the necessary steps for the establishment of the WUO.[17] In other places the decision to establish a Water User Association is itself through petitions supported by farmers and landowners themselves.[18]

The fact that these processes were not thought about and never adopted in any of the State laws in India has also meant that the farmers know little abut the law and rights created in their favour through these laws. This last point is especially important for the purposes of the present chapter as it points to limits of rights-based approach in developing WUAs across the country.

Specific Rights of WUAs and the Water User: The Position in State Laws

The structure of the WUAs in various state laws has been discussed above. Amongst these there is no single legislation which specifically

---

[16] The implications of this nature of law making is in itself a critical area of enquiry though outside the purview of the present paper.
[17] A similar procedure is foreseen by the Bulgarian legislation.
[18] Take for example California where the process of establishing an irrigation district is initiated by petition. Such a petition must be supported by a majority of land owners or at least 500 land owners who hold title to not less than 20 per cent of the value of the land to be included within the proposed district.

talks about the rights of the WUAs or of the individual water user. However, there are two Rules made under two different laws which explicitly mention the rights and responsibilities of the WUAs. These two Rules are – The Andhra Pradesh Farmers' Management of Irrigation Systems Rules, 2003 and the recently notified Chhattisgarh Sinchai Prabandhan Me Krishkon Ki Bhagidari Niyam, 2006. These Rules explicitly talk about the right and responsibilities of the WUAs and the right and responsibilities of each of the water users in these Associations. This gives the impression that under these Rules there is an equal emphasis on both the individual and the group's rights. This aspect will be closely seen in the following paragraphs. Besides, conceptually the rights with the WUAs can also be seen as internal and external rights of the WUAs. As explained in the first section of the chapter above, apart from developing an understanding on the external water rights of the group, which it can use to its advantage against every one outside the group, there is a need for better appreciation for internal water rights laying down the right of the group member's vis-à-vis each other. An elaboration of the rights specifically vested by the two Rules in AP and Chhatisgarh within the above typologies is attempted below.

Specific Rights with the Water User Association

The First Set of Rights with WUAs:

The Andhra Pradesh Farmers' Management of Irrigation Systems Rules, 2003 and the Chhattisgarh Sinchai Prabandhan Me Krishkon Ki Bhagidari Niyam, 2006 both make clear that the Water Users' Association has (a) Right to obtain information in time about water availability, opening/closing of main canal, periods of supply and quantity of supply, closure of canals, etc. (b) Right to receive water in bulk from the irrigation department for distribution among the water users on agreed terms of equity and social justice; and also (c) Right to receive water according to an approved time schedule.

Clearly all of the above rights are well meaning rights with the WUA and are critical for its survival simply because if there is no water there would be no point having a WUA. Both the Rules, however, do not make clear that if the right to receive water in bulk from the irrigation department is not honoured what remedies might lay with the WUA. In other words, whilst there is a generally worded right, there is no accountability of the department that has been established through this provision. Besides, merely saying that these rights exist will not be enough if the irrigations systems are not properly rehabilitated to be in such a condition where minimum water flow could be maintained. Many of the Water User Associations today are paper entities because this minimum condition necessary for their existence is just not present. The short point here is that without assessing the water resource and ascertaining its availability, assigning water rights is the last thing that will improve irrigation management in the country.

The Second Set of Rights with WUAs:

The above mentioned rights are followed by another set of rights under the two Rules. These are mentioned as – (a) Right to allocate water to non-members on agreed terms and conditions (b) Right to levy separate fees for maintenance of the system (c) Right to levy any other fee or service charges, to meet management costs and any other expenses (d) Right to obtain the latest information about new crop varieties, and their pattern, package of practices, weed control, etc., for agriculture extension service and purchase inputs such as seeds, fertilisers and pesticides for use of its members (e) Right to have full freedom to grow any crop other than those expressly prohibited by a law and adjust crop areas within the total water allocated without causing injury to neighbouring lands.

These above set of Rights relate to some 'second generation' concerns and acquires meaning only if the basic requirement – that of availability of water is met. In light of the fact that in most of the WUA there is a scarcity of water, and especially so with the tail-end members, one can't visualise a scenario where the water is allocated by the WUA to non-members. Likewise, both the willingness and the ability to charge additional fees/water charges are questionable and this points to a larger problem. Even the existing water charges, which are not paid in many circumstances, are far less than the expenditure needed for proper operation and maintenance of the system. All across the country the irrigation fees are a small fraction of the operation and maintenance costs of the systems and an even smaller fraction of the actual costs of private lift irrigation with diesel pumps. The highly subsidised irrigation fee structure has helped establish a low-level equilibrium. Farmers are unwilling to demand improved maintenance and service from the irrigation department lest it might result in higher irrigation fees. In turn, the department staff justifies lack of maintenance and poor operation and maintenance by citing low irrigation fees.[19] In such a scenario, an active search by the WUA for information on crop varieties and agriculture extension service is some years away notwithstanding the right that the WUA in Andhra Pradesh and Chhatisgarh has been granted today. These set of rights provokes one to admit that vesting of a substantive right is one thing while having a capacity to claim it is another.

The Third Set of Rights with WUAs:

The third set of Rights with the WUA under the two Rules are worded as follows: (a) Right to participate in planning, and designing of micro irrigation system; (b) Right to suggest improvements/modifications in the layout of Field Channels/Field Drains to supply water to all the farmers in the command, (c) Right to plan and promote use of the ground water; (d) Right to carry out other agro-based activities for economic upliftment of its members; (e) Right to utilise the canal bunds – as long as such use is not obstructive, or destructive to hydraulic structures – by planting timber, fuel, or fruit trees or grass for augmenting the income of the farmers' organisation; (f) Right to engage in any activity of common interest of members in the command area related to irrigation and agriculture and supplementary businesses for self sufficiency and sustainability of the Organisation; and (g) Right to receive funds and support from various development programmes of the central and state government and other development organisations.

A closer look at the listing above can easily show to the reader that all these rights are essentially management and planning functions and are not rights in the strict sense of the term.

---

[19] V. Upadhyay, 'Command Area Development: Restructured Guidelines', 40(29) *Economic and Political Weekly* 3119 (2005). Perhaps the biggest reason that perpetuates this equilibrium is the fact that there has been absence of a link between the payment of service charges and the prospect of improvement in services provided by the system in response to the user requirements and this has provided a fertile ground for the politicisation of the cost recovery processes.

Specific Rights of the individual Water User

Both the Andhra Pradesh Farmers' Management of Irrigation Systems Rules, 2003 and the Chhattisgarh Sinchai Prabandhan Me Krishkon Ki Bhagidari Niyam, 2006 not only talk about the rights of the WUA but also try and pin down the right of its members. The right vested with the individual water user specifically includes: (a) Right to suggest improvements/modifications in water deliveries; (b) Right to get information relating to water availabilities, allocations, opening/closing of canals and outlets, period of supply, frequency, etc.; (c) Right to receive water as per specified quota for use; (d) Right to sell or transfer the water share to any other water user within the operational area of Water User Association with the permission of the concerned Water User Association and without affecting the rights of the other members of the Association; (e) Right to participate in the General body meeting and receive annual reports; and (f) Right to receive equitable benefits from the activities of the organisation.

The first three rights mentioned above can suggest that these rights of the water users are directly dependent on the availability of water with the Water User Association. Thus access to irrigation waters is at the base of the realisation of all these rights both for the Association and also its members. The right mentioned in (d) above, i.e., the 'right to sell or transfer the water share..' opens up a useful space for tradeable water rights but except in isolated pockets it is a provision that is futuristic and this is especially true for the State of Chhatisgarh. When it comes to 'right to participate...' suffice it to say here that right to participation does not ensure participation. The 'right to receive equitable benefits....' is a right with the members that is delightfully vague for the Water User Association. It will not have much meaning unless the bye laws of the WUAs give meaningful content to it.

The Specific Internal and External Water Rights of the WUAs

Under the two State Rules mentioned above, the *rights of the WUA that it can exercise with its members* specifically include: (a) Right to allocate water to non-members on agreed terms and conditions; (b) Right to levy separate fees for maintenance of the system; and (c) Right to levy any other fee or service charges, to meet management costs and any other expenses. As distinguished from this the specific *'external'* rights of WUA which it can use to its advantage against every one outside the Association include the following: Right to obtain information in time about water availability, opening/closing of main canal, periods of supply and quantity of supply, closure of canals, etc.; Right to receive water in bulk from the irrigation department for distribution among the water users on agreed terms of equity and social justice; Right to receive water according to an approved time schedule; Right to obtain the latest information about new crop varieties and their pattern, package of practices, weed control, etc., for agriculture extension service and purchase inputs such as seeds, fertilisers and pesticides for use of its members. Right to have full freedom to grow any crop other than those expressly prohibited by a law and adjust crop areas within the total water allocated without causing injury to neighbouring lands.

The limits of these rights have been indicated in the preceding paragraphs. Given the state of the laws and the state of the irrigation systems, it is a moot question as to how much the WUAs are empowered through the mere vesting of these rights.

IV: Locating Rights in Irrigation Systems: Water Entitlements as the Way Ahead

The analysis of the rights of the WUAs and the individual water user should make clear that while everyone agrees that India should evolve a formal water rights system, this is simply a starting point. The recognition of rights has to move beyond this commonly agreed, indeed axiomatic, proposition. We must identify the precise nature of the water rights we are discussing, and also how to evolve them in the specific social and legal context of the country. Without this, the mere accordance or recognition of rights will be an illusory gain. The fact that water rights with WUAs has been more of an illusory gain even where the State Rules talk explicitly about them provoked an insightful commentator to say that a striking aspect of India's PIM programmes is the little attention that is given to water rights. It has meant that the governments' rights to water are unchallenged, while its obligations to deliver water to WUAs are rarely legally binding.[20]

---

[20] David Mosse, *The Rule of Water: Statecraft, Ecology, and Collective Action in South India* (New Delhi: Oxford University Press, 2003).

The points made in the above analysis have also shown that the critical concerns on increasing the access to the water resource apart from larger questions relating to who controls the resource as well as ownership rights on them have never been addressed. As pointed out in the first section of the chapter it may be dangerous to treat these as mere abstract enquiries. For example, in Mexico the large-scale irrigation management transfer programme was accompanied by a revision in the water rights law, and water users' organisations are even demanding rights to water at the headwork of irrigation systems.[21] The fact that this has not happened in India and the argument that this needs to happen fast can be more strongly put in a historical context. Notably, medieval inscriptions of South India have revealed various functions relating to irrigation, which were exercised by the village assemblies. These included ownership of water resources, powers to arrange for construction, repair and maintenance of tanks, powers regarding land transactions relating to irrigation, levy and collection of cess, powers to engage and remunerate local functionaries, maintenance of records, disputes settlement and relations with the Central Government.[22] The range of power with the village assemblies at that time is in sharp contrast with the restrictive functions largely including only water distribution, management and local monitoring, that has been vested with the Water User Associations under the new State laws. It may also be mentioned here that State officials may not necessarily disagree that the functions with WUA are restricted but they tend to attribute the restrictive range to the lack of capacity with the farmers and the water users.

Quite apart from whether and how much the lack of capacity with the farmers is true – and notwithstanding the fact that such an argument mocks at history – the present author feels that given the state of irrigation systems there are at least two minimum conditions that need to be specifically put down as essential first steps in the laws as the way ahead from here.

---

[21] This suggested to Ruth Meinzen-Dick that some kind of control rights over water to user groups can be an effective part of PIM programs. *See* Ashok Gulati, R. Meinzen-Dick and R. V. Raju, *Institutional Reforms in Indian Irrigation* (New Delhi: Sage Publications, 2005).

[22] *See* Agarwal and Narain, note 15 above.

The first mandatory condition is that with the existing WUAs the irrigation departments across the states need to carry out time-bound joint inspection of the irrigation canals followed by identification and execution of priority works for rehabilitation of the existing canal systems. This needs to be put down as an essential non-negotiable right of the WUAs because without these talking about their water rights is really putting the cart before the horse. All the rights are located in a system and for rights to be effective the system needs to work. Secondly, to ensure that a fully functioning turned – over system maintains the water flow in it the minimum water entitlement of the WUA needs to be built in to the laws so that a total volume of water is guaranteed to be supplied to a Water Users Association at agreed points of supply. If this is put down as part of the law, water from the canal system shall be supplied to the Water User Association at various levels, from tail to head on bulk basis measured volumetrically as per their water entitlements. In other words, before talking about the *water rights* of the WUAs and the water users their *right to water* needs to be honoured. The state of Maharashtra has already taken a lead in this regard in the recently enacted Maharashtra Management of Irrigation System by Farmers Act 2005 by building in such water entitlements in the Act. There is no good reason – although there can be many excuses – as to why the other states can't follow their example.

6

Customary Rights and their Relevance in Modern Tank Management: Selected Cases in Tamil Nadu

A. GURUNATHAN AND C.R. SHANMUGHAM

Introduction

In India, the use of natural resources and their associated technologies and laws have their origin in a base of traditional jurisprudence. The governance of important natural resources such as village water resources was decentralised, having its legal basis almost entirely in custom. A custom is law not written but established by long usage and consent of our ancestors. In a legal sense, a custom means a long-established practice considered as an unwritten law. In another sense, a custom depicts a long practice or usage having the force of law. Customs mostly take the place of law and regulate the conduct of men in the most important concerns of life. At times, customs die away or are abolished or suspended by statutory law. Nevertheless, customs have been a source of law independent from known sources, namely religious or ethical doctrine, texts or royal decrees, as far as traditional Indian jurisprudence is concerned. We can observe that many of these customary practices are even now in vogue in land-holding patterns, traditional water technologies, forest use, agriculture and fisheries. The legal frameworks based on customs provide a wealth of information on sustainable resource use and management.[1]

Food security plays a crucial role in addressing the needs of a growing population and it is inextricably linked to poverty alleviation. Water is a crucial input for enhancing crop production and providing food security. Minor irrigation tanks seen in plenty across the nation and especially in the Deccan Plateau have been supplying rain water for agriculture by effectively harvesting monsoon rains. Indeed, they have been traditionally managed by the local communities who have, over the years, evolved certain regulations for distribution and integrated management of water. Those regulations adapted by the community to suit the changing situations over the years have become the customary rights in tank management.

The Government of India introduced the 73rd and 74th amendments to the Constitution in 1992, thereby requiring the state governments to create a statutory three tier local self-government structure down to the village level. Several natural resources including tanks and ponds were brought under the jurisdiction of these bodies. The Indian government also passed a Panchayat Raj (Extension to Scheduled Areas) Act (PESA) in 1996, which empowered the *gram sabhas* (village general bodies) in the fifth scheduled areas to have the right to decide upon or veto development projects within their jurisdiction.[2]

Therefore, the practices followed by the community from time immemorial over water bodies fall under the scope of custom and customary practices.

DHAN Foundation which works towards conservation and development of small-scale water bodies like tanks and ponds through community institutions, sought to examine the customary

---

[1] M.S. Vani, Customary Law and Modern Governance of Natural Resources in India – Conflicts, Prospects for Accord and Strategies (Paper submitted to the Commission on Folk Law and Legal Pluralism; XIII International Congress, Chiang Mai University, Thailand, April 2002).

[2] Sharachchandra Lele, A.K. Kiran Kumar and Pravin Shivashankar, Joint Forest Planning and Management in the Eastern Plains Region of Karnataka: A Rapid Assessment (Bangalore: Centre for Interdisciplinary Studies in Environment and Development, Technical Report, 2005).

practices and rights traditionally held by the users of tanks, as a research study with the guidance and support of Development Centre for Alternative Polices (DCAP), New Delhi.[3] The authors present the findings on the customary rights and their relevance in tank management by reviewing select cases in Tamil Nadu.

Customary Tank Management in India

Irrigation tanks, one of the very important water resources for the rural community in India, occupy a significant position in agricultural economy. They support the livelihood of farmers and cover 3.0 million hectares of gross irrigated area in the country. They account for more than one-third of total gross irrigated area of three South Indian states of Andhra Pradesh, Karnataka and Tamil Nadu. They have also played a crucial role in safeguarding the local ecosystems. In the context of management of tanks in India, one could notice that by and large the local management systems developed and practiced for centuries have served the multiple needs of the rural community.

Local autonomy was a characteristic feature of tank management in India. Maintenance of tanks was one of the main community activities. There were well-defined customary practices regarding water allocation, operation of sluice outlets and sharing of usufructs. There were village committees and functionaries specifically assigned the task of handling tank related matters, as described below in greater detail under customary rights and practices.

Nevertheless, after independence, the continuous neglect of these unique indigenous tank systems due to various reasons has resulted in their degradation and several small-scale water resources have even become extinct. The decline in tank fed agriculture has been rapid in the past four to five decades. This situation has led the affluent farmers in the tank command areas to go in for wells,

---

[3] DHAN Foundation, 2004, Study on Customary Rights and their Relation to Modern Tank Management in Tamil Nadu, India; K. Sivasubramanian, Irrigation Institutions in two Large Multi-village Tanks of Tamil Nadu (Chennai: Madras Institute of Development Studies, PhD Thesis 1995), and M.S. Vani, note 1 above.

leaving the small and marginal farmers in the lurch. Many of the ayacut (irrigated) lands of tanks situated along the outskirts of large towns and cities have been converted into house sites due to urbanisation. Consequently the tanks were further neglected. They became dumping grounds for wastes and lost their storage capacity. This situation led to encroachments in the common lands of the tank complex, particularly in the tank bed and along the feeder channels. Tank system has a special significance to the marginal and small farmers as most of them are dependent on the tanks for their livelihood through irrigation, domestic and livestock water use and inland fishing.

A. Customary Rights and Practices in Tanks

Customary Rights to tank water and other associated usufructs have been exercised from time immemorial by farming as well as non-farming villagers, according to the norms evolved with their consensus. It was felt necessary to understand the customary rights and practices indigenously developed and traditionally practiced by the community, how over a period of time other interventions have changed them and the implications of such changes on the community as well as on the resources themselves. The study of customary rights made by DHAN during 2003–04, was based on the available records as well as through intensive field studies, mainly to document the present pattern of intra- and inter-tank management systems. The study undertaken with the support of DCAP had the following objectives:

(i) To investigate historical and still existing customary rights in tank systems in Tamil Nadu and their relation to past and present customary management of tanks.
(ii) To review the current irrigation law and policy of the State in relation to institutions and management processes, including review of the institutionalisation of irrigators under the official modern tank management strategies and through initiatives of non-government organisations.

The study was conducted in tanks situated in the southern districts of Tamil Nadu. Archival and public records and other literature, government orders and court verdicts were reviewed for a proper understanding of the problems in general and specific to the study areas. Selected individual farmers were interviewed through a standardised interview schedule.

1. Traditional customary practices in vogue

Village communities had their own norms inherited from their ancestors regarding the management of irrigation tanks and various other related issues. The water management by and large still remains with the villagers. Their informal/formal associations take care of such functions. More important is the water acquisition in chains of tanks which is dealt with by the villagers. Government authorities stay far away from this activity.

2. Collective action

In most of the villages studied, the people had their village bodies in the name of village committees which were not registered or formalised by external authorities. But those bodies were enforcing, monitoring, and regulating various issues related to tank management. Such committees had full powers to do these activities and the people respected them. They were also able to resolve conflicts that occurred in the community in an effective manner. The term 'community' used in this paper refers to the village community – a group of people having a common interest.

3. Composition of committees and their general functions

Village committees formed with due representation comprised all the caste groups of the village, thus demonstrating a democratic approach in their decision-making process. Among the sample villages both formal and informal associations exist. The study could not find any issue of domination or control by any single group in the matter of water acquisition or its distribution. The dominant groups are high caste groups and influential groups of farmers. The possible tensions that could happen would be in the matter of allocation and distribution of water between well-owning farmers and others or head-end and tail-end farmers or farmers served by high-level and low-level sluice outlets.

The village committee decided on various works like closing of breaches in the channel banks and tank *bunds*, and took action on those who worked against the interests of the tanks and their supply sources. Other activities included resolving conflicts related to water allocation and distribution among the water users.

4. Routine system operation and maintenance work

In most cases, the villagers appointed a person to manage the irrigation tanks and the irrigation activities. Such persons called *neerkatties* (water distributors) regulated the flow of tank water through sluice outlets according to the needs of the standing crops and distributed it to individual land holdings in an effective and equitable manner. They were paid in kind by all the farmers at the time of harvest for the services rendered by them.

Villagers themselves carried out the tank maintenance work like cleaning the feeder and main irrigation channels, strengthening of the *bunds* and desilting the tanks through *Kudimaramath* (community participation), either by contributing labour or money as decided by the village committee. It is seen in most cases that the Water Users Associations had been instrumental in undertaking annual maintenance work. However, of late, the farmers have reported that the routine system maintenance work and employment of *Neerkatties* are slowly disappearing, due to lack of adequate finance, the non-cooperation of farmers and breakdown of unity and collective action in the villages.

5. Water management functions

The entire tank management affairs were primarily undertaken by the committee. To ensure the proper use and control of the water management functions, the villagers themselves decided the area to be irrigated, generally as a proportion of the total area cultivated by each farmer. The government or any external agency was not involved in such functions in any manner at any time.

In a few villages, during the scarcity period when a tank had a registered ayacut of say 50 acres and water was available in the tank only for 30 acres during a particular cropping season, the water was allocated equally to every cultivator irrespective of the size of his land holding in the tank ayacut. All landholders could raise irrigated crop only up to that limit and raise rain-fed crops in the remaining area. Thus the small and marginal landowners stood to benefit more than the large landowners in terms of percentage of land holdings irrigated.

The villagers were allowed to rear cattle like sheep and goats and they were also allowed to use the channel banks and adjoining areas for grazing their animals. Duck rearing, inland fish culture and growing aquatic flowers like water lily and lotus were the other activities promoted in a few village tanks.[4]

The village communities were enjoying their freedom to decide the priorities according to the needs of the farmers and water availability. For example, a village would fill the drinking water pond once in the beginning of the rainy season and again at the end of the season from the irrigation tank, even when water was required for irrigating the crops. This practice was seen as an activity for maximising the utility of the available water. Such a regulation needed a lot of cooperation from the farmers and the rest of the villagers. Usually drinking water was given top priority (which conforms to the National Water Policy) and only thereafter crop water and other requirements were considered. This practice was found to be common in most of the villages studied.

6. Water sale and regulation

The community was enjoying the power to transfer any of the rights related to the irrigation tanks. In a few places in Ramanathapuram district, like Mudukulathur, a big tank which is located in a cascade of tanks, the villagers worked hard to bring water by cleaning the long-winding feeder channels, and fixing prices for the lower down tank beneficiaries for letting water into their tanks. This was happening because of the non-cooperation of the lower down tank villagers to bring the water through a joint

---

[4] C.R. Shanmugham and A. Gurunathan, Customary Water Rights and Jurisprudence in Tank Management: Case Studies from a Community Livelihood Perspective (Lead paper presented at the Workshop entitled 'Law and Administrative Practices in Water and Related Issues addressing Access to Safe and Adequate Water for the Poor in South India', organised by the National Law School of India University, Bangalore, 2006).

effort. The buyers used to see the cost of water as their price for non-cooperation during the water acquisition works. In a particular village, (Thoori in Ramanathapuram district) the farmers exerted social pressure by stopping the irrigation water to the villages downstream, when they refused to contribute either labour or cash for carrying out maintenance works in the irrigation tanks. By this method, they induced the people from other villages to cooperate with them in such common endeavours.

7. Unregistered ayacuts

In some villages, the farmers permitted their fellow farmers to irrigate crops grown in non ayacut areas when there was plenty of water in the tank, after collecting some fee from them and adding the same to the village common fund. This was a good proposition, considering the frugal and efficient use of available water and maximising the collective returns. It is more equitable than fixing up the ayacuts in a rigid format. However this was reported only in Ramanathapuram district and not in the other places. Thus they ensured equitable distribution of irrigation water among all the village communities.

8. Revenue from usufructs

In all the selected villages, the farmers reported that they were enjoying full rights over the irrigation water and they had power to utilise the usufructs as desired by them. However, in recent years, such use of the usufruct revenues by them is objected to by the government authorities, mainly the revenue department, and not by the Water Resources Organisation (WRO) or the local *Panchayats*. Government is even now the owner of the land in which the tanks are built and used. The Revenue department is considered as the owner of all government lands and so it collects land revenue and usufruct revenue from these lands. The tanks are only vested with Water Resource Organisations or Panchayats and so they are not recognised as owners. The funds raised from usufructs were generally used for temple festivals and for maintenance of tank systems. In general, the villagers had the right to rear fish and grow trees on the *bunds* or in the water spread area of the tank and enjoy the proceeds.

B. Synopsis of Select Court Cases on Customs and Customary Rights in Modern Tank Management in Tamil Nadu[5]

The rights and obligations as between the state and ryots in India in the matter of irrigation rest largely on unrecorded customs and practices. Though generally customary rights are not recorded, there are cases in which such rights are meticulously recorded.

1. Customary irrigation rights: A recorded Mamul Nama in Vellore district

It is quite interesting to observe the recorded irrigation rights of *pattadars* (command area landowners) of 188 tanks of Vellore taluk in Tamil Nadu in 1815 under the heading 'Water *Mamul Namas*'. Again these were printed by the British in the year 1907. The *Mamul Namas* have been written in Tamil and signed or attested with thumb impression by the *Karnam* (Accountant in Village) and by the important farmers of the village.[6] It is astonishing to note how meticulously the *mamul namas* have been written, recording the period in which the tanks received water supply, the quantity of water available during particular months, the area that could be cultivated, when the tanks got full supply, the mode of irrigation to be adopted during the distress period, the permissible number of wells that could be sunk in the command area and the crops that could be cultivated in the area.

Even though the irrigation rights and practices were not recorded in many instances, they were meticulously observed by the ryots and the community from time immemorial. However, some customary rights could be ascertained from the 'A' register maintained by the revenue department and the old settlement records. These customary rights along with *Kudimaramath* systems were followed with high dedication and vigil by the ryots and villagers during the zamindari system and even under the rule of East India company for some time. But after the introduction of Ryotwari settlements by the middle of nineteenth century, the

---

[5] For more such cases, *see* DHAN Foundation, note 3 above.
[6] English version of the Mamulnama extracted from GO No. 660 I (8 February 1918), as cited in K. Sivasubramanian, note 3 above.

effectiveness of the traditional system deteriorated progressively, with the result that the tanks were not maintained well in the country.

2. Select court cases

Courts have been proactive in protecting customary rights and have granted compensation to farmers who have been denied their rights. As early as in 1892,[7] prior to independence, the division bench of the Madras (now Chennai) High Court addressing an issue of violation of customary rights, laid down some basic principles of law. In a case where the irrigation water supplies diminished due to the Government excavating a diversion channel, the farmer approached the Court asserting his right to a certain quantum of water from the irrigation source. The court held that Government diversion under the orders of the collector, 'causing a material diminution in the supply for the cultivation of plaintiff's lands was in excess of the powers possessed by him for the regulation of the supply of water for irrigation purposes'. The Court however, also acknowledged the right of the government to divert the waters. The Court ordered recovery of damages resulting from such diversions.

In another instance, the land that was customarily irrigated from a channel that flowed downstream was blocked at a higher point. The aggrieved farmer approached the Court but the defendants contended that the channel never irrigated the lands either of the fourth defendant or the plaintiff, and that it stopped somewhere near their own lands. It was then found that the channel in question continued as a definite water course up till the fourth defendant's land and that the water flowed over the lands of the fourth defendant's field and joined another channel which irrigated the plaintiff's lands. The Court held that the obstruction to flowing water at a place where it exists as a regular water course is an 'actionable wrong'.[8] The Court also granted an injunction against the defendants from interfering with the plaintiff's rights.

In another case, the diversion of water from Cholavaram tank and Redhills tank to Madras (Chennai) city for drinking purposes

---

[7] *Ramachandra Iyer v. Narayanasami*, 1892(2) MLJ 279; 1893 ILR Mad. 333.
[8] *Valluri Adinarayana v. Ramudu*, SA No 931 of 1911, XXIV MLJ 17.

was detrimental to the existing ayacut of over 5000 acres. The question which arose for decision was whether the government was entitled to supply water to Madras without regard to the rights of the cultivators in the old ayacut as they existed in 1860. The Court held that in the Madras presidency the ryot was entitled to receive the water which his lands had been accustomed to for irrigation purposes without interference by the Government or anyone else.[9] However, the Government cannot be required to supply water when none was available but it has an obligation of conserving and distributing the water available in the interest of the particular ayacut, during years of shortage. The only obligation of the Government was to make an equitable distribution of water. But the ryot has a claim against the Government when it withholds from him the water which he has a right to demand taking into consideration the supply available.

A ryot holding land under the old ayacut in a tank could successfully sue the Secretary of State for a declaration that he was entitled to a sufficient supply of water for the cultivation of the crop per annum subject to the power of the Government to control the distribution of the available water in the interest of the land holders, whose lands comprised of the old ayacut was well recognised. It was pointed out that the rights and obligations as between the state and the ryot in this country, so far as supplying water for irrigation purposes was concerned, vested largely on unrecorded custom and practice. In this particular case the Court awarded damages of Rs 450 for the loss suffered by the ryot. This is a verdict which has been given in recognition of the custom that prevailed then.

The Madras High Court held in the Naicken case that it was not open to the Government to interfere with the customary method of supply of water to the field.[10] The Court observed that a ryot in respect of his Ryotwari land is entitled to receive from the Government the water necessary and sufficient for irrigating his registered wet lands. The Government has the right to regulate the method and manner of supply. It can therefore indicate to him the source or method of supply and he is bound to accept those

---

[9] *C.N. Marudhanayagam Pillai v. Secretary of State for India*, 1939 MLJ 176.
[10] *Annaswami Naicken and Others v. C. Manicka Mudaliar and Others*, AIR 1937 Madras 957.

indicated. But as an incident to the tenure, the ryot has the right to receive the water, whether it is called contractual or proprietary. In this case the High Court upheld the right of the Government as only to regulate the method and manner of water supply to the individual land.

3. Customary irrigation rights – legal position – inroads made in the customary rights due to statutory law and the Constitution of India

Customary rights on the use of water have always been recognised by law; but this customary right is not an absolute right and is subject to the paramount right of the state to regulate and control the supply of water for irrigation purpose. The customary right of the ryots has also undergone a change after the enactment of the Madras Irrigation Tanks (Improvement) Act 1949, and the Constitution of India.

In the Nageswara Iyer case, the dispute relates to the question of customary rights enjoyed by a particular village which was affected by digging a channel to feed another village.[11] The diversion channel proposed by the government, the villagers feared, would cause loss to the village and hence they preferred a suit for a declaration of this exclusive right to the water flowing in the channel and for an injunction restraining the government from carrying out the proposed diversion based on their claim to customary right and prescription. The villagers were reportedly enjoying the water from the channel exclusively for over one hundred years. The Court held that claim based on prescription was precluded by the relationship between the plaintiff and defendant and though the plaintiffs are entitled to the accustomed supply of water for irrigation of their lands, yet they could not acquire any exclusive right to challenge the paramount right of the state to regulate and control supply of water in public streams and channels. The Court also held that the extent of right of the plaintiff villages to take water from the channel for irrigation of their lands should be determined with reference to the accustomed uses of the water by the plaintiff and not with reference to the entries in the registry. The Court laid down the dictum that custom cannot give exclusive right to the detriment of the state.

---

[11] *Secretary of State v. P.S. Nageswara Iyer & others*, AIR 1936 Madras 1923.

When the Manjalar dam was constructed in Madurai district of Tamil Nadu, the lower-down ayacutdars affected by the proposed construction approached the court for injunction against the construction. They succeeded in getting the injunction in the lower court. But on appeal, a division bench of the Madras High Court held that no injunction or stay against the project devised in public interest and welfare of the people could be granted. They also held by interpreting Article 39 (b) of the Constitution of India, that in the interest of social and distributive justice, such injunction against public oriented schemes has to be denied and basic tenets of the principle of equality have to be upheld. The court also held interpreting section 4 of the Tamil Nadu Irrigation Tanks (Improvement) Act 1949, that the State Government has power to regulate and distribute water for effective irrigation of agricultural lands. The court held it was unjust and inequitable to deny others water for one crop while agreeing to provide water to the plaintiffs for raising second crop, merely because their lands were registered as 'double crop', wet lands.

In another recent judgment, the Madras High Court has held that the request for permanent injunction restraining the state from interfering with supply of water to a tank could not be granted. The court also held that the suit was barred because of Section 4 of the Tamil Nadu Irrigation Tanks (Improvement) Act 1949.[12]

From the above discussions it is clear that the customary right against the state is no longer absolute. The state can make regulations and make inroads into the customary rights of individuals in the larger public interest. It has also been held that with the coming into force of the Constitution of India, the principle of social and distributive justice has to be upheld. Article 39 (b) of the Constitution of India enjoins the state to direct its policy to secure that the ownership and control of the material resources of the community are so distributed as to best sub-serve the common good and the words 'material resources' have been assigned a broad meaning to include not only natural but physical resources also.

4. Fishery rights in tanks

As per Section 133 of the Tamil Nadu Panchayats Act (Section 85 of the Tamil Nadu Panchayats Act 1958) maintenance of irrigation

---

[12] *State of Tamil Nadu v. Sudalli Pothinadar*, 1999–1–L. W. 129.

works (Minor Irrigation Tanks) vests with the Panchayat Union and village panchayat. Section 133 (3) declares that when the maintenance of any irrigation work is transferred under the section, the fishery right of such tanks shall be transferred to the village panchayat or the panchayat union, subject to such conditions as may be prescribed. The important conditions laid down by the government are:

(i) If by custom the fishery right is vested within the community, it would be continued.
(ii) If fish *patta* had been granted in favour of any person, the panchayat and panchayat union should not interfere with it till such *patta* was cancelled.

These rights were sought to be transferred to fish farmers' society in the 1980s by the government and Collector through executive orders, contrary to law and statutory provision. Because the Collector happened to be the Chairman of the fish farmers' society in many cases, the transfer of fishery right to fish farmers' society was not challenged by the panchayat or any other person/body.

The Andhra Pradesh High Court in the Ipur Grama Panchayat case contended that the taking away of the fishery right from the panchayats and giving it to the fisherman cooperative society by issuing executive orders under Article 162 of the Constitution cannot have over-riding effect over the statutory provision. The court held that the orders were illegal and without any authority.[13]

In a fishery dispute arising between an individual and panchayat, the Madras High Court held that only the panchayat had the fishery right in view of the statutory vesting of fishery right in the panchayat.[14]

Fishery right will also accrue to the panchayats under Section 83, of the 1958 Act (Section 132 of the 1994 Act). If by custom such right belongs to the inhabitants of the village from a particular source of irrigation and if the collector makes a declaration that such rights are vested with the Panchayat, the income from the fishery will then be administered by the panchayat for the benefit of the inhabitants. It is very important that the declaration to this

---

[13] *Ipur Grama Panchayat v. Government of Andhra Pradesh*, 2000 (4) ALT 678.
[14] *Natarajan v. Ramapuram Panchayat, Thanjavur District*, 1999 (1) MLJ 598.

effect has to be issued by the collector. Until such a declaration is issued the panchayat will not have any fishery right.

The Madras High Court has recently held that Section 83 of the 1958 Act (Section 132 of the 1994 Act), clearly contemplates a declaration by the government, vesting the community property or income to Panchayats. There is no automatic vesting under Section 83 without the declaration by the government (delegated to the Collector). In this case, in the absence of such a declaration, the court held that the fisheries were owned and enjoyed by the *Karaiswans* of 43 *karas* and the income derived was utilised as per custom and usage for common good. Fishery *Pattas* for the suit properties were granted to the *Karaiswans* as far back as 1912. The *kist* (tax) relating to fishery rights was being paid to the government regularly. In such circumstances the beneficiaries (*Karaiswans*) would have the fishery rights and the Panchayat Union or the village panchayat would not have any right to interfere with the fish *patta*.[15]

Earlier instructions regarding fishery rights have been completely modified after the issue of statutory rules by the Government.[16] Under this rule, the Panchayat Union Commissioner is authorised to auction the fishery right in all tanks vested with the Panchayat Union and all PWD tanks other than the provincialised sources in the Panchayat Union area. Panchayats will auction the fishery rights in all water bodies vested with them.

5. Custom that prevailed in water scarce area in tank and drinking water pond

Ramanathapuram district in South Tamil Nadu is renowned for customs in the management of tanks and ponds. Being a water scarce district in a drought prone region, coupled with saline ground water, the surface water bodies remained lifelines and as it is well understood by the people, the customs are strictly adhered to and any change in this led to conflicts and communal disharmony.

Mudukulathur tank is located in Mudukulathur taluk of Ramanathapuram district. The tank irrigates an ayacut area of more than 40 ha and the farmers who live in the surrounding villages

---

[15] *Alagar Iyengar and 12 Others v. State of Tamil Nadu*, 2002–4–L. W 498.
[16] G.O. (R.T.) No. 169 RD Department, 16 August 1999.

of Thoori, Ettiseri, Kadambankulam and Selvavinayagapuram own the land. Traditionally, Thoori villagers were maintaining and managing the Mudukulathur tank. Till mid 1980's, the villagers from Thoori used to invite ayacutdhars from the other remaining three villages for mobilising voluntary labour to clean up the feeder channel from its original source Ragunatha Cauvery which is a tributary of Gundar river.

After 1980s, the practice has been converted into mobilising money rather than mobilising labour from the same villages for the cost equal to their labour. This has happened because of the behaviour of one or two villagers who did not send adequate number of labourers. This practice had also collapsed in the mid 1990s. During 1999, Thoori villagers had spent Rs 25,000 to clean the supply channels and fill the big tank at Mudukulathur. They vehemently refused to release any water even after the Public Works department engineers tried to open the sluices. Thoori farmers put forth the argument 'No payment for the clearing of channel and hence no water'. After a lot of tension and arguments, two villagers paid Rs 10,000 and Rs 6,000 respectively and got their share of water. These types of custom enforcing tank management issues are common in such drought prone arid plains of South Tamil Nadu.

The alluvial formations in a few pockets in the Ramanathapuram district and in the proximity of the Gulf of Mannar coast are attributed to salinity in ground water. The villagers in many parts of the district used to fill their *Ooranis* (drinking water ponds) from the tanks as a custom. This happens at the beginning of the rainy season (September), at the end of the season (December) and once again during summer (June). This has been the way of life and the source of their drinking water for many years. They could not separate drinking water from irrigation tanks for ages. It is also enforced and practiced that nobody should pump or bail the water below the sill level of the sluice outlets of the irrigation tanks.

6. Conflict in sharing usufructory rights from tanks in Dindigul district

Athoor is a traditional zamin village bound by its heritage and cultural practices of a multi-caste village in Southern Tamil Nadu.

It is situated twenty kilometres south west of the district head quarter, Dindigul. Athoor Village Committee was established even before 1900 with a view to help the village gain certain benefits from the then government. Late Savarimuthu Pillai was active in the welfare of Athoor and Sempatti villages and he was claimed to be a charismatic leader. He is reported to have laid the foundation for the Athoor Pattadhars' Committee (APC). It was registered in the year 1993. The Executive Committee consisted of four office bearers namely President, Vice President, Secretary and the Treasurer and thirteen Executive Committee members who constituted the apex body in the decision making process.

Athoor village has a series of tanks, namely Pulvettikulam, Karunkulam and Pagadaikulam. These tanks are all situated in a single line from east to west of the village. They receive water supply from the rainfed non-perennial river named Gundar. The ayacut area commanded by these tanks is given in Table 1.

These lands belong to 703 farmers. Of them about 73 per cent belong to marginal farmers' category and only 1.5 per cent belong to big farmers while the remaining are small farmers.

Water had to be distributed by the agreed (customary) rules formed by the Athoor Pattadhar Committee (APC). They include,

- *Maniams* (water distributors) have to distribute the water in an orderly manner sequentially (Head to Tail end)
- If any one needs water beyond the requirement they have to request the APC only, which in turn will suitably instruct the *maniams*.
- During the periods of scarcity, water delivery time will be fixed on the basis of availability and certain prefixed norms on the basis of equity.

Table 1: Tanks in Athoor Village.

Tank	Water Spread Area (ha)	Ayacut Area (ha)	Cultivated Area (ha)
Pulvettikulam	68.750	165.505	156.005
Karunkulam	20.030	34.075	31.520
Pagadaikulam	33.085	88.480	81.580
Total	**121.865**	**288.06**	**269.105**

Fishing rights from these tanks are with the villagers as per custom, under which the villagers auction the fishing rights. The returns from the auction are used for temple and tank related purposes only. All the religions get their share of revenue for their respective religious festivals and it is made known to all the villagers. They have been adhering to this norm for more than forty years.

The customary rights that followed in a consensus based decision making process of APC were:

- Irrigation rights as per the (customary) rules framed
- Appointment of *Maniams* for irrigation
- Fishing rights
- Segment (*Kandam*) based Watch and Ward system through appointment of guards men.
- Cattle rearing and other Recreational Activities
- Auctioning right over the use of threshing floor (*Kalam*) at the time of harvesting.

7. Dispute on fishery usufructs: Loss of rights

The tank fishery rights of the villagers were taken away by the government agencies by way of collecting a tax *Meen Pasy* (fish tax) in order to recognise the rights of the villagers to have fishery under their control. Way back in 1946, the government tried to cancel the fishery rights of the APC. But the then president, Thiru I. Savarimuthu Pillai fought against it in courts and finally a stay was awarded by the Madras High Court stopping the take over of the tank fishery right from the villagers.

Again in mid 1980s, the then Tamil Nadu Government brought the tanks under the Fish Farmer Development Agency Act and declared the tank as part of the pilot tanks where fishery was proposed to be promoted. The Assistant Director of the Fisheries Department, Dindigul requested the Tehsildar to cancel the APC's customary rights to fishery. The APC put up more than ten years of legal battle in the court of law. But in the year 1998 the High Court vacated the stay and announced that the right to fishing from the tank has been vested with the Assistant Director, Fisheries Department, Dindigul. The suit was dismissed as not pressed by the plaintiffs. The plaintiffs could not continue to fight legally as the District Collector happened to be the Chairman of FFDA also and there was no leader to continue the battle. So the APC lost its enjoyment of fishing rights from 1998 onwards.

Like Athoor, Sithayankottai Town Panchayat situated twenty kilometres southwest of Dindigul lost its customary fishing rights enjoyed by Village Farmers' Protection *Sangham* over five decades to fishery department which issued a letter during 1998. In this village even now the mainstay of people – agriculture is practiced under two rain-fed tanks namely, Thamaraikulam and Puliyankulam and also in the direct ayacut area of Thamaraikulam Rajavaikkal. The direct ayacut of Rajavaikkal and two tanks command 471.065 ha.

In this village, Mr N. Abdul Khadar (who was later elected as Rajyasabha MP) organised the farmers and started a formal association namely Sithayankottai *Grama Vivasaigal Pathukappu Sangam* (village farmers' protection association). This sangam undertook the following tank related activities.

- Efforts to clean Rajavaikkal and the two other tanks every year.
- Regulate water distribution.
- Purchase land for the Puliyankulam Tank Farmers' Association building construction.
- Fish rearing activities in the tanks.

Such a well performing *Sangam* which has been traditionally enjoying all the usufructory rights including fishery in the tanks witnessed problems with the fishery department. The association approached the Madras High Court to reserve order in favour of the sangam due to their customary practices since ages which the court conceded. While the case against the fishery department was pending with the High Court, they continuously enjoyed their rights using the injunction until 1998. In 1998, the fishery department invited contractors for fishing in the tank, but no one came forward to apply for the contract fearing the *Sangam* and the fact that villagers may not allow any fishing which is against the customary practice. Then the case was dismissed as it was not pressed for a decision and the merit of the case was also not contested. Annex 1 provides the legal points on which the *Sangam* fought the case.

8. Encroachments and the rights of cultivators: A case of Rasingapuram village in Theni district[17]

Rasingapuram is one of the village panchayats in Bodinaickanur block of Theni district. It is a multicaste village wherein more than twelve castes are residing with traditional and cultural bondage. This village is situated 23 kilometres away from Theni in the south west direction. Total geographical area of the village panchayat is 2618.28 ha with around 1640 households. The total population of the village is 6426 (Male 3272 and Female 3154). The main village Rasingapuram is surrounded by four hamlets within the Panchayat jurisdiction. *Kurumba goundar* is the dominant caste in the village.

It is said that the village was prosperous in early 50s and 60s due to the irrigated cotton cultivation when the people used to move a large number of cotton bales from the village to Theni Cotton market. Goundankulam is a tank fed by Suthagangai Odai, a non-perennial wild stream emerging from the Western Ghats. In addition the village used to have several ponds.

This village was one of the front liners in getting electricity in the late 50s. This combined with free electricity and agricultural credit to sink wells in the 1960s led the villagers to sink more than 250 wells. Ruthless mining of ground water from the wells made the farmers dig 100 feet deep bore wells inside the open wells of 80–100 feet depth. The over dependence on wells coupled with state ownership of tanks, led the farmers to neglect the tank. Using this opportunity, a few power-centric as well as greedy farmers encroached the feeder channel, ploughed the tank bed, sunk two wells, got electricity supply by unfair means as well as raised hundreds of coconut trees. They enjoyed the benefits over twenty long years. The continued efforts of the villagers to vacate the encroachments failed to yield any positive result in their favour. By the year 1997 the total water spread area of 5.17 ha had been reduced to around 1.46 ha beside complete dismantling of the tank bund. The villagers who owned lands in the ayacut as well as the others who tried to protect the water spread since 1985, failed in their effort. In total, ten farmers had encroached the land as given in Table 2.

---

[17] Asian Development Bank, *Rehabilitation and Management of Tanks in India – A Study of Select States*, Publication Stock No. 122605 (Philippines: ADB, 2006).

Table 2: The encroachments declared as legitimate *patta*.

Sl. No.	Name of the encroacher	SF No.	Patta No.	Extent of encroachment (ha)
1	Krishnasamy.S	346/1	45	0.445
2	Ramuthai.K	346/2	1553	0.515
3	Kariappan.C	346/3A1	139	0.230
4	Srinivasan.S	346/3A2	2148	0.040
5	Keppammal.S	346/3B1	2149	0.035
		346/3B2	348	0.220
6	Malarkodi.S	346/4	-	0.230
		346/5	-	0.295
7	Ondiveeran	346/6	-	0.300
8	Thangamani	346/6	-	0.300
9	Perumal.O	346/6	-	0.300
10	Subramani.P	346/6	-	0.800
	Total			3.710

During the year 1996, the farmers approached DHAN Foundation for help to remove encroachments and revive the tank. The farmers were interested in restoration and reclaiming the tank through eviction. They felt that their efforts so far had not been successful and so an organisation like DHAN could guide them properly in resolving the problem. They formed a formal Tank Farmers' Association and arrived at a consensus for making a contribution to the tank rehabilitation works.

The villagers approached the District Collector for funding the project and they got the approval. The work to the estimated value of Rs 88,000 was allotted to the Tank Farmers' Association (TFA) under *Namakku Namae* (Self Help) Scheme. After a great deal of struggle, a land survey was organised by the Tehsildar and the boundary was established for the tank at least on paper. The villagers took up the construction of the tank bund after evicting the total area of 0.485 ha under the SF No 346/3A1, 346/3B1 and 346/1. However, the encroachers were continuously making threats as well as taking legal steps to stop the tank restoration work through every possible mean. Since a part of the tank was revived, many wells in the vicinity got rejuvenated by next year (during 1998) and many villagers started pressing for complete eviction of the encroachers.

The villagers again tried to get funds from the Panchayat Union for reviving the rest of the tank. This time they evicted around 1.00 ha of land using force and coercion and spent Rs 1.80 lakhs. Then the encroachers consulted lawyers and filed a case against the Collector for illegal eviction from their lands. The village farmers agitated a lot and jointly decided to evict all the encroachments at any cost and collected Rs 25,500 from the village. Using this as their contribution, they got a sanction order for water harvesting work for an amount of Rs 1.02 lakhs under the village self sufficiency scheme. This time the villagers formed a stable and big bund around the revived water spread area. They also completely evicted the encroachers from the supply channel using coercive means. By this exercise, they have encircled the entire area of the tank bed. A lone encroacher sitting in the middle of the tank bed went on an all out offensive against the villagers. He was successful in getting an interim injunction to the works sanctioned by the Government. As of March 2008, the case is still pending a decision in the High Court, Chennai.

Presently the villagers are confronted with a question of whether their retrieved land will survive in the court battle. In case the court upholds the *Patta* given to the encroacher in the eighties what would be the fate of the tank? Their efforts to get impleaded in the court case also did not meet with success because of the Government Pleader's assertion that it was not necessary for them to get impleaded in the case.

Annexure 2 contains the time line of the encroachments and the efforts taken by the villagers.

C. Learnings

1. Customary practice in vogue

The communities had their own norms inherited from their ancestors regarding the management of the irrigation tanks and various other related issues. The tank management by and large still remains with the villagers. Their informal/formal associations take care of such functions. More important is the water acquisition in the chain of tanks which is dealt with by the villagers, and the Government authorities stay away from this activity.

The above cases also reiterate the collective action of the villagers by composition of committees with stipulated roles and responsibilities. They also follow the routine system of operation and maintenance work, by appointing water guides/*Maniams* for irrigation and contributing labour or money for cleaning irrigation channels and water courses.

2. Revenue from usufructs

In all the selected villages, the farmers reported that they were enjoying full rights over irrigation water from tanks and they had power to utilise the usufructs as desired by them. Usufructs from tank systems include trees, shrub and grass grown on tank *bunds* and water spread areas, tank silt and fishing in tanks. However in recent years, such use of the usufruct revenues by them is objected to by the government authorities – mainly the Revenue department and not by the Public Works department or the local panchayats.

The Revenue department collects the tax for 2 C *patta* based on the type of trees planted in tank complexes and recognises the right of individuals who planted and guarded the trees and allows them to get monetary benefit from them. However, the tank users of present times want to generate some form of revenue from the tanks as a matter of right. The villagers as a forum demand that the customary rights to usufructs which they were enjoying earlier be restored to them and the panchayats can oversee that the funds are utilised for the maintenance of the village tanks so that some of the prevailing illegal practices could be prevented. In many villages the Water User Associations have expressed their willingness to share the revenue generated from usufructs with the local panchayats. In a few villages they have also entered into a memorandum of understanding with the panchayats and are already enjoying a share of the revenue from the usufructs.

3. Encroachments in tank system

One of the challenges faced in storing rainwater in the tanks up to their designed capacity is encroachments that are made along the supply and surplus channels, and tank water spread areas. Such encroachments constrain the carrying capacity of the channels resulting in only partial inflow of runoff into tanks from their catchment areas and reduced outflow of the surplus water from the tank. The encroachments also induce the encroachers to willfully break the surplus weirs or tank *bunds* in order to protect their crops in the encroached tank bed area from water logging and damage. The low storage of tanks caused by such encroachments deprives the poor from having full use of the tank water. The existing laws to evict the encroachments are long drawn and ineffective. There exists a rule in the government office of revenue[18] that no water body could be encroached upon by any individual organisation nor can *patta* right be given to any one to use such land for any purpose other than for conservation of the water body. This rule has also been, in recent times, upheld both by Madras High Court and by the Supreme Court of India. Yet this is not strictly followed in all cases. Responding to the appeal made by an informal body namely 'Conservation Council for Small Scale Water Resources' and a Policy Brief submitted by DHAN Foundation, the Tamil Nadu Government enacted the 'Tamil Nadu Protection of Tanks and Eviction of Encroachment Act, 2007' which has come into force during the year 2007. It simplifies as well as hastens the procedure for eviction of encroachments. As this Act does not cover the smaller panchayat tanks and ponds, representations have been made to the Government to include all water bodies under the purview of the Act.

4. Water rights and access to use

In the modern era of legal jurisprudence, the courts are indeed willing to understand and/or accept new 'rights' such as right to water, right to food, right to work and right to clean environment. As per the 73rd and 74th amendments of the Constitution of India, water management is one among the subjects listed in Schedules 11 and 12 for devolution to the Panchayats and Nagarpalikas. This needs to be put into practice, as presently it is the Panchayat Unions (Rural Development Blocks)[19] which are exercising these powers.

---

[18] Standing Orders of the Board of Revenue in the erstwhile Madras Province (presently Tamil Nadu Government).

[19] The Rural Development Department of Government of Tamil Nadu functions as a three tier administrative setup at the district level. These include the local panchayat of elected representatives, the Panchayat Union (comprising the Panchayat presidents) known as the Development Block, and the Zilla Parishad (District Council) comprising elected representatives and chaired by the District Collector.

D. Way Forward

The view that water is a common property resource is virtually universal among civil societies. Even the former Prime Minister of India, Mr. A.B. Vajpayee, in his address to National Water Resources Council on 1 April 2002 proclaimed that the community is the rightful custodian of water. This statement calls for a paradigm shift to local management and customary practices of rural water resources.

The existing laws need a thorough review in order to make them much more stringent so that customary rights as well as small village water resources, namely tanks and ponds could be preserved before they become extinct. Like the Reserved Forest Protection Act, the framing of enactments to conserve all traditional village water bodies from all social evils have to be introduced in the Parliament by law and approved by policy makers.

While there are some local customary regulations and practices adopted by the rural people for the use of surface water harvested and stored in tanks and ponds treating them as common property resources, there are no effective regulations for the development and use of ground water, either by custom or by legal provisions. Some Bills and Acts have been passed by a few state governments but they are not strictly put into effect. This has resulted in unrestricted withdrawal of ground water, mostly by the rich well owners who could afford to have their own wells either for irrigation or for industries. But the poor people who could not afford to sink private wells are the sufferers, as they do not have access to ground water. If community wells are provided for their use, they will be able to operate, maintain and manage this water resource and thus have access to it at an affordable cost. There are large contradictions in the use of ground water. While there is a restriction on spacing of wells to be constructed in order to conserve this precious resource, there is no restriction on the withdrawal of ground water in terms of either time or quantity. It is well known that withdrawal of ground water in excess of its recharge is dangerous and it will lead to the ruin of the present and the future generations. The government's encouragement to excessive withdrawal of ground water is through its subsidised or free electric supply to farmers. These contradictions have to be addressed immediately by suitable and effective regulations and their strict adherence through effective implementation. If the policy of the government is to support farmers pumping ground water for increased crop production, it must also ensure that the withdrawal of that much water is made good by recharge. Tanks and ponds would meet this requirement to serve as recharge basins. Desilting of tank water spread area which is a component of tank rehabilitation will enhance the recharge of ground water. How effectively the laws or legislations can be framed to prevent the exploitation of ground water resources and induce their recharge through appropriate measures, is a matter for serious consideration.

In the globalisation era, for achieving Millennium Development Goals with water as a tool to alleviate poverty, the government has the onus to take up legal as well as policy reforms in favour of community managed Natural Resources Management. It is beyond doubt that these native water wisdoms exist over many decades, surviving and attributing to the village's economic growth. The water bodies need rehabilitation suited to their design standards to ensure water and food security in the coming years. Adequate resource allocation and policies empowering village communities to own and manage these systems similar to 'Kudimaramath' and/or 'syndicate agricole' (followed earlier by the French in Pondicherry) made as a form of new law, are required. The Central government may, if need be, undertake to protect water resources in the nation by bringing them under 'Concurrent list' of the Constitution from the present 'state' list. In addition, the financial resources allocated for the revival of these vital village water resources to harvest the rain water and manage the demand of water by multiple stakeholders effectively, have to be increased. There is need for action-based grassroots research studies to identify successful customary practices across India and how they can be dovetailed to fit into modern law.

In conclusion it is suggested that, the shortage of fresh water that every successive generation is going to encounter and the compelling need to produce more crops with less water so as to ensure food security to the poor, are addressed immediately by all concerned. This has to be done by enacting appropriate Acts and Rules, and by developing stringent legal framework for the conservation of water. More importantly, there is an urgent need for a campaign to enhance mass awareness, to highlight that conservation of all the existing surface and ground water resources is everybody's business. Last but not the least, let us all recognise the traditional customary rights and restore them to the local communities through appropriate legal validation.

Annexes

Annex I

The issues raised in the Case are as follows: (*See* Section B.7)
The petitioner *Sangam* consists of the farmers making a livelihood from the tanks for ages.

1. The farmers are the rightful holders of the water and the government is only collecting tax in recognition of our rights. The petitioner is having the fishing rights for and spending the proceeds for the benefit of the tanks and the village. No single individual gets the benefits of the fishery. The revenue resettlements have also confirmed the rights of the petitioner.
2. The proceeds from the fishery is only marginal compared to the stakes on water for agriculture. Therefore the petitioner will give priority to farming and will use the entire water for agricultural production in the village even if the proceeds from fishery is going to be fully affected for want of water. This cannot be the case if the Fish Farmers Development Agency's (FFDA) appointed contractors come into the picture.
3. FFDA has only been in existence to promote inland fishery in the district from water resources. Modern fishery may affect the customary practice of fish farming and is not suitable for agricultural areas and tanks because the use of chemicals and others may affect agriculture, sanitation and hygiene.
4. The right to fishery is a natural or common law right vested with wetland owners and cannot be taken away by the State. Such taking away of the natural and common rights infringes the fundamental rights of the petitioners.
5. Right to fishery is vested with the petitioner from time immemorial and is inseparable from agrarian and irrigation rights. Such a right of occupation, trade or business cannot be taken away from the citizens through executive orders without due process of law.
6. The fishery contractor may not provide the water at times of water scarcity foregoing his losses in scarcity years. Considering the paddy production and its value, fishery production is not worth comparing. No data is made

available of such losses to the villagers in such eventualities. No consultation is made before taking away the rights.
7. When the Government is collecting *Meen Pasy* (fishery cess) from the Petitioner, how can it give the same rights to the FFDA for the same activity?
8. The common interest of the village will be affected and the unity, integrity and communal harmony of the villagers will be affected by such action of the government.
9. When the Government still focuses on agricultural production for basic need fulfillment, how can an enterprise like fishery be made out at the cost of agriculture, since the fishery production sets out certain quantity of water at the cost of agriculture.

Annex 2

The timeline of the encroachments and efforts taken by villagers. (*See* Section B.8)

1. During the 1800s (under the *Zamindari* system), the Goundankulam tank was maintained by the ryots with the support of the zamindar. The map published by the British in the early 1800s clearly depicts the size of the tank and its sources. The tank was laid in SF No. 776 of the village and the total extent of tank water spread was marked as 5.17 ha. The map had been recovered from the archives and used as evidence in the court case by the villagers. However, the subsequent maps published in 1980 do not show the extent fully because of the *patta* issued in the middle of the tank.
2. During the time of the Zamindars the land tax was collected by *Avildhar* from the farmers and much of it was used for maintenance activities. These are narrated and remembered by some of the villagers. (However, these need verification).
3. After the *Zamindari* system was abolished, the water body was brought under the list of Panchayat Union tanks of the Bodinaickanur Panchayat Union. So this tank is presently maintained by the Block Development Officer (BDO) of the Panchayat Union. For all practical purposes

the BDO's office has to represent the case but it hardly does so because of its preoccupations!

4. This tank was used for ground water recharge after the 1950s and hence more than 60 wells sprung up in the command area benefiting around 200 acres. The functioning of the tank was reduced from the direct source to an indirect source of irrigation, rendering the tank bed vacant most of the time because of lack of maintenance of the inlet channel to the tank. The encroachers had started their act because of the farmers' indifference and neglect. Complete neglect by the villagers had resulted after the intense well irrigation in the 1960s. The works attended by the villagers such as cleaning of the supply channels, and protecting the tank from the encroachments had almost been given up.

5. More number of wells were dug for irrigation in later years and the depth of the tank has been continuously increased. This reached around 120 feet making the well-irrigated agriculture economically unviable.

6. Most of the wells dried in the late 1980s due to over-exploitation of ground water resulting from their poor recharge capacity. By then the tank water spread area was gradually encroached upon by the foreshore farmers. The encroachers started to cultivate the land and paid land tax with penalty which is a meager amount compared to the realisation. There were ten encroachers cultivating 3.71 ha of land in an intense manner. They also dug wells inside the tank for irrigation purposes and got electric connections for them using illegal means.

7. The survey number 776 of the older settlement of the nineteenth century got changed into SF No. 346 in the year 1981 after the re-survey. The fragmentation of the tank bed was marked without giving any notice to villagers and hiding it from the knowledge of the ayacutdars. This has resulted in the encroachers getting *patta* (ownership).

8. The new realisation by the farmers had come after a series of failures of their wells. They started working for eviction of encroachments and sending petitions to various officers and none of them worked. They organised as a formal group by themselves into Tank Farmers' Association in

the year 1997, got it registered through Theni District Tank Farmers' Federation and approached it with a militant attitude.
9. They secured funds from District Rural Development Agency (DRDA), Panchayat Union and other sources and evicted the encroachers by themselves without any formal support from the lower level bureaucracy. In the same way they also cleared the supply channel encroachment also through coercion, and threats.
10. Presently the last and final battle to retain what they have restored is being held at the High Court against the encroachers. No one is sure about what would happen there.

7

Ground Water – Legal Aspects of the Plachimada Dispute

SUJITH KOONAN

1. Introduction

Plachimada is a small village in the Palakkad district of the state of Kerala. 'Plachimada' has become synonymous with debate on legal regime of control and use of ground water after the Hindustan Coca-Cola Beverages Private Limited (hereafter the Company) started a plant in Plachimada. The plant was commissioned in March 2000 to produce its popular brands such as Coca-Cola, Fanta, Sprite, Limca, Kinley Soda, Maaza and Thumps Up.[1]

The local people in Plachimada started their protest against the Company within two years after the Company started production.[2] The local people complained that the quality and quantity of ground water in the area has deteriorated due to over-exploitation of ground water by the Company.[3] While the public protest against the company was growing, the Perumatty Grama Panchayat (hereafter the Panchayat) refused to renew the license of the Company in 2003.[4]

---

[1] C.R. Bijoy, 'Kerala's Plachimada Struggle: A Narrative on Water and Governance Right', 41/41 *Economic and Political Weekly* 4332, 4333 (2006).
[2] *See* the special volume of *Keraleeyam* on Plachimada issue, Vol. 90, January 2005. *See also* Bijoy, note 1 above at 4334.
[3] Ibid.
[4] Ibid., at 4335.

The refusal to renew the licence of the Company by the Panchayat was the beginning of the legal battle.[5] The issue reached the Department of Local Self Government, Government of Kerala for 'appropriate orders' as per the direction of the Kerala High Court.[6] However, the legal battle did not end at the level of the Department of Local Self Government. The issue of ground water depletion and the refusal of the Panchayat to renew the license of the Company came before the Single Judge and subsequently before the Division Bench of the Kerala High Court as appeal. The matter is now pending before the Supreme Court of India.

The major legal issues discussed by the Kerala High Court in the Plachimada case was the right of a landowner to extract ground water from his land and the power of the Panchayat (or Local Bodies in general) to regulate the use of ground water by private individuals.[7] Apart from this, the legal framework regulating the quality of ground water (pollution control laws) also forms part of the legal regime. Even though the issue of pollution and its impacts on public health and local economy were raised in the protest against the Company, pollution control laws have not been a major focus in the Plachimada case.

In this background, the first part of the paper briefly describes the factual background leading to the Plachimada case. The second part analyses the legal and institutional framework addressing the issue of ground water depletion and pollution. The third part discusses the Plachimada case as decided by the Kerala High Court. The Kerala government enacted the Kerala Ground Water (Control and Regulation) Act in 2002. The Act was notified in 2003. By this time the matter had already come before the Kerala High Court and therefore, this Act has not been applied in this case. Since the Act is the major statutory framework to address situations like Plachimada in future, an analysis of the Act is included in the

---

[5] The Company challenged the cancellation of license in the Kerala High Court. See *Hindustan Coca-Cola Beverages Private Limited v Perumatty Grama Panchayat and Anr*. The Kerala High Court, Original Petition No. 13513 of 2003, Judgement of 16 May 2003.

[6] Ibid.

[7] The word 'Plachimada case' is used in this paper to indicate cases decided by the single judge and division bench of the Kerala High Court. These decisions are discussed in detail in the later part of this paper.

fourth part. The fourth part also describes the major contentions raised in the pending appeal in the Supreme Court of India.

2. The Factual Background

Palakkad district in the state of Kerala, where the Coca-Cola plant is situated, is an important agricultural region and is popularly known as the 'rice bowl of Kerala'.[8] Majority of the people in this district depend upon agriculture for their livelihood.[9] Plachimada depends on ground water and canal irrigation for agricultural and domestic purposes.[10] Plachimada is also home to several scheduled castes and scheduled tribes.[11] The villagers are predominantly landless and agricultural labourers.[12] The site of the plant is surrounded by a number of water reservoirs and canals built for irrigation.[13] Palakkad district is in the rain shadow area of the Western Ghats and is thus a drought-prone area.[14]

The local people started their agitation against the Company within a year after the Company set up its plant in Plachimada.[15] The major demand of the protest was the immediate closure of the Company.[16] Later, several non-governmental and political organisations joined the agitation. Several study reports have been

---

[8] Jananeethi, Report on the Amplitude of Environmental and Human Rights Ramifications by the HCCBPL at Plachimada 1 (Thrissur: Jananeethi, July 2002).

[9] Ibid.

[10] *See* Bijoy, note 1 above at 4333.

[11] The area, which is said to have affected due to the working of the Company, consists of thirty to forty per cent tribals and ten per cent dalits. *See* C.R. Bijoy, note 1 above at 4333.

[12] *See* Jananeethi, note 8 above at 1.

[13] The site is located barely three kilometers to the north of the Meenakkara Dam reservoir and a few hundred meters west of the Kambalathara and Venkalakkayam water storage reservoirs. The Moolathara main canal of the Moolathara barrage passes less than ten metres north of the factory compound and the main Chittoor River runs very close to the Coca-Cola plant. *See* Jananeethi, note 8 above at 1.

[14] R.N. Athavale, Water Management at the Coca-Cola Plant at Moolathara Village, Palakkad District, Kerala State, India (on file with the author, 2002).

[15] *See* Keraleeyam, note 2 above.

[16] *See* Bijoy, note 1 above at 4334.

published explaining the causes and effects of the deterioration of ground water quality and quantity in Plachimada. They give different explanations regarding the causes and effects of ground water problems in Plachimada.

Keraleeyam, a Malayalam journal which actively supports public protests against the Company, reported that the people started facing adversities within six months after the Company started its production.[17] A study conducted by Jananeethi, an NGO based in Kerala, reported that salinity and hardness of water had risen after the Company started its manufacturing process.[18] A study conducted by Dr Sathish Chandran revealed that water from some open wells and shallow bore wells in the nearby area has an extremely unpleasant, strong, bitter taste.[19] The people who used this water complained of a variety of illnesses such as burning sensation in the skin, greasy, sticky hair; stomach disorders and skin deformities.[20] It was also reported that a few wells in the nearby area had become dry after the Company started ground water extraction.[21] The insufficiency of water had also resulted in the decline of agricultural production. Consequently, local economy and life in the area was alleged to have been ruined.[22]

R.N. Athavale, Emeritus Scientist in National Geophysical Research Institute, Hyderabad, conducted a study on the issue of water problems in August 2002 at the request of the Company. The Athavale Report concluded that there was no 'field evidence available to indicate over-exploitation of ground water from the premises of the Coca-Cola Plant'.[23] The report observed that the depletion of ground water in Plachimada could be due to the

---

[17] *See* Keraleeyam, note 2 above. *See also* Yuvajanavedi, Report on the Environmental and Social Problems Raised due to Coca-Cola and Pepsi in Palakkad District (Thiruvananthapuram: Yuvajanavedi, November 2002).

[18] *See* Jananeethi, note 8 above.

[19] Sathish Chandran, Adverse Environmental Impact of the Hindustan Coca-Cola Beverages Pvt. Ltd. located in the Plachimada Area in the Perumatty Panchayat in the Chittur Taluk of the Palakkad District (report on file with the author, 2002).

[20] Ibid.

[21] *See* Athavale, note 14 above.

[22] *See* Yuvajanavedi, note 17 above.

[23] *See* Athavale, note 14 above.

deficit in rainfall and consequent insufficiency in the recharge or replenishment of ground water. The report further concluded that 'water quality deterioration of any well in the neighbourhood cannot be considered as due to the pumping activity in the plant area'.[24]

The problem of pollution due to solid wastes in Plachimada came to popular attention through a British Broadcasting Corporation (BBC) report. On 25 July 2003, the BBC reported the presence of heavy metals – lead and cadmium – in quantities higher than the approved limit in the sludge supplied by the Company as fertiliser.[25] The BBC report has also alleged that the Company had clandestinely dumped the sludge in the nearby river-bed. The BBC study shows that the sludge supplied by the Company is dangerous to health and it had no value as manure. The heavy dumping of the sludge in agricultural fields has also been reported by Jananeethi in 2002.[26]

The Kerala Pollution Control Board (KPCB) examined the sludge samples from the factory premises and found cadmium in higher concentration than the approved maximum limit under Schedule 2, Class A of the Hazardous Waste (Management and Handling) Rules, 1989 as amended in 2003 and therefore directed that it should be treated as hazardous waste. The KPCB, thereafter, directed the Company to 'take immediate action to stop the supply of this waste to external agencies and also internal use as manure'.[27] The Company was also directed to recover the sludge that has been already transported outside and store the same in a secured site within the factory premise.[28]

The KPCB conducted a further study on this matter and concluded that: 'the concentration of cadmium and other metals were found to be below the limit prescribed under the Schedule 2 of the Hazardous Waste (Management and Handling) Rules, 1989

---

[24] Ibid.

[25] A transcribed version of the BBC report is on file with the author.

[26] *See* Jananeethi, note 8 above; Yuvajanavedi, note 17 above.

[27] Letter issued by the KPCB to the Hindustan Coca-Cola Beverages Private Limited, Letter No. PCB/HO/HWM/CC-PLT/2003 dated 7 August 2003.

[28] Letter issued by the KPCB to the Hindustan Coca-Cola Beverages Private Limited, Letter No. PCB/HO/HWM/CC-PLT/2003 dated 7 August 2003.

as amended in 2003, and hence the solid wastes generated in the Company will not come under the said rules'.[29] But it was also stated in the report that the presence of cadmium in the common Panchayat well is double the permissible limit and touches the permissible upper limit in another well. The KPCB's comment about this was that: '... in the common Panchayat well could a small quantity of cadmium be detected'.[30] Later, the KPCB sent a letter to the President of the Perumatty Grama Panchayat informing them that water in the Panchayat well should not be used for drinking purposes.[31]

Another study conducted by the Central Pollution Control Board (CPCB) two months after the KPCB study concluded that the sludge from the Effluent Treatment Plant (ETP) and the sludge supplied by the Company to farmers for using as fertiliser contain heavy metals like lead and cadmium in more than permissible limits. The CPCB report warrants the sludge to be treated as per the Hazardous Waste (Management and Handling Rules) 1989, as amended in 2003.[32]

The Supreme Court Monitoring Committee (SCMC), constituted by the Supreme Court of India under Writ Petition No. 657/95, visited the Company and nearby areas in August 2004.[33] The SCMC noticed that the drinking water source adjacent to the Company was contaminated due to the illegal dumping of wastes by the Company.[34] By taking note of the SCMC findings, the KPCB

---

[29] Kerala State Pollution Control Board, A Study Report on the Presence of Heavy Metals in Sludge Generated in the Factory of M/s Hindustan Coca-Cola Beverages Pvt. Ltd., Palakkad (Thiruvananthapuram: Kerala State Pollution Control Board, September 2003).

[30] Ibid.

[31] Letter No. PCB/PLKD/W-217/2001 dated 31 October 2003.

[32] Central Pollution Control Board, Report on Heavy Metals and Pesticides in Beverages Industries (Delhi: Central Pollution Control Board, November 2003).

[33] *See Research Foundation Science Technology Natural Resource Policy v Union of India*, Supreme Court of India, Writ Petition No. 657 of 1995, Order dated 14 October 2004, para. 7.

[34] Supreme Court Monitoring Committee On Hazardous Wastes (SCMC), Report of the visit of the SCMC to Kerala with recommendations, 14 August 2004, Source: http://www.thesouthasian.org/archives/2006/pdf_docs/SCMC_Report_on_Kerala_Visit%5B1%5D%20August%202004.pdf.

directed the Company to close the factory until it complies with the provisions of the Hazardous Waste (Management and Handling Rules), 1989 as amended in 2003.[35]

The legal battle related to the ground water issue in Plachimada began when the Perumatty Grama Panchayat passed a resolution on 7 April 2003 refusing to renew the license given to the Company.[36] The Company challenged the action taken by the Panchayat in the Kerala High Court.[37] The Kerala High Court directed the Company to approach the appropriate forum, that is, the Department of Local Self-Government.[38] The Department of Local Self-Government directed the Panchayat to constitute an expert group to study the matter and decide accordingly.[39] Having felt aggrieved by the direction of the Department of Local Self-Government, the Panchayat approached the Kerala High Court.

3. Legal and Institutional Framework

The Coca-Cola Company started their operation in the year 2000 and the people's agitation against the Company began in 2002. Meanwhile, the Kerala legal system underwent a major change in 2002 through the enactment of the Kerala Ground Water (Control and Regulation) Act. But the said Act was not applicable as it was notified only in 2003. In the absence of a specific statutory framework, principles such as the public trust doctrine and the common law rule regarding the right of the landowner over ground water have been discussed in the Plachimada case. Apart from these principles, there are a number of environmental laws that could have been applied in the Plachimada case.

---

[35] The KPCB's direction to the Hindustan Coca-Cola Beverages Private limited, vide order No. PCB/HO/H&R/485/04 dated 23 August 2004.

[36] *See Perumatty Grama Panchayat v State of Kerala and Ors.*, The Kerala High Court, Original Petition (Civil) No. 34292 of 2003, Judgement of 16 December 2003, para. 2.

[37] *Hindustan Coca-Cola Beverages Private Limited v Perumatty Grama Panchayat and Anr.*, The Kerala High Court, Original Petition No. 13513 of 2003, Judgement of 16 May 2003.

[38] Ibid.

[39] *See Perumatty Grama Panchayat*, note 36 above, para. 3.

3.1 Principles

One of the basic issues in water law is that of rights, that is, what kind of rights do the people have, or ought to have, and what are the rights of the state.[40] This implies that the legal relationship between and among the state, individuals and water resources is one of the basic issues that need to be defined legally. Even though there have been a number of enactments defining rights of the state and individuals regarding surface water resources, there was no such specific law(s) regarding ground water.[41] Hence, the regime mainly consisted of principles of which two important principles discussed in the Plachimada case are the Public Trust Doctrine and the common law principle on ground water rights.

3.1.1. Public Trust Doctrine

The Public Trust Doctrine (PTD) describes the state's relationship with water resources and citizens of the state.[42] The doctrine is based not on the notion of rights but that of duties.[43] It means, state has a duty to protect, preserve, manage and use the trust property in the public interest. The trust has been reposed in the state by the public as a fundamental contract that the state will act in public interest and in the interest of ecology.[44] The meaning of the doctrine can be expressed as: 'the state which holds the natural waters as a trustee, is duty- bound to distribute or utilise the waters in such a way, that it does not violate the natural right to water of any individual or group and safeguards the interest of the public and of ecology (or nature)'.[45]

Tracing the origin of the public trust concept, most scholars look to the Institutes of Justinian, a body of Roman civil law

---

[40] Chhatrapati Singh, 'Water Rights in India', *in* Chhatrapati Singh ed., *Water Law in India* (Delhi: Indian Law Institute, 1992).

[41] Chhatrapati Singh, *Water Rights and Principles of Water Resource Management* 39 (Bombay: Tripathi, 1991).

[42] Melissa Kwaterski Scanlan, 'The Evolution of the Public Trust Doctrine and the Degradation of Trust Resources: Courts, Trustees and Political Power in Wisconsin', 27 *Ecology Law Quarterly* 135 (2000).

[43] *See* Singh, note 41 above.

[44] *See* Singh, note 41 above at 76.

[45] Ibid., at 76.

assembled approximately in 530 CE.[46] This text articulated the universal notion that water courses should be protected from complete private acquisition in order to preserve the lifelines of communal existence.[47] Ancient Roman law recognised public right in water and the seashore which were unrestricted and common to all. These rights were considered as part of natural law. It was considered as 'common to mankind by the law of nature. No one is forbidden provided he respects habitations, monuments and buildings'.[48]

The roots of public trust doctrine can also be seen in *dharmasatra*, as the king was the upholder and protector of natural resources for and on behalf of the people.[49] The 'public trust values in water' can also be found in many other ancient legal systems. For instance, in Chinese water law of 249–207 BC, in ancient and traditional customs of people of Nigeria, in Islamic water law, in the laws of medieval Spain and France, in the Mexican laws, etc.[50]

The concept of public trust was originated to provide public access to the waterways for commercial benefit, and their preservation was viewed as a factor to facilitate trade and establish communication lines.[51] The scope of the doctrine has been widened, in the course of time, from 'access to all' to 'preservation of all natural resources'.[52] The widening of the scope of the doctrine can

---

[46] Joseph L. Sax, 'The Public Trust Doctrine in Natural Resource Law: Effective Judicial Intervention', 68 *Michigan Law Review* 471 (1970); Patricia Kameri-Mbote, 'The Use of the Public Trust Doctrine in Environmental Law', 3/2 *Law, Environment and Development Journal* 197 (2007).

[47] George Smith and Michael Sweeny, 'Public Trust Doctrine and Natural Law: Emanations within a Penumbra', 33 *Boston College Environmental Affairs Law Review* 307–344 (2006).

[48] *See* Scanlan, note 12 above at 140.

[49] *See* Singh, note 41 above at 76.

[50] *See* Sax, note 46 above.

[51] *See* Smith and Sweeny, note 47 above.

[52] *See* Joseph L. Sax, 'The Public Trust Doctrine in Natural Resource Law: Effective Judicial Intervention', 68 *Michigan Law Review* 471 (1970); Patricia Kameri-Mbote, 'The Use of the Public Trust Doctrine in Environmental Law', 3/2 *Law, Environment and Development Journal* 197 (2007). *See also* Cynthia L. Koehler, 'Water Rights and the Public Trust Doctrine: Resolution of the Mono Lake Controversy', 22 *Ecology Law Quarterly* 541–589 (1995).

be mainly attributed to American courts.⁵³ Hence, it could be seen that the PTD is not new to the legal system. It has been in existence for a long time and has been widened in scope over the years to respond to the changing needs.

In the contemporary context, the PTD seems to cast a duty upon the state to protect and preserve natural resources for and on behalf of beneficiaries, that is, the people. The beneficiary not only includes the present generation but future generations also.⁵⁴ The duty of the state also includes the duty to furnish information regarding the trust property to the beneficiaries.⁵⁵

There is no statute in India which makes the PTD a part of the Indian legal system. The doctrine has been incorporated into the Indian legal system by the Supreme Court of India. The Supreme Court in *M.C. Mehta v Union of India (1997)* took note of the development of the doctrine through American case laws and scholarly writings.⁵⁶ By quoting relevant American case laws and scholarly articles, the Supreme Court seems to have recognised and accepted the PTD as a valid legal principle having contemporary relevance in the Indian context. In the said case, the Supreme Court declared that 'our legal system – based on English common law – includes the public trust doctrine as part of its jurisprudence'.⁵⁷ It was further held that the state is the trustee of all natural resources which are by nature meant for public use and enjoyment. Public at large is the beneficiary of the sea-shore, running waters, airs, forests and ecologically fragile lands.⁵⁸

The PTD, as explained by the Supreme Court in *M.C. Mehta Case*, has been recognised in several subsequent case laws.⁵⁹ The legal validity and contemporary relevance of the doctrine has also

---

⁵³ *M.C. Mehta v Union of India*, Supreme Court of India, (1997)1 SCC 388, para. 33.
⁵⁴ *See* Kameri-Mbote, note 46 above.
⁵⁵ Ibid.
⁵⁶ *M.C. Mehta v Union of India*, (1997)1 SCC 388, paras 24–33.
⁵⁷ Ibid., para. 34.
⁵⁸ Ibid., para. 34.
⁵⁹ *See M. I. Builders Pvt. Ltd. v Radhey Shyam Sahu* (1999) 6 SCC 464; *Intellectual Forum v State of Andhra Pradesh* (2006) 3 SCC 549; *Karnataka Industrial Area Development Board v Kenchappan* (2006) 6 SCC 371.

been recognised in a report of the law commission of India.[60] Moreover, as per the Constitution of India, 'the law declared by the Supreme Court shall be binding on all courts within the territory of India'.[61] Supreme Court decisions are to be regarded as law of the land unless and until changes have been made through a subsequent Supreme Court decision or an express statutory provision. Therefore, the PTD can be considered as a part of environmental law in India and should be followed mandatorily.[62]

The public trust doctrine, in principle, can be a basis of the power of the state to control the use of ground water by private individuals. It can also be a theoretical basis to explain the duty of the state to take measures for the protection and preservation of natural resources (ground water in the present case) for present and future generations. Inaction on the part of the state would amount to violation of the trust and cannot be justified in law.

3.1.2 Common Law Rule on Ground water

One of the peculiar facts in the history of water law is the separate development of law governing surface water sources – such as lakes and rivers – and that governing ground water. British Common law recognised rights of riparians, that is, the usufructuary right subject to state control. Principles evolved under common law mainly addressed rights in surface water. Ground water was dealt with under a different regime, that is, as part of an individual's right to enjoy property. The government's control was not applicable to water sources like wells, tanks, tube wells existing in private land. Early irrigation laws and the Indian Easements Act establish the existence of a different set of principles for surface water and ground water.[63]

Common law considered ground water as part of the soil in which it exists. Ground water was considered as a chattel attached to the land without having a distinctive character of ownership

---

[60] *See* Law Commission of India, Report on Proposal to Constitute Environmental Courts, 186th Report (New Delhi: Law Commission of India, 2003).
[61] *See* Article 141 of the Constitution of India *in* P.M. Bakshi, *The Constitution of India* (Delhi: Universal Law Publishing Co., 2006).
[62] *See* Bakshi, note 61 above at 134–135.
[63] *See* Singh, note 41 above.

from the earth.⁶⁴ Common law rule permitted the landowner to extract any extent of ground water, even though it is dangerous to his neighbours or may diminish or take away the water from neighbouring wells.⁶⁵ Common law dismisses such a problem with the curt observation that such a result is *damnum absque injuria* (not actionable under law).⁶⁶

Law of easements is not applicable to ground water.⁶⁷ Right to draw water from a well to irrigate the field cannot be acquired as an easement right by prescription.⁶⁸ Section 17 (d) of the Easements Act lays down that there could not be any prescriptive right in ground water, not passing in a defined channel. The law on this question is discussed at great length in the case of *Acton* v. *Blundell* that the easement right cannot be obtained over the percolating ground water flowing in undefined channels.⁶⁹

This common law position has been followed in India also. It is very clear from the words of Chandra Shekhara Aiyar J. that:

> The general rule is that the owner of a land has got a natural right to all the water that percolates or flows in undefined channels within his land and that even if his object in digging a well or a pond be to cause damage to his neighbour by abstracting water from his field or land it does not matter in the least because it is the act and not the motive which must be regarded. No action lies for the obstruction or diversion of percolating water even if the result of such abstraction be to diminish or take away the water from a neighbouring well in an adjoining land.⁷⁰

---

⁶⁴ *See* Singh, note 41 above at 39.
⁶⁵ V. Sitararama Rao, *Law Relating to Water Rights* 185–186 (Hyderabad: Asian Law House, 1996).
⁶⁶ Frazier v Brown, 12 Ohio st. 294 (1861) as cited in Robert Emmet Clark ed., *Water and Water Rights*, Vol. I, 71 (Indiana: The Allen Smith Company Publishers, 1967).
⁶⁷ *See* Rao, note 65 above at 185.
⁶⁸ *Het Singh v Anar Singh*, A.I.R. 1982 All. 468.
⁶⁹ *Manturabai v Ithal Chiman*, A.I.R. 1954 Nag. 103 [A.I.R. 1951 Nag. 447 reversed] as cited in G.C. Mathur ed., *Amin and Sastry's Law of Easements* 434 (Lucknow: Eastern Book Company, 1984).
⁷⁰ *Kesava Bhatta v Krishna Bhatta*, AIR 1946 Mad. 334, 335.

Common law recognised it as a right of the land owner to divert or appropriate ground water from his land. The control which common law applies over this right is that when exercising this right, a landowner shall not cause any damage to water flowing in a defined channel. The same idea has been reiterated by Lord Hatherley as 'if you cannot get at the underground water without touching the water in a defined surface channel you cannot get at it at all. You are not by your operation or by any act of yours, to diminish the water which runs in a defined channel'.[71]

The historical reason for the evolution of these rules seems to be the lack of knowledge about ground water hydrology which prompts one to leave it out of control.[72] Since the mechanisms for tapping ground water was not much improved, the chance of extraction of too much water was not in existence and as such it was unlikely to cause any serious social problem which requires mediation through law. Both these reasons have now become obsolete. The science of hydrology developed fast and now the processes involved in the recharge and discharge of ground water and the quantity of water available in a region are matters within the human knowledge. The behaviour of ground water is no longer a mystery. Availability of powerful mechanical devices for drawing ground water has also resulted in tilting the balance.[73] The quality and quantity of ground water have deteriorated due to indiscriminate exploitation. Another implication of the common law rule is that it leaves out all landless people and tribals who may have group (community) rights over the land but not private ownership. These situations necessitate the evolution of a new jurisprudence to ensure access to water for all and the protection and preservation of the resource.

---

[71] *Grand Junction Canal Co. v Shugar*, (1871) 6 Ch. A. 483 as cited by Chandra Sekhara Aiyar J. in *Kesava Bhatta v Krishna Bhatta*, AIR 1946 Mad. 334.

[72] *See* Lawrence J. MacDonnell, 'Rules Guiding Groundwater Use in the United States', 1 *Indian Juridical Review* 43, 46 (2005); Sanjiv Phansalkar and Vivek Kher, 'A Decade of the Maharashtra Groundwater Legislation: Analysis of the Implementation Process', 2/1 *Law, Environment and Development Journal* 67 (2006).

[73] A. Narayanamoorthy and R.S. Deshpande, *Where Water Seeps! Towards a New Phase in India's Irrigation Reforms* 36 (New Delhi: Academic Foundation, 2005).

3.2 Environmental Laws

Environmental laws in India have developed in the last three decades. Since the 1970s, a number of statutes have been enacted to address various aspects of the environment such as prevention and control of pollution, conservation of forest and protection of wildlife.[74] One of the important features of this development is the emphasis on protection and preservation of the environment. Prior to this phase, Indian environmental law mainly consisted of claims made against tortious actions such as nuisance and negligence.[75] The Indian judiciary, particularly the higher judiciary, also made remarkable contributions to the development of environmental laws in this country.[76] Thus it could be said that environmental law in India has been developed through legislative and judicial initiatives.[77]

The Water (Prevention and Control of Pollution) Act, 1974 (hereafter 'The Water Act'), the Environment Protection Act, 1986 (hereafter 'The EP Act') and the Hazardous Wastes (Management and Handling) Rules, 1989 as amended in 2003 (hereafter 'The Rule') are the major legal frameworks that have been in force since the beginning of the Plachimada problem. These enactments

---

[74] Some of the major environmental legislations are: Water (Prevention and Control of Pollution) Act, 1974; Air (Prevention and Control of Pollution) Act, 1981; Environment Protection Act, 1986; Forest (Conservation) Act, 1980; Wild Life (Protection) Act, 1972; Biological Diversity Act, 2002; Public Liability Insurance Act, 1991; National Environment Tribunal Act, 1995 and National Environment Appellate Authority Act, 1997. For an analysis of environmental legislation in India, *see*, Shyam Divan and Armin Rosencranz, *Environmental Law and Policy in India: Cases, Materials and Statutes* (Oxford: Oxford University Press, 2001).

[75] B.N. Kirpal, Developments in India Relating to Environmental Justice, Paper presented in Global Judges Symposium on Sustainable Development and the Role of Law in Johannesburg, South Africa, on 18–20 August 2002. *Source:* http://www.unep.org/law/Symposium/Documents/Country_papers/INDIA%20.doc.

[76] Major principles and doctrines of environmental law have been incorporated as part of Indian law by the Indian judiciary through case laws. Some of the important case laws are discussed later in this paper. For further references, *see* Divan and Rosencranz, note 74 above.

[77] *See* Kirpal, note 75 above.

provide legal and institutional mechanisms to address various aspects of ground water quality and quantity issues in Plachimada.

The central government enacted the Water Act with the object of 'prevention and control' of water pollution and to 'maintain or restore' the 'wholesomeness' of water. The preamble to the Water Act gives an indication that the Act is meant for protecting and preserving water in the larger interest of living and non-living organisms. The EP Act also sets forth the same philosophy in a comprehensive manner to cover the whole ecosystem. It is expressly stated that the object of the EP Act is the 'protection and improvement' of the environment. Hence, these laws provide the framework for the protection and preservation of the environment.

The word 'pollution' under the Water Act is defined broadly to include all direct and indirect actions, which can render water harmful or injurious to public health, safety or to the life of other organisms.[78] The authority constituted under the Water Act, Pollution Control Board, is empowered to carry out the objectives of the Act, that is, prevention and control of water pollution. The Water Act prescribes a two-tier institutional mechanism, one at the central level (hereafter 'Central Board') and the other at the state level (hereafter 'State Board'). The Pollution Control Board also has the responsibility to implement the EP Act. Therefore, powers and responsibilities of the Pollution Control Board are very wide and it is the primary agency responsible to take care of the quality of the environment as a whole.

The State Board under the Water Act is empowered to enter and inspect any premises, conduct investigation and advise the state government with regard to the prevention, control or abatement of water pollution.[79] Moreover, the State Board is also empowered to issue any order, which includes the order requiring any person concerned to construct sufficient mechanisms for the disposal of sewage and trade effluents or to modify, alter or extend any such existing system or to adopt such remedial measures necessary to prevent, control or abate water pollution. It also has the power to issue an order of closure, prohibition or regulation of

---

[78] The Water (Prevention and Control of Pollution) Act, 1974, Section 2 (e). *Source:* http://www.ielrc.org/content/e7402.pdf.
[79] Ibid., Section 16 (h).

industries.[80] The Water Act also makes it mandatory for any industry that is likely to pollute water to obtain a license under the Act.[81]

The Water Act also gives some special and overriding powers to the Central Government. The Central Government is empowered to give directions to any person, officer or authority in exercise of its powers under the Act.[82] This includes power to direct closure of any industry, prohibition or regulation of any industry and stoppage or regulation of supply of water, electricity or any other services.[83] The person or authority against whom such directions have been given is bound to comply with them.[84]

The Rule has been enacted under the EP Act to specifically address the alarming problem of hazardous wastes.[85] The Rule lays down detailed schedules, which consists of lists of hazardous wastes to be treated as per the Rule. Hazardous wastes are classified into different categories depending upon toxicity; prohibition or restriction is prescribed accordingly.

The Rule requires only authorised dealers to deal with hazardous wastes. The generator of hazardous wastes is duty bound to give the authority (the Pollution Control Board) all details about the waste.[86] The generator is also required to obtain authorisation from the authority to handle, treat, transport and dispose of the waste.[87] The authority will grant permission after examining whether facilities are in compliance with the Rule or not. It is the duty of the authority to make sure that the concerned industry has sufficient mechanisms to treat hazardous wastes so as to avoid implications upon public health, public safety and the environment. In the event that pollution does occur, the rule expressly places the

---

[80] Ibid., Sections 16 (1) and 33A.

[81] Ibid., Section 25.

[82] *See* The Water Act, Section 33A.

[83] Ibid., Section 33A.

[84] Ibid., Section 33A.

[85] The legal regime regulating toxic substances in India has been developed largely as a response to the Bhopal Tragedy that occurred in December 2004. For details, *see* Divan and Rosencranz, note 74 above at 514–562.

[86] The Hazardous Wastes (Handling and Management) Rule, 1989, Rule 5 (2). *Source:* http://envfor.nic.in/legis/hsm/hsm1.html, Rule 4.

[87] Ibid., Rule 5 (2).

liability upon the polluter to reinstate or restore the damaged element(s) of the environment.[88] If the polluter fails, the authority has the power to order the polluter to deposit an estimated amount that will be adjusted towards the expenses incurred to restore the environment.[89]

The normative contents of environmental laws in India have been widened through the interpretative role played by the Indian higher judiciary. Some of the cardinal principles developed as part of international environmental law are now part of Indian law. Of which, important principles relevant to the Plachimada case are the precautionary principle, polluter pays principle, public trust doctrine and the principle of absolute liability.[90]

After discussing the constitutional and statutory provisions related to the environment, the Supreme Court of India in *Vellore Citizens' Welfare Forum case* held that '...we have no hesitation in holding that the precautionary principle and polluter pays principle are part of the environmental law of the country'.[91] The Court further defines the precautionary principle. The precautionary principle casts duty upon the state to take measures to '...anticipate, prevent and attack the causes of environmental degradation'.[92] As per the precautionary principle, the precautionary measures shall not be postponed because of scientific uncertainty.[93]

The polluter pays principle has been recognised as part of Indian law by the Supreme Court in the *Bichhri case*.[94] It was held that '...once the activity carried on is hazardous or inherently

---

[88] Ibid., Rule 16 (2).

[89] Ibid., Rule 16 (2). The polluter pays principle has also been incorporated expressly in the National Environment Tribunal Act of 1995. *See* Section 3 of the National Environment Tribunal Act of 1995.

[90] The Law Commission of India in its 186th report has recognised these principles and recommended it to be applied. For details, *see* Law Commission of India, Report on Proposal to Constitute Environmental Courts, 186th Report (New Delhi: Law Commission of India, 2003), Chapter VII.

[91] *Vellore Citizen's Welfare Forum v Union of India,* Supreme Court of India, (1996) 5 SCC 647, para. 14.

[92] Ibid., para. 11.

[93] Ibid.

[94] *Indian Council for Enviro-Legal Action and Ors. v Union of India,* Supreme Court of India, (1996)3 SCC 212.

dangerous, the person carrying on such activity is liable to make good the loss caused to any other person by his activity irrespective of the fact whether he took reasonable care while carrying on his activity...'[95] The polluter pays principle as understood and illustrated by the Supreme Court of India seems to be linked with the liability of the polluter in the case of hazardous substances.[96] The Court seems to have used the principle as a basis of the liability of the polluter vis-à-vis damages caused to individuals and the environment. The liability as illustrated by the Supreme Court is absolute in nature.[97] The Supreme Court further held that sections 3 and 5 of the EP Act empower the Central Government to give directions and take measures for giving effect to this principle.[98]

Hence, it appears that environmental laws in India provide legal framework to deal with pollution related problems. The legal framework also provides institutional mechanism to implement laws. The scope of environmental laws has been widened through judicial initiatives, that is, by incorporating some of the cardinal principles as part of environmental laws in India. The Government and courts in the country are bound to follow and apply these principles.

This legal and institutional framework seems to be relevant in the ground water related issues in Plachimada. The powers under the abovementioned legal framework have been invoked more than once. This is clear from various investigations and directions given by the KPCB to the Company including the closure order.

However, it is to be noted that investigations and actions taken by the KPCB and other government bodies such as Central Ground Water Board have been criticised by the Plachimada Struggle Committee. After analysing the CGWB's report, *Keraleeyam*, a Malayalam magazine pointed to the presence of high TDS (Total

---

[95] Ibid., para. 65; *see also M.C. Mehta v Union of India*, AIR 1987 SC 1086 (Oleum Gas Leak Case).

[96] For details regarding the origin and development of the polluter principle, *see* Nicolas de Sadeleer, *Environmental Principles: From Political Slogans to Legal Rules* (Oxford: Oxford University Press, 2002).

[97] Ibid., para. 66; *see also M.C. Mehta v Union of India*, AIR 1987 SC 1086 (Oleum Gas Leak Case).

[98] Ibid., para. 67.

Dissolved Substances), hardness, EC (electrical conductivity) and high chloride content in the wells situated within hundred meter circumference from the Company. This element should have been an important one to decide the link between the Company and the ground water pollution in Plachimada. But CGWB has neglected this fact.[99] It has been further criticised that the report by the Kerala Ground Water Department, though identified pollution problem in some of the wells, neglected it as marginal.[100]

It could be seen that most of the reports recognise depletion and pollution of ground water in Plachimada but investigations conducted by government agencies do not find a link between the depletion of ground water in Plachimada and the Company.[101] However, it is to be noted that the environmental jurisprudence in India, particularly the precautionary principle, requires the Government to take adequate measures even in the absence of sufficient scientific evidence. At the same time a report by the KPCB confirms the pollution (including ground water pollution) caused by the dumping of hazardous waste by the Company. This fact seems to be sufficient for the KPCB to invoke the polluter pays principle. The Central Government is also empowered to give effect to the polluter pays principle under the EP Act. No such action is reported to have been taken by the KPCB or the Central Government.

---

[99] *See* Keraleeyam, note 2 above. *See also* Central Ground Water Board (CGWB), A Report on the Ground Water Conditions in and Around Coca-Cola Beverages Private Limited Company, Plachimada Village, Palakkad District, Kerala (Thiruvananthapuram: CGWB, 2003).

[100] *See* Keraleeyam, note 2 above.

[101] *See* Kerala Ground Water Department, Report on the Monitoring of Wells in and Around the Coca-Cola Factory in Plachimada, Kannimari, Palakkad district (Kerala Ground Water Department, September, 2003). A later report by the Kerala Ground Water Department recognised that the depletion of groundwater could be '...the combined effect of lower than normal rainfall and groundwater draft, especially by the wells in the factory'. *See* Kerala Ground Water Department, Report on the Monitoring of Water Levels and Water Quality in Wells in and Around the Hindustan Coca-Cola Factory at Plachimada, Palakkad District (Thiruvananthapuram: Kerala Ground Water Department, 2006).

3.3 The Role of the Panchayat

The role and powers of the Panchayat to regulate the use of ground water requires a special mention in relation to the Plachimada case. This was the major issue before the Kerala High Court in the Plachimada case.

The decentralisation policy, as it stands now, has been introduced as a result of 73rd and 74th Constitutional Amendment in 1992.[102] It envisages the constitution of the panchayat and devolution of power by the State Government to enable the panchayat to act as a micro level unit of local self-governance. However, the constitutional provision does not transfer powers and authority to panchayats. Powers and authority of the panchayat is required to be devolved through state legislation. Hence, the role of the panchayat necessarily depends upon the concerned state legislation. Most of the states have enacted laws to implement the constitutional norm envisaged in the 73rd and 74th amendment.[103] The Kerala State Government enacted the Kerala Panchayat Raj Act in 1994 (hereafter the PR Act).

The power of the panchayat over water resources in its jurisdiction is recognised in the Constitution and the PR Act. The subjects 'minor irrigation, water management and water shed development' and 'drinking water' have been included in the Schedule of the powers and functions of the panchayat in the Constitution of India.[104] The PR Act provides that all water resources, except the one passing through more than one panchayat, shall be considered as 'transferred to and absolutely vested' in the panchayat.[105] It means the panchayat has the power to control the use of drinking water resources in its jurisdiction. The PR Act requires factories and industries to obtain a license from the

---

[102] The 73rd and 74th Amendment deals with the Panchayat and Municipality respectively. *See* P. M. Bakshi, *The Constitution of India* (Delhi: Universal Law Publishing Co., 2006).

[103] Government of India, Annual Report of the Ministry of Rural Development (New Delhi: Ministry of Rural Development, Government of India, 2002–2003).

[104] *See* Bakshi, note 102 above, Article 243G, Eleventh Schedule, Entry 3 and 11.

[105] The Kerala Panchayat Raj Act, 1994, Section 218.

panchayat to establish factories.[106] Further, the PR Act gives responsibility to the panchayat to abate the nuisance created by any factory or industries in its jurisdiction.[107]

A combined reading of all these provisions indicates that the panchayat has the responsibility to maintain water resources and to take necessary measures to abate pollution problems in its jurisdiction. Public health and welfare seem to be the rationale for granting these powers to the panchayat. Therefore, the panchayat has the power to take necessary actions to protect the right of the people to clean and safe drinking water. However, it has been observed that though the political decentralisation has been successful, there is minimum administrative and financial decentralisation. Administrative and financial powers generally remain with the State Government.[108] The lack of administrative and financial powers tends to weaken the capability of the panchayat to carry out its responsibilities as envisaged in the Constitution and the PR Act.

The decentralisation principle has been introduced to constitute and enable panchayat raj institutions to function as units of local self-governance. The Kerala Government has implemented this principle by enacting the PR Act. Having acknowledged this legal development and its meaning and spirit, measures or actions taken by the panchayat to carry out its responsibilities need to be respected and facilitated. The curtailment of powers of the panchayat through administrative or judicial action would be against the meaning and spirit of the decentralisation principle.

One of the important responsibilities of the panchayat, under the PR Act, is to protect and preserve drinking water resources in its jurisdiction. The Perumatty Grama Panchayat, by invoking this provision, has taken action against the Company by refusing to renew the licence of the Company. However, this action of the Panchayat triggered the legal battle.

4. Plachimada Case

It is already stated that the cancellation of the licence of the company by the Panchayat was the major bone of contention in the writ

---

[106] Ibid., Section 233A.
[107] Ibid., Section 233.
[108] *See* Government of India, note 103 above.

petition filed before the Kerala High Court. Hence, the major question before the High Court was the legal validity of the Panchayat's action vis-à-vis the right of the Company to extract ground water from the land owned by the Company.

4.1 Background of the Case

The legal battle began when Perumatty Grama Panchayat (hereafter the Panchayat) passed a resolution deciding not to renew the licence of the factory on 7 April 2003. The Panchayat issued a show cause notice in this regard to the Company. The rationale for the Panchayat's action was stated as to stop the heavy usage of ground water by the Company which has caused depletion of ground water, heavy drought and drinking water scarcity and to avoid other environmental issues.[109]

In response to the show cause notice, the Company denied the allegations stated in the notice. It was claimed by the Company that it runs with all statutory clearances and the Panchayat can cancel its licence only if there is any violation of conditions of licence.[110] The Panchayat rejected the Company's claim for being 'against the facts' and decided to cancel the licence and directed the Company to stop production by a Resolution dated 5 May 2003.[111]

The Company challenged the order of the Panchayat in the Kerala High Court.[112] The Company was directed to approach the

---

[109] See Perumatty Grama Panchayat v State of Kerala and Ors., The Kerala High Court, Original Petition (Civil) No. 34292 of 2003, Judgement of 16 December 2003.

[110] Ibid. The licence issued by the Panchayat consisted of eight conditions. Out of which two conditions are relevant to the Plachimada case. Condition No. 3 provided that: 'the water using for the manufacturing should be tested periodically and the source of water should be kept sanitary without causing any pollution'. Condition No. 7 provided that: 'All wastes should be disposed off (sic) properly'. Conditions in the licence have not been discussed further by the Kerala High Court. See Perumatty Grama Panchayat, Proceedings of the Special Grade Secretary, 27 January 2000 (a copy of the proceedings on file with the author).

[111] Ibid.

[112] See Hindustan Coca-Cola Beverages Private Limited, note 5 above.

appropriate authority, that is, the Local Self Government Department (LSGD).[113] On appeal, the government ordered the Panchayat to constitute a team of experts from the departments of Ground Water and Public Health and the State Pollution Control Board to conduct a detailed investigation into the allegation and to take a decision based on the investigation report.[114] The Panchayat filed a writ petition, the major case in relation to Plachimada, against the order of the Government on the ground that 'protection and preservation of water resources is the exclusive domain of the Panchayat. When the Panchayat takes a decision based on relevant materials, the Government cannot interfere with it and dictate, how the Panchayat should act in the matter'.[115]

4.2 Case Law Analysis

The Plachimada Coca-Cola case came before the High Court of Kerala questioning the authority of the Panchayat to order the closure of the factory on the grounds that over-exploitation of ground water by the Company has resulted in acute shortage of drinking water. The major question addressed by the Court was, whether the Grama Panchayat has the power to regulate the right of a private individual or a company to extract ground water from their land or not and whether the Panchayat has the power to issue closure order against a company on account of over-exploitation of ground water or not. The writ petition, at the first instance, was decided by the single judge of the Kerala High Court and the appeal was decided by the division bench.

4.2.1 Single Judge Decision

The question considered by the single judge of the Kerala High Court was whether the decision of the Panchayat to cancel the

---

[113] Section 276 of the PR Act says that an appeal from the decision taken by the Panchayat would lie to the tribunal constituted under Section 271 (s) of the Act. But the tribunal was not constituted by the government at the time of the litigation. In the absence of the Tribunal, the LSGD used to exercise the function of the appellate body. See *Hindustan Coca-Cola Beverages Private Limited*, note 5 above.

[114] See *Perumatty Grama Panchayat*, note 109 above.

[115] Ibid.

licence of the Company and order its closure on the ground of excessive extraction of ground water was legal and whether the intervention of the government through its appellate jurisdiction was legally sustainable.[116]

It was argued on behalf of the Panchayat that the Panchayat is authorised to preserve water resources in its jurisdiction as per the Kerala Panchayat Raj Act. Therefore, the closure order issued by the Panchayat was legitimately in the interest of the general public. Further, it was argued that the Government could not dictate to a licencing authority as to how it should work. On the whole, the Panchayat argued mainly on the basis of the discretionary and exclusive power of the Panchayat under the Constitution of India and the Kerala Panchayat Raj Act.[117]

The Company argued that the Government was the appellate authority under the Kerala Panchayat Raj Act and therefore the Government has the authority to cancel the order of the Panchayat. It is not proper for the Panchayat to challenge it. The company also justified the Government's decision by arguing that the order against the company was a non-speaking order. The order was not supported by any authoritative scientific report or investigation. It was argued further that there was no statutory prohibition on digging of bore-wells at the time when the Company started production. Therefore, legally there was no restriction upon the Company to extract ground water from its land.[118]

The Court invalidated the closure order issued by the Panchayat. It was held that the Panchayat was not authorised to issue a closure order on the ground of excessive extraction of ground water by the Company.[119] The Court further held that 'the Panchayat can at best, say, no more extraction of ground water

---

[116] It has been stated by the single judge that '…in this case, the notice was issued *only* on the ground of excessive exploitation of ground water and the decision to cancel the license was taken *only* on the basis of that ground. Therefore the Panchayat fairly submitted that the validity of its decision and that of the Government on this point *alone* need be considered by this Court in this case' (emphasis added). *See Perumatty Grama Panchayat v State of Kerala*, High Court of Kerala, India, W.P. (C) No. 34292 of 2003, Judgement dated 16 December 2003, para. 8.

[117] Ibid., para. 5.

[118] Ibid.

[119] Ibid., para. 12

will be permitted and ask the company to find alternative sources for its water requirements'.[120] The Single Judge seems to have recognised the power of the panchayat to restrict or prohibit the use of ground water in its jurisdiction. At the same time, to issue an order of closure on the ground of protection of drinking water sources was held as beyond the authority of the panchayat.

This legal proposition implies that the panchayat has powers to take action in proportionate to what its responsibility requires. In this instant, the Single Judge appears to have considered the closure order as not in conformity with the proportionality test. The Court considered prohibition on the use of ground water as an adequate measure to discharge the responsibility of protection of drinking water sources. If the Court had an opportunity to discuss pollution issues, the decision would have been different. This is significant given the fact that 'proper waste disposal' has been included as a condition in the licence and the Company is proved to have violated this condition.[121]

At the same time the Court answered the second question affirmatively, that is, whether the Panchayat has the power to restrict or prohibit the extraction of ground water. The Court disapproved the argument made by the Company that in the absence of law the Company can extract any quantity of ground water from its land. The contentions of the Company were held incompatible with the emerging environmental jurisprudence under Article 21 of the Indian Constitution.[122] It was held that:

> Even in the absence of any law governing the ground water, I am of the view that the Panchayat and the State are bound to protect the ground water from excessive exploitation. In other words the ground water under the land of second respondent

---

[120] Ibid.
[121] *See* Perumatty Grama Panchayat, Proceedings of the Special Grade Secretary, 27 January 2000 (a copy of the proceedings on file with the author). For reports of Pollution Control Board confirming the unauthorised disposal of wastes, *see* Kerala State Pollution Control Board, A Study Report on the Presence of Heavy Metals in Sludge Generated in the Factory of M/s Hindustan Coca-Cola Beverages Pvt. Ltd., Palakkad (Thiruvananthapuram: Kerala State Pollution Control Board, September 2003), and Central Pollution Control Board, Report on Heavy Metals and Pesticides in Beverages Industries (Delhi: Central Pollution Control Board, November 2003).
[122] Ibid., para. 13.

(the Company) does not belong to him. Normally, every landowner can draw a reasonable amount of water, which is necessary for his domestic use and also to meet agricultural requirements. It is a customary right.[123]

The Court appears to recognise the right of the Company to exploit ground water from its land in a 'reasonable quantity'. The Court further gives explanation as to what amounts to reasonable quantity, that is, 'the quantity that is necessary for his domestic use and also to meet agricultural requirements'.[124]

The Single Judge relied upon the Public Trust Doctrine as recognised by the Supreme Court of India in the *M.C. Mehta* case.[125] It was held that being the trustee of natural resources, it is the duty of the state to protect ground water resources against over-exploitation.[126] The inaction of the state in this regard will tantamount to the infringement of the constitutionally guaranteed right to life under Article 21.[127] The Court also found basis in the Kerala Panchayat Raj Act. It was held that 'the duty of the panchayat can be correlated with its mandatory function No. 3 under the third schedule to the Panchayat Raj Act namely, 'maintenance of traditional drinking water resources'.[128]

The common law rule on ground water was held as outdated and incompatible with the emerging environmental jurisprudence. It was stated that:

> The principles applied in those decisions cannot be applied now, in view of the sophisticated methods used for extraction like bore-wells, heavy duty pumps etc. ...are incompatible with the emerging environmental jurisprudence developed around Article 21 of the Constitution of India.[129]

Based upon the above findings, it was decided that the Company should be restrained from excessive extraction of ground

---

[123] Ibid., para. 13.
[124] Ibid., para. 13.
[125] *M.C. Mehta v Kamal Nath* (1997) 1 SCC 388.
[126] *Perumatty Grama Panchayat v State of Kerala*, High Court of Kerala, India, W.P. (C) No. 34292 of 2003, Judgement dated 16 December 2003, para. 13.
[127] Ibid., para. 13.
[128] Ibid., para. 13.
[129] Ibid., para. 13.

water from its land. It was further held that the Company, like any other landowner, should be permitted to extract ground water, which must be equivalent to the water normally required for irrigating crops in 34 acres of plot. The Panchayat was given the power to decide the quantity of water that can be legitimately extracted by the Company. The Panchayat was also given the power to monitor and inspect the ground water consumption of the Company.

To sum up, ground water was held as a national wealth and as a resource that belongs to the entire society, that is, a subject of public trust.[130] The panchayat and the state in general were held to be custodians of ground water resources in its jurisdiction. The right of an individual to use ground water was made subject to the restrictions imposed by the state. In result, the decision is in tune with the present water law reforms through which ground water is being shifted from the individual to Government control.[131] The Single Judge decision also recognises the fundamental right of individuals under Article 21 of the Constitution of India. It was held that the over-extraction of ground water by a person or a company is likely to infringe the fundamental right of others guaranteed under the Constitution and therefore the state is duty bound to take actions to prevent it.

4.2.2 Division Bench Decision

Being aggrieved by the Single Judge decision, both the Panchayat and the Company filed appeals. Apart from that there were other appeals in connection with the licence issuing power of the Panchayat. Since all these matters were interlinked, the division bench considered and decided all the appeals together.[132]

---

[130] Ibid.

[131] Philippe Cullet, 'Water Law Reforms: Analysis of Recent Developments', 48(2) *Journal of the Indian law Institute* 206 (2006). *Source:* http://www.ielrc.org/content/a0603.pdf.

[132] Four appeals have been filed in the Kerala High Court, they are: *Hindustan Coca-Cola Beverages Private Limited v The Perumatty Grama Panchayat and Ors.*, W.A. No. 2125 of 2003; *The Perumatty Grama Panchayat v. State of Kerala and Ors*, W.A. No. 215 of 2004; *The Perumatty Grama Panchayat v State of Kerala and Ors.*, W.A. 1962 of 2003 and *The Perumatty Grama Panchayat v Secretary to Government, Local Self Governance, Government of Kerala*, W.A. No. 12600 of 2004.

In the appeal, the Panchayat presented that it had no issues with the Company and was merely anxious about the miseries of the people. It was presented on behalf of the Panchayat that, if there are proper solutions for the scarcity of water and other environmental problems, the Panchayat would never object to an industry capable of providing employment and other development. At the same time the Company argued that the Single Judge had been wrong in saying that ground water in a piece of land does not belong to the owner of the land but to the public.

The division bench stated that in the absence of a specific statute prohibiting the extraction of ground water, a person has the right to extract ground water from his land. Such an extraction could not be considered illegal. In this context, the division bench stated that 'we do not find justification for upholding the findings of the learned judge (Single Judge) that the extraction of ground water is illegal...we cannot endorse the finding that the company has no legal right to extract his wealth'.[133] The division bench also disapproved the reasoning of the Single Judge based on the Public Trust Doctrine and said that 'abstract principles could not be the basis for the Court to deny basic rights unless they are curbed by valid legislation'.[134] The Court further held that 'the reliance placed...in Kamal Nath's case is not sufficient to dislodge the claim'.[135]

The division bench also rejected the reasoning of the Single Judge on the basis of powers of the panchayat under the PR Act. It was said that: '...reference to the mandatory function referred to in the third schedule of the Panchayat Raj Act, namely "maintenance of traditional drinking water resources" could not have been envisaged as preventing an owner of a well from extracting water from there as he wishes'.[136] The Division Bench appears to have recognised ground water as a 'private water resource' and accepted the proposition of law that the landowner has 'proprietary right' over it.

Based upon this premise it was held that 'the Panchayat had no ownership over such *private water resources* and in effect denying

---

[133] Ibid., para 35.
[134] Ibid., para. 35.
[135] Ibid., para. 43.
[136] Ibid., para. 35.

the *proprietary rights* of the occupier and the proposition of law laid down by the learned judge (single judge) is too wide for unqualified acceptance' (emphasis added).[137] The division bench asserted that '...ordinarily a person has a right to draw water, in reasonable limits, without waiting for permission from the Panchayat and the Government. This alone could be the rule and the restriction an exception'.[138]

The Court reaffirmed this proposition of law and said that 'it always will be permissible for an occupier to draw water out of his holding'.[139] In the opinion of the Court, the permissible restriction, in public interest, can only be to compel the occupier of the land to ensure that his conduct does not bring about a drought or imbalance in the water table.[140] Having said so, the division bench rejected the proposition of law as observed by the Single Judge. It was held that 'ground water under the land of the Respondent (the Company) does not belong to it may not be a correct proposition in law'.[141] The Court appears to have affirmed the common law principle and consequently rejected the power of the panchayat to restrict or prohibit this right.

The division bench rejected the allegation of pollution and the quality problem of the products of the Company. It was held that the Panchayat was ill-equipped to examine technical matters like that of pollution and the purity of the products of the Company.[142] The division bench also rejected the Joint Parliamentary Committee (JPC) report on the purity of the products of the Company on the ground that the JPC report had not referred to any samples collected from the factory in Plachimada.[143]

The division bench accepted the decision of the Government regarding the constitution of an expert committee to investigate the matter. As a result, an expert committee was constituted to

---

[137] Ibid., para. 35.
[138] Ibid., para. 43.
[139] Ibid., para. 49.
[140] Ibid., para. 49.
[141] Ibid., para. 43.
[142] It is to be noted that the Court has not declared it as beyond the authority of the panchayat. It suggests that if the panchayat is technically equipped, it can go into the matter of the purity of the products.
[143] Ibid., para. 50.

study and investigate the issue. The expert committee submitted an interim and a final report in the Court. By accepting the fact of water scarcity in the area, the expert committee concluded that the reason could be the declining rainfall in the last several years. The Committee had recorded the opinion that the unregulated withdrawal of ground water from wells within the factory complex and also outside had aggravated the water shortage. The report concluded that the annual ground water requirement of the Company, at the average rate of five lakh litres per day, could be allowed, if average rainfall was available. The report also suggested that the consumption should be reduced proportionately to the decrease in rain fall, for example, if rainfall was less by ten per cent, the exploitation of water was to be reduced to four lakh litres per day. The expert committee report has been accepted as such by the division bench by saying that 'it appears to be authentic, based on data collected, mature and therefore acceptable'.[144]

To sum up, major proposition of law propounded by the division bench was that the landowner or the occupier of the land has the right to withdraw ground water from his land. This is part of his proprietary right. Any restriction on this right is an exception and should be supported by express statutory authorisation. The Court expressly rejected the PTD on the ground that this principle does not find expression in statutes.

It appears that the division bench, in principle, reversed the Single Judge's decision. The Single Judge Decision was premised on the PTD and considered ground water as 'national wealth' and therefore belongs to the society. Whereas the division bench rejected this proposition and asserted that right to draw ground water was part of proprietary rights and any restriction on this right was an exception. The division bench appears to have relied upon the common law rule on ground water. Having premised on public law concepts such as the PTD and Constitutional rights, the single judge upheld the power of the panchayat while the division bench

---

[144] Ibid., para. 46. The expert committee report has been criticised on the ground that it had relied upon unrealistic and unscientific data. The Committee has also been criticised for being unrepresentative of the interests of the local community and the Panchayat. *See* K. Ravi Raman, 'Corporate Violence, Legal Nuances and Political Ecology: Cola War in Plachimada', 40 (25) *Economic and Political Weekly* 2481 (2005).

Ground Water – Legal Aspects of the Plachimada Dispute 189

upheld private property rights and disapproved the role of the panchayat to regulate ground water use.

5. The Future: An Analysis

The future course in the context of the Plachimada case consists of two important areas. First, the general legal framework of ground water developed in Kerala after the Plachimada case. The legal regime of ground water in Kerala is not the same as it was in the Plachimada case. The ground water legal regime has undergone reforms. The Kerala Ground Water (Control and Regulation Act), 2002 is the major result of legal reforms. This is the major legal framework expected to address Plachimada like situations in future. Secondly, the pending appeal in the Supreme Court of India. This is another area through which a development in the ground water legal regime is likely to occur.

5.1 Plachimada in the Supreme Court [145]

The root cause of the Plachimada case was the Panchayat's refusal to renew the licence of the Company on the grounds that the Company's over-exploitation of ground water has caused an acute shortage of drinking water and other environmental problems in the Panchayat. Therefore, all major arguments presented in appeal before the Supreme Court seeks to justify the Panchayat's action against the Company.

The Panchayat sought to justify its action on the grounds that there had been insufficient water for agricultural and drinking purposes and this shortage had resulted in popular protests in the Panchayat. The Panchayat argued that the action taken against the Company was its duty under the Kerala Panchayat Raj Act and in conformity with the underlying spirit of the 73rd Amendment to the Constitution.

The Panchayat contended that the power to control or restrict the ground water extraction comes under the mandatory duty of the Panchayat. Objectives sought to be achieved through all these provisions or powers are public safety and public welfare. By

---
[145] The analysis in this part is based on the Special Leave Appeal filed in the Supreme Court on behalf of the Panchayat.

relying upon this legal background, it has been strongly contented in the Special Leave Petition (SLP) that the High Court was wrong in directing the Panchayat to renew the licence. The High Court did not consider the powers of the Panchayat as envisaged under the PR Act and the Constitution of India. It was also submitted that the High Court has no power under article 226 to give such a direction to a licencing authority.

It was submitted that the over-exploitation of ground water by the Company has resulted in drying up of wells in the Plachimada area and also contamination of water resources. These 'ground realities' have been ignored by the division bench of the High Court. The Panchayat has also submitted its arguments based upon the 'right to life' jurisprudence. It has been presented that there was an acute scarcity of drinking water in the area and therefore the action taken by the Panchayat was in the larger interest of public health and safety. Otherwise it would have been a violation of the right to life and the right to livelihood under Article 21 of the Constitution.

The priority principle, the duty of the state to protect and preserve the environment and the right to livelihood are the arguments presented by the Panchayat to support its part.[146] The SLP has also relied upon on liability principles under the tort law. It was argued that property rights vested in the Company does not extent beyond the four boundaries of its property. Any activity, even though carried out in their property, if adversely affecting the life as well as the proprietary right of the owner of the adjoining property, then it is the duty of the authority to interfere with such activity and to ensure the maintenance of rights and basic amenities to its citizens. It has also been argued that when the enjoyment of property by one person causes harm to the life and property rights of the adjoining owner, the liability under tort arise and the victim is entitled to compensation.

All the abovementioned submissions tend to establish the power of the panchayat to manage and develop water resources in its

---

[146] The priority principle requires the government to set priorities in the allocation of water for various competing uses such as domestic, agricultural, industrial and commercial. This principle has already found expression in the National Water Policy, 2002. *See* National Water Policy (2002), Para 5. *Source:* http://www.ielrc.org/content/e0210.pdf. *See also* Cullet, note 131 above.

jurisdiction. In a way, arguments presented in the Supreme Court are an attempt to establish the state's control over natural resources.

The Plachimada case addresses primarily two issues, the pollution problem and the question of control over the private person/company's ground water extraction from their property. The first issue would help to provide a specific remedy to the Plachimada crisis and the second one may clear the way to arrive at a balance between public interest and the right of the Company and to lay down basic principles underlying the ground water legal regime. The second issue is closely related to the decentralisation principle, that is, the role of the panchayat vis-à-vis regulation of ground water use.

The pollution problem, from the very beginning, has not been an issue in the case. The Kerala High Court disposed of the issue by saying that pollution was not the main question to be decided in the case brought before the Court. In the present SLP too, the pollution problem has not been highlighted. The liability issue (of the Company) has been argued mainly on the basis of principles under tort law. Given the fact that pollution caused due to solid wastes from the Company was confirmed, the polluter pays principle could be a legal basis of compensation claims.

The second issue seeks to address the broader question of balance between the power of the state (particularly local bodies) to regulate ground water use and the right of the landowner to draw ground water from his land. The discussion on this issue is likely to be centered on the PTD and the common law rule. The subsequent changes in the ground water legal regime, that is, the Kerala Ground Water (Control and Regulation Act), 2002 and the decentralisation principle could also be an influencing factor. The said Act seeks to empower the government to regulate the ground water exploitation and provides legal and institutional framework for that purpose and the decentralisation principle tends to give regulatory powers to local bodies and encourage public participation in governance.

5.2 The Ground Water Act

The reforms in ground water laws is said to have begun when the Ministry of Water Resources circulated a Model Bill for Regulation of Ground Water to all States and Union Territories in 1970, which was revised in 1992 and 1996. The Model Bill was again reviewed in 2005. Several states have enacted their own ground water laws, mostly following the Model Bill.[147] The Kerala Government enacted the Ground Water Act in 2002. The Act was not in force when the legal battle was started. Although the Act was notified later, it was too late to apply the regulatory framework envisaged in the Act in the Plachimada case. Since this major legal framework is supposed to manage similar issues in the future, the scheme of the Act is explained and analysed here.

5.2.1 Introducing the Act

The Kerala Government has enacted the Kerala Ground Water (Control and Regulation Act), 2002 for the conservation of ground water and for regulation and control of its extraction and use.[148] The Act explicitly considers the fact that ground water is a critical resource of the state and the undesired environmental impacts of the indiscriminate extraction of ground water in the state. Hence, the State Government considered it necessary to regulate the use of ground water in the interest of the public.[149] The schemes envisaged in the Act aims to control and regulate the extraction and use of ground water by private individuals and companies.

The Act provides for !the constitution of a State Ground Water Authority as an institutional mechanism for implementing the Act.[150] The authority is responsible and empowered to fulfill the objectives of the Act. This is the competent body to advise the Government to initiate policy actions to protect and preserve ground water resources in the state.

The Act is not applicable to all users of ground water or to all geographical areas in the state. The application of the Act is limited by quantitative and geographical restrictions. First, the term 'user of ground water' includes only persons using ground water from

---

[147] *See* the Kerala Ground Water (Control and Regulation Act), 2002; The Goa Ground Water Regulation Act, 2002; The Himachal Ground Water (Regulation and Control of Development and Management) Act, 2005; The West Bengal Ground Water Resources (Management, Control and Regulation) Act, 2005.
[148] Hereafter referred to as 'the Act'.
[149] *See* the preamble of the Act.
[150] Hereafter referred to as the 'Authority'.

a pumping well.¹⁵¹ The definition of the term 'pumping well, expressly excludes open wells fitted with pumps driven by an engine or motor of horse power up to 1.5 and bore wells and dug-cum bore wells fitted with pumps driven by an engine or motor of Horse Power up to three.¹⁵² This provision tends to exclude small-scale users, most likely the domestic users. Secondly, the Act is only applicable to notified areas. The Government, on recommendation of the Authority, is entrusted with the power to declare a particular area as a notified area, if it is necessary in the public interest to regulate the ground water use in that area.¹⁵³ The notification process is the discretion of the Government and the role of the Authority is only advisory in nature.

The Act provides permit and registration system as a tool for regulating ground water use. The Act makes it mandatory for every person who desires to dig a well or to convert an existing well into a pumping well to seek permission from the Authority.¹⁵⁴ The Act further gives guidelines for the Authority to consider before accepting or rejecting the permit applications. It includes purposes like digging wells, quality and quantity of ground water in the area, potential danger to existing users, distance from the existing wells, etc.¹⁵⁵ The rules made under the Act makes it mandatory that a ground water scientist deputed by the Authority should visit the concerned place and after studying the geology and existing ground water conditions of the area give an investigation report with recommendations. If necessary, geophysical survey may also be done in addition to the hydrogeological survey.¹⁵⁶ The Act also requires the existing users of ground water in a notified area to register their wells.¹⁵⁷ Here also the Authority can accept or reject the application on reasonable grounds. The guidelines for

---

[151] Kerala Ground Water (Control and Regulation Act), 2002, Section 2 (h). *Source:* http://www.ielrc.org/content/e0208.pdf.
[152] Ibid., Section 2 (f).
[153] Ibid., Section 6.
[154] Ibid., Section 7 (1).
[155] Ibid., Section 7 (7).
[156] Kerala Ground Water (Control and Regulation) Rules, 2004. *Source:* http://www.kerala.gov.in/dept_irrigation/gopno.17-2004.pdf.
[157] Ibid., Section 8 (1).

accepting or rejecting the application for registration are more or less same as that required for the permit.[158]

The Act contains provisions to protect public drinking water resources. The Act requires permission from the Authority to dig wells within 30 meters of any public drinking water resources.[159] The Authority is authorised to grant permission for the purposes of drinking or agriculture, if the digging of the well is not likely to affect public water resources. The power of the Authority to grant permission is restricted by an express term 'drinking purpose or for agriculture'. This means there is no question of other competing uses like commercial or industrial purposes within 30 meters from a public drinking water resource. In the absence of express provisions dealing with priorities, this provision can be used as a guideline for the Authority to set priorities before granting a permit or certificate of registration or to put conditions in the permit or certificate of registration.

The Authority may grant the permit or certificate of registration upon conditions necessary for the implementation of the Act.[160] The Authority may also change conditions in the permit or certificate of registration.[161] The Authority can cancel the permit or certificate of registration on grounds such as non-compliance with conditions or procurement of permit/certificate based on false facts. The Authority can also use this power if the ground water situation in the area demands a higher degree of restriction.[162]

5.2.2 Critical Analysis of the Act

The object of the Act is to promote the conservation of ground water and regulate the use of ground water. The Act recognises the existing indiscriminate exploitation of ground water in some areas of the state and its negative environmental impacts. However, the legal framework as envisaged in the Act seems to be insufficient to achieve the stated objectives.

First of all, regulatory tools are applicable only to 'notified areas'. The power to notify a particular area is vested with the

---

[158] Ibid., Section 8 (5).
[159] Ibid., Section 10.
[160] Ibid., Sections 7 (4) and 8 (3).
[161] Ibid., Section 11.
[162] Ibid., Section 12.

government on the recommendation of the authority constituted under the Act. It would be proper if areas being commercially utilised by water-based industries are deemed to be 'notified areas'.

The Act does not incorporate the principle of prioritisation as envisaged in the National Water Policy of 2002.[163] However, the prioritisation principle could be seen as implied in the provision for the protection of public drinking water resources.[164] The provision which permits only drinking water and agriculture in areas within 30 meters from a drinking water source conveys the second priority given to agriculture. The Act does not differentiate between the competing uses of ground water. Different grades of regulation should have been envisaged under the statute for different uses like drinking water and other domestic purposes, agricultural purposes and commercial uses. Since agricultural and commercial users are the big users of ground water, the grade of the regulation and penalties for violations ought to have been prescribed separately. But the Act has drawn up a single procedure, framework and penalty for all uses. Moreover, the penalty as prescribed under the statute appears to be insufficient.[165] The penalty prescribed in the Act is not sufficient to restrain big companies from over-exploiting ground water. The cancellation of the permit or registration should have been made a punishment in addition to the fine or imprisonment for the second offence.

The polluter pays principle is considered to be an important part of environmental jurisprudence. The Supreme Court of India has incorporated the polluter pays principle as a part of the Indian legal system.[166] The principle requires the polluter to pay compensation for damages caused to the people and ecology. The Act prescribes only penalties of nominal fines and imprisonment. Despite the repeated recognition of this principle by the Supreme Court, it has not found expression in the Act.[167]

---

[163] *See* National Water Policy, 2002, Section 5. *Source:* http://www.ielrc.org/content/e0210.pdf.

[164] *See* the Kerala Ground Water (Control and Regulation Act), 2002, Section 8.

[165] Kerala Ground Water (Control and Regulation Act), 2002, Section 21. *Source:* http://www.ielrc.org/content/e0208.pdf.

[166] *Vellore Citizen's Welfare Forum* v *Union of India* (1996) 5 SCC 647; *Indian Council for Enviro-Legal Action and Ors.* v *Union of India*, (1996)3 SCC 212.

[167] A better approach, in this regard, can be seen in the Andhra Pradesh Water, Land and Trees Act, 2002. *Source:* http://www.ielrc.org/content/e0202.pdf.

The Act prescribes a centralised planning and management strategy. The role of local communities and local bodies has not found expression in the Act. It is most unlikely that the protection and preservation of natural resources like ground water become effective without participation at the local level. Therefore, the decentralisation principle needs to be one of the bases of legal framework for ground water regulation and management. Though the Act points out conservation as a major objective, the Act does not prescribe any measures for resource augmentation. In this regard, it would be beneficial to note some features of other ground water laws in the country. For instance, The Andhra Pradesh Water, Land and Trees Act of 2002 has included rain water harvesting structures as a requirement for building constructions.[168]

Absence of base data regarding the quantity and quality of ground water in Plachimada aquifer was the major obstacle faced by government agencies in examining the causes and effects of ground water depletion in Plachimada. Therefore, the preparation of data on ground water in the state and periodic monitoring of the quality and quantity need to be an essential part of the legal framework. In fact, some states have incorporated provisions in this regard in their ground water laws. For instance, the West Bengal Ground Water Law contains provisions requiring periodic preparation of district-wise ground water data.[169]

6. Conclusion

Separate statutory framework for ground water in India has been developed since 1990s. Prior to this period, the legal regime mainly consisted of legal principles linked, to a large extent, to land ownership.[170] The development of a separate ground water legal regime was initiated by the Ministry of Water Resources through its Model Bills. A statutory framework for regulating ground water use was enacted in the state of Kerala in the year 2002, which came into force in 2003.

---

[168] *See* the Andhra Pradesh Water, Land and Trees Act, 2002, Section 17 (2).

[169] *See* the West Bengal Ground Water Resources (Management, Control and Regulation) Act of 2005, Section 9 (a). *Source:* http://www.ielrc.org/content/e0502.pdf.

[170] *See* Cullet in this volume.

The issue of ground water depletion and pollution in Plachimada has arisen prior to the 2002 Act. Therefore, the reforms in ground water laws did not find place in the Plachimada case. Legal principles – common law rule and the PTD – were the major point of discussion in the Plachimada case. Both these principles were relied upon by the Kerala High Court in the Plachimada case. Given the differences in basic premises and outcome of these principles, finality was not achieved in the Plachimada case. The matter is now pending in the Supreme Court of India.

The Plachimada case, as decided by the Kerala High Court, addresses the issue of ground water depletion allegedly caused due to the over-exploitation of ground water by the Company and the consequent cancellation of licence of the Company by the Panchayat. Therefore, the focus of the case was on the power of the Panchayat to regulate ground water use in its jurisdiction and the right of the landowner to draw ground water from his land. The Kerala High Court decided the matter differently in principle. The Single Judge relied upon the PTD and decided in favour of the power of the Panchayat. Whereas, the division bench relied upon the common law rule and decided in favour of the Company's right to draw ground water from its land.

Apart from the issue of ground water depletion, pollution was another major issue raised in Plachimada. The issue of pollution due to effluents discharged from the Company was confirmed by various government agencies in their reports. The Company was directed to stop its production on this ground by the Pollution Control Board. This situation makes it viable for the Government to invoke the precautionary principle and polluter pays principle because these principles are part of environmental laws in India as per the Supreme Court decisions.[171] Moreover, the environmental jurisprudence developed under Article 21 of the Constitution of India requires the government to take appropriate actions to mitigate the problem of ground water depletion and pollution.[172]

---

[171] *Vellore Citizen's Welfare Forum v Union of India* (1996) 5 SCC 647; *Indian Council for Enviro-Legal Action and Ors. v Union of India*, (1996)3 SCC 212.

[172] *Subhash Kumar v State of Bihar*, AIR 1991 SC 420; *F.K. Hussain v Union of India*, AIR 1990 Ker 321; *Venkatagiriyappa v Karnataka Electricity Board & Others*, 1999 (4) Kar. L.J. 482.

A case study on Plachimada, which largely symbolises the discussion on ground water legal regime, reveals the multiplicity of laws and institutions. The ground water laws and institutions constituted under it, the pollution control laws and PCBs, the Panchayat and the Central Ground Water Board are the major components of ground water legal regime in India. The coherence and cooperation among and between these components would be a major challenge for the legal system in the coming days.

UNIT III
Perspectives on Privatisation

8

Tirupur Water Supply and Sanitation Project: A Revolution in Water Resource Management?

ROOPA MADHAV

1. Introduction

The Tirupur Water Supply and Sanitation Project (henceforth the Project) typifies an incongruity in water resource management and prioritisation. It is paradoxical that the textile industry which has over-extracted and polluted the water bodies in and around Tirupur for a decade now, has been given first priority in designing a water supply scheme. Also ironically, the scheme draws clean raw water (nearly 125 million litres per day) from a river source, but does not conceive of a design for the discharge and treatment of the same. Pollution caused due to industrial waste continues to plague Tirupur, even as state of the art technology is being installed to extract and transport water from distant sources. The scheme, in effect, has a direct bearing on the efforts to ensure recycling of waste water and 'zero effluents discharge' and in turn, the broader agenda of sustainable water management and conservation.

The Tirupur Water Supply and Sewerage Project has many firsts to its credit – 'it is the first project to be structured on a *commercial format*; the first *project-specific public limited company* for water and sewerage with equity participation of major beneficiaries; first *concession* by a state government to a public limited company to draw raw water for domestic and industrial uses and to collect revenues; the *first index-based user charges* and direct *cost recovery* for urban environmental services; *construction, operations and maintenance* of infrastructure and related services by experienced domestic and international operators'.[1] (emphasis added) It is also the first investment of the International Financial Corporation (the 'IFC') in the water sector in the country.[2] The Project reportedly also has a technical sophistication that is unmatched in the country. Notably, the bulk of the water supplied under this project is for industrial purposes – 115 million litres per day (MLD) of the 185 MLD extracted supplies water 24 × 7, to nearly 900 industrial units. Though not representative of the standard 'privatisation of water supply' model, this case study throws light on the new and emerging legal arrangements in promoting Public-Private Partnerships (PPPs), in the water sector.

After the implementation of the Project in 2005, there have been significant changes in the water service delivery in the town. The directives of the Madras High Court, in the pollution cases,[3] have impacted the implementation of the Project, raising several issues around the sustainability and viability of this PPP experiment. This case study examines the reasons behind the new project – the institutional, financial and legal aspects of the Tirupur PPP. It also examines important legal issues such as the right to water, competing interests in water, financing of projects, waste water management and the environmental consequences of the PPP. More particularly, it questions the wisdom of planning a water supply project that seeks to prioritise the needs of a polluting industry over the basic water needs of the region.

---

[1] Chetan Vaidya, The Tiruppur Area Development Program – Focus on Urban Infrastructure and Private Sector Participation (Indo-US Financial Institutions Reform and Expansion Project – Debt Market Component FIRE (D), Note No.13, 1999). *Source:* http://www.niua.org/indiaurbaninfo/fire-D/Project No.13.pdf.

[2] *See generally* IFC Project summary. *Source:* http://www.ifc.org/ifcext/ southasia.nsf/Content/SelectedProject?OpenDocument&UNID =68985B 0E317547B785256A03007C23CE.

[3] *Noyyal River Ayacutdars Protection Association and Others* v. *Government of Tamil Nadu*, Writ Petition Nos. 29791 and 39368, decided on 22 December 2006 by the Madras High Court.

2. Background to Water Privatisation and PPP in India

Like in most developing countries, drinking water needs in India are largely met by the public sector. The entire cycle of water treatment, production, distribution and maintenance is the responsibility of the State. Large-scale neglect of the system and low investment in the maintenance and upgrading of existing facilities has necessitated urgent reforms in the sector. Fresh infusion of funds from the Government is meagre, as the larger policy design is to roll back the State in favour of the private sector in all sectors, including water services.

Despite being backed by a clear policy thrust towards privatisation of water utilities for domestic and industrial purposes in India,[4] progress has been slow, with less than one-tenth of the urban centres being privatised.[5] There are nearly 32 projects, either in the pipeline or being implemented in the water and sanitation sector, with the modes of privatisation ranging from operation and maintenance contracts, management contracts, concession agreements, BOOT (build-own-operate-transfer) to BOT (build, operate, transfer).[6]

The wholly privately owned BOT model has now fallen by the wayside, superseded by a distinctive new model for Public Sector Partnership (PSP), the joint venture (JV) or time-bound divestiture model. This new privatisation model does not fit neatly into any of the categories usually used to classify PSPs, such as divestitures, concessions or leases. Instead, it combines the sale of a minority stake in the utility assets under a Public-Private JV ownership structure with a limited contract lifespan, usually between fifteen and fifty years, with shorter contracts for smaller scale rehabilitation or extension projects.

---

[4] Section 13, National Water Policy, 2002. *Source:* http://www.ielrc.org/content/e0210.pdf.

[5] *See generally* National Institute of Urban Affairs (NIUA), Status of Water Supply, Sanitation and Solid Waste Management in Urban Areas (New Delhi: NIUA, June 2005).

[6] *See generally* Gaurav Dwivedi, Rehmat and Shripad Dharmadhikari, *Water: Private, Limited: Issues in Privatisation, Corporatisation and Commercialisation of Water Sector in India* (Badwani: Manthan Adhyayan Kendra, 2006).

The Government of India (GoI) defines a 'Public Private Partnership' project to mean a project based on a contract or concession agreement, between a Government or statutory entity on the one side and a private sector company on the other side, for delivering an infrastructure service on payment of user charges. According to Alison Mohr, the emergence of Public Private Partnerships heralds a new era of collaboration between public-private actors across different policy fields including transport, housing, and healthcare to name a few. PPPs first emerged in the United States in the late 1970s and early 1980s in response to what was seen (mainly by neoliberals) as the poor performance of the public sector, and the view that the State had reached its financial limits as far as the provision of public services were concerned. PPPs were initially introduced as a more publicly acceptable alternative to privatisation and hence as a tentative first step towards full privatisation agreements.[7] Greater participation by the private sector corporations and the international funding agencies has transformed the privatisation debate in the country.

It is widely believed that the privatisation of the water sector would ensure more efficient delivery of services, bring in much needed finances to cash-strapped government utilities and free up precious resources for allocation in other important social and welfare sectors. However, proponents against water privatisation argue that water is essentially a public and social good and to ensure the welfare of all citizens, it cannot be left solely in the hands of private actors. Treating water as a commodity, based on market principles, will, they argue, adversely impact the human right to water and the right to life. To fully realise these rights, the state's obligation to deliver this essential service is vital.

Project designs have since evolved from wholly private sector participation to a joint venture model of public-private partnership. The Tirupur Project is the first experiment in PPP in the water sector in the country.

---

[7] Alison Mohr, Governance through 'Public Private Partnerships': Gaining Efficiency at the Cost of Public Accountability? (International Summer Academy on Technology Studies – Urban Infrastructure in Transition, 2004). *Source:* http://www.ifz.tugraz.at/index_en.php/filemanager/download/311/Mohr_SA% 202004.pdf.

3. Water Supply and Sanitation in Tirupur

A. History of Water Supply and Sanitation Infrastructure in Tirupur

Supply of drinking water, both in urban and rural areas, has been the responsibility of the local bodies. To assist the local bodies to effectively source, monitor and implement the distribution of water, statutory boards were set up by the state governments. In Tamil Nadu, the Metrowater Board (for the city of Chennai and surrounding areas) and the Tamil Nadu Water Supply and Drainage Board (TWAD Board or the Board), were set up. The TWAD Board, set up in 1971, seeks to assist local bodies in the planning and execution of water supply and drainage schemes all over the state. In its operation, it acts as the executing agency for the local bodies who have to pay the Board for the capital works. Once installed, the maintenance of the system is the responsibility of the local body, but in several instances, the TWAD Board is also acting as the maintenance and operating agency.

Prior to the New Tirupur Area Development Corporation Ltd. (NTADCL), there was no piped water supply to the industry. The town received water from two schemes, for which the primary source is the River Bhavani situated at Mettupalayam, 54 km away from the city. A total of 28.50 MLD per day is supplied by the two schemes from Mettupalayam head works. The Tirupur Municipality (TM) was vested with statutory powers for the operation and maintenance of the water supply system within municipal limits. Accordingly, Scheme I, maintained by the Tirupur Municipality, with a capacity of 7 MLD, was commissioned in the year 1965 to supply water to the Tirupur Municipality and seven wayside villages. Scheme II, which is outside the Tirupur Municipality limits, was commissioned in the year 1992. It was operated and maintained by the TWAD Board and had a design capacity of 32 MLD. It supplied 24 MLD of water to the Tirupur Municipality and four town panchayats and the remaining water was supplied to the 44 wayside villages.

While the second scheme was designed for an ultimate population of 3 lakhs by the year 2011, it had, by the year 2006, already hit the 3.5 lakhs mark. Water supplied by the two schemes is unable to meet the water demands of the city. Likewise, the panchayats in the surrounding Tirupur area to which water is being supplied from the NTADCL scheme (popularly called the 'third scheme'), have expressed grievances about the projected water needs and the population figures proposed under the scheme.[8]

The industries in the Tirupur Local Planning Area (TLPA), including the TM area, do not have access to reliable piped water supply. Water is an essential commodity in the cotton knitwear production process. The existing water supply system does not provide water to the dyeing and bleaching industries. The industries have largely relied on their own resources to access water supply. Private water suppliers abstract ground water and supply it to the industries through tankers. Prior to the setting up of NTADCL, the demand for water by the industries was so high that several agriculturists resorted to selling ground water to the industries, adversely impacting the water source for agricultural purposes.

There is no underground drainage system in Tirupur. Disposal of night soil is normally by way of individual facilities and disposal of liquid waste (sullage and kitchen waste) is through the open drains. The main mode of individual disposal in the town is through septic tanks, low cost sanitation units and through public conveniences. About 62 per cent of the total population is covered by sanitation facilities comprising of septic tanks and soak pits. Community facilities in the form of 36 community toilets have been provided by the Tirupur Municipality for about 8,113 households belonging mainly to the economically weaker sections and low income groups. The slum area within the Tirupur Municipality currently does not have access to any sanitation facilities including toilets.

On the industrial front, with over 700 industries, the contribution of industrial discharge in Tirupur is significant. This wastewater, to a large extent, is untreated and is discharged into the River Noyyal, dry wells or onto open lands. While there have been efforts of late to set up common effluent treatment plants and individual treatment plants, the technology used by several units is not advanced enough to effectively deal with the effluents.

---

[8] Interview with panchayat representatives at Chettipalayam on 10 March 2007.

B. NTADCL: A Water Revolution in Tirupur?

The Tirupur Water Supply and Sanitation Project originated in the context of domestic water supply being limited to a few hours on alternate days; industries not having access to piped water supply and relying heavily on water tankers for water supply; available water sources – both surface and ground water – being polluted heavily by the textile industry and the fast depleting ground water in the region.

The initiative for the formation of NTADCL came from the Tirupur Exporters Association (TEA). TEA, a registered society of owners of the industrial units manufacturing textiles in Tirupur, supported a plan for the development of infrastructural facilities, particularly those relating to water treatment and supply and sewage treatment for the enhancement of the productivity and export potential of the industrial units in Tirupur. The Government of Tamil Nadu (GoTN) mandated the Tamil Nadu Corporation for Industrial Infrastructure Development Limited (TACID) to identify infrastructure projects so as to enhance Tirupur's export and industrial potential.

In 1993–94, TACID formulated an integrated Tirupur Development Plan (TADP) for the TLPA, which envisaged several schemes including those relating to services of treatment and supply of potable water in the service area and the offtake, treatment and disposal of sewage in the Tirupur Municipality. The Government of Tamil Nadu along with TACID and TEA, with a view to leveraging the resources, approached the Infrastructure Leasing and Financial Services (IL&FS), a non-banking financial services company, for assistance in raising finances for the Project.

A Memorandum of Understanding was signed on 25 August 1994, between the Government of India, TACID, TEA and IL&FS (henceforth the MoU). The Concessionaire contract (Build-Own-Operate-Transfer) has been granted jointly by the Government of Tamil Nadu and the Tirupur Municipality to the NTADCL for a period of thirty years. In furtherance of this MoU, NTADCL has been incorporated on 24 February 1995 as a public limited company under the Indian Companies Act, 1956 with initial equity participation from the GoI, TACID (representing GoTN), TEA and IL&FS. The GoTN and the Tirupur Municipality have agreed to grant a concession to NTADCL, to develop, finance, design, construct, operate, maintain and transfer on strictly commercial principles, on an integrated basis, the water treatment and supply facilities and sewage treatment facilities including the right to draw water from the River Cauvery (the concession agreement).

Under the concession agreement, NTADCL undertakes, either by itself or through its subsidiaries, to implement the project, on strictly commercial principles on an integrated basis, to:

(a) provide a water abstraction, treatment and distribution service by undertaking to develop, finance, design, construct, own, operate, maintain and transfer to GoTN or its nominee, the Water Treatment Facility for the purpose of supply of Potable Water to TM and other purchasers, outside the jurisdiction of TM at the Water offtake points;

(b) provide sewage offtake, treatment and disposal service by undertaking to develop, finance, design, construct, own, operate, maintain and transfer to GoTN or its Nominee, the Sewage Treatment Facility, for the purpose of offtaking, treating and disposing Sewage, delivered by TM at the Sewage Offtake points;

(c) develop, design, finance and construct the Water Distribution System and the Sewerage system, which would be transferred to TM by NTADCL upon the issuance of the Construction Completion Certificate.

The agreement provides for recovery of costs and operation and maintenance through a composite water and sewerage charges. The tariff structure provides for annual revision linked to indexation and any unusual increases are to be approved by the Price Review Committee. The base project return is 20 per cent per annum on 185 MLD project cost.

The EPC1 contractor (River intake well and pumping station; water treatment plant and booster pumping station; transmission main – 56 kms; master balancing reservoir) is Hindustan Construction Co. Ltd. and EPC2 Contractors (3 Feeder Mains – 93 kms; water distribution stations; distribution network; distribution network to wayside villages; sewerage system – 124 kms; low cost sanitation) are Mahindra & Mahindra and L&T Ltd. The O&M contract has been awarded to United Utilities, UK and Mahindra & Mahindra.

The other related contracts, apart from the concession agreement, are the Bulk Water Supply Agreement between NTADCL and the Tirupur Municipality, Shareholders' Agreement, Common Loan Agreement, Engineer Procure Construct Contract, and Operation & Maintenance Contracts.

Bidding process

NTADCL selected a consortium for the design and construction of the project facilities and their operation and maintenance over the concession period, through an international competitive bidding process. Global tenders were called for, and after selection the O&M contract was awarded to a consortium led by the Mahindra Group, with Bechtel Enterprises and United Utilities International, UK, being the other members of the Consortium. While ownership of the project assets lay exclusively with NTADCL in its capacity as the concessionaire, the consortium has an equity share in NTADCL.

Project capacity

The key features of the project are: 56 k pipeline from the Cauvery river; a water distribution network of about 350 km; raw water and sewerage treatment plants; pumping stations; and conveyance facilities. The intake of water is at the river Bhavani and a water treatment plant is located at the water source. The clean water is then transported 56 kms (through steel pipes which have 1400 and 1200 mm dia) to the Master Balancing Reservoir. The villages located alongside the pipeline – Kanchikoil (2.7 MLD), Perundurai Chennimalai (8.4 MLD), Uthukali (3.9 MLD) – and the Netaji Apparel Park (1.0 MLD) are being supplied water. The water from the Master Balancing Reservoir is then supplied to the distribution stations through three feeder mains. Feeder Main I supplied to the Tirupur Municipality and Feeder Mains II and III supplies to the Tirupur Local Planning Area.

When fully operational, this water supply and sewerage system will supply 185 MLD of water to about 900 textile firms and over 1.6 million residents in Tirupur, Tamil Nadu and the surrounding areas. About 125 MLD of water is to be supplied to the knitwear dyeing and bleaching industry, 25 MLD to the residents of Tirupur including 60,000 slum dwellers and 35 MLD to the region's remaining rural towns, villages and settlements.

C. Institutional and Financial Aspects

It is believed that the lack of budgetary resources prompted the local industry to spearhead the implementation of TADP on a commercial format. NTADCL, a special purpose vehicle (SPV) and project sponsor, was formed to access commercial funding. The Project itself was split into three separate contracts – two were awarded on an engineer, procure and construct (EPC) basis and the third was to operate and construct (O&M) the finished facility. Hindustan Construction Company Ltd. was the EPC1 (river intake well and pumping station; water treatment plant and booster pumping station; transmission main – 56 kms; master balancing reservoir) contractor and the EPC2 contract (3 Feeder Mains – 93 kms; water distribution stations; distribution network; distribution network to wayside villages; sewerage system – 124 Kms; low cost sanitation) was awarded to the Mahindra & Mahindra/ Larsen and Toubro joint venture.

Project funding was a mixture of debt and equity, an approach which involved a number of sources, including public money, various commercial interests, financial institutions and international funding agencies. The financing structure is as follows:

Equity: Rs 322.7 Crores (US $ 69 million); senior debt Rs 613.8 Crores (US $ 132 million).

Subordinate debt: Rs 86.5 Crores (US $ 18 million). The total amount of Rs 1023 Crores (US $ 220 million) consists of EPC Cost (Rs 650 Crores), owners cost (Rs 127 Crores), contingency cost (Rs 93 Crores), interest during construction (Rs 142 Crores) and initial working capital (Rs 11 Crores). The project cost is to be financed through a debt-equity ratio of 1.5:1, which includes equity (Rs 322.70 Crores), Subordinate debt (Rs 86.50 Crores), and Debt (Rs 613.80 Crores).

Assistance came from the IL&FS and from the US Agency for International Development (USAID) with loan guarantees over 30 years for US $ 25 million. The World Bank provided a line of credit to IL&FS and the ADB, through its private arm, ADIQUA Holdings, has a 27 per cent stake in the Project.[9]

---

[9] The equity ownership is as under: TWICL – 31.56%; ADIQUA – 27.05%; LIC – 6.01%; GIC – 6.01%; New India Assurance – 1.13%; UTI – 1.13%; NICL – 0.90%; OICL – 0.68%; TIDC – 0.68%; Mahindra Infrastructure Developers Ltd – 4.51%; Mahindra Construction – 2.25%; Mahindra Holdings – 2.25%; WSA Engineers – 4.51%; and Others – 11.33%.

4. Context and Objectives of the PPP

In 1995, a special purpose vehicle, NTADCL, was set up as a public limited company, with equity holders consisting of the Government of Tamil Nadu, TACID, TEA and IL&FS. Floated as the first public-private partnership in the water sector, this BOOT experiment has been operational since August 2005. It is responsible for the offtake, treatment and transmission of water, distribution of water to industries and the municipality for domestic consumption, and treatment of the collected sewage, and maintenance of the sewage treatment plants. The Project primarily seeks to address the water needs of the industrial area in Tirupur, with the bulk of the water being supplied to the industry.

The impetus for the PPP Project can be traced to the severe water crisis faced by the industries in Tirupur in the mid-1990s. Faced with a booming textile industry and bracing itself for take off in the post MFA (Multi Fibre Agreement) phase, the government was compelled to examine the water situation in the industrial area. Though the TWAD Board drew up a project plan, the Project ultimately took shape as a Public-Private Partnership, the cost of which is estimated at Rs 1023 crores. The PPP must also be examined in the light of the severe environmental damage caused by the textile industry discharging effluents into the surrounding water bodies and extensively contaminating ground water in the region. While the Project focuses on the supply of water to the industry, the needs of a burgeoning township and its surrounding villages have been given short shrift. Likewise, the Project envisages low-cost sanitation programs but effluent treatment of industrial discharge, which is crucial to preventing further damage to the water bodies and the environment, has not been integrated into the Project. It is worthwhile to place the entire project in its larger socio-economic and political context.

Clean water is crucial for the dyeing and bleaching units as the quality of water impacts the process of dyeing and bleaching. There is a severe water crisis in Tirupur due to over- extraction of ground water and contamination of ground water and other surface water bodies. Prior to the setting up of NTADCL, the water needs of the industry were being met by the supply of water through tankers, drawing water from open and bore wells in the surrounding taluks such as Avanashi, Palladam, Annur, Kangeyam and other taluks of Erode district. Roughly between two to three thousand lorries with a capacity to transport 10,000 to 12,000 litres/trip were plying around seven to ten trips daily to supply clean water for the textile wet processing.[10] NTADCL was planned in this context, to primarily meet the demands of the industry.

A. Location and Population

Tirupur, a municipal town located about 50 km east of the city of Coimbatore in Tamil Nadu, is located on the banks of the River Noyyal, a tributary of the River Cauvery. In addition to the River Noyyal, waterways such as Jamunai Pallam, Rajavaikkal and other local water bodies and ponds exist in Tirupur. The town is renowned as India's leading cotton knitwear centre, accounting for over 90 per cent of the country's exports in this sector.

According to the census, the total population of Tirupur was 2,35,666 in 1991 and increased to 2,98,853 in 2001.[11] The migrant population in Tirupur is very high, given the high employment opportunities in the hosiery industry, and this directly impacts the projections for infrastructure needs based on the resident population. The total slum population of Tirupur town as per the 1991 census was 57,780 (about 25 per cent of the Tirupur population) distributed in 88 notified slum areas.

B. Socio-economic and Demographic Profile

The textile industry in Tirupur has transformed the socio-economic profile of the region. The industries were largely owned by the

---

[10] S. Eswaramoorthi *et al.*, 'Zero Discharge – Treatment Options for Textile Dye Effluent: A Case Study at Manickapurampudur Common Effluent Treatment Plant, Tirupur, Tamil Nadu' (Paper presented at the International Conference on Soil and Groundwater Contamination: Risk Assessment and Remedial Measure organised by Environment with People's Involvement and Co-ordination in India, Hyderabad 8–11 December 2004). *Source:* http://www.epicin.org/documents/zero_discharge.zip.

[11] Prakash Nelliyat, 'Public-Private Partnership in Urban Water Management: The Case of Tirupur', *in* Barun Mitra, Kendra Okonski and Mohit Satyanand eds., *Keeping the Water Flowing* 149, 154 (New Delhi: Academic Foundation, 2007).

Chettiars in the early days but the Gounder community has increasingly begun investing in the smaller units. It is a story of the emergence of the Gounder peasant workers into a working class on the factory shop floor, morphing further to becoming owners of small units and now, controlling a large part of the industrial activity in Tirupur and the surrounding districts of Coimbatore. Alongside the growth of the Gounder presence, there has been a huge influx of migrant labour from surrounding taluks and districts and also from the state of Kerala. Therefore, the caste composition of the town is forever evolving and changing, but the dominant castes remain the Brahmins, Gounders, Naidus and Chettiars.

The political profile has largely been influenced by the strong labour movement in Tirupur. CPM and CPI have a large presence alongside the regional parties of AIADMK, DMK and the others, while CITU and AITUC dominate the labour unions in the textile industry.

The feminisation of labour is a more visible recent trend, resulting from globalisation and the transformation of labour practices around the country. This is clearly evident in Tirupur too. Despite a conscious and sustained attempt at abolishing child labour in the industry, several units continue to employ child labour.

C. Profile of the Textile Industry

The first hosiery unit in Tirupur was set up in 1893. The district of Coimbatore grows good quality cotton essential for the yarn and knitting industry. Around the 1950s, South India, especially the Coimbatore district, grew into a major powerloom centre. The high concentration of ginning, weaving and spinning mills in the area provided the impetus for the growth of the knitwear industry in Tirupur. Till the 1960s, the industry in Tirupur produced mainly gray yarn and bleached vests for the domestic market. The industry changed profile and gathered momentum in the mid-1980s, with a surge in textile exports. In 1984, the first export consignment was shipped. The 1990s brought in greater demand for finished garments and this expansion saw the feminisation of labour in Tirupur. It must also be noted that rapid expansion has led to the unplanned and haphazard growth of the industry.

The production process comprises of four important operations, decentralised and sub-contracted to various firms and units. These are:

(i) knitting of cotton yarn and making grey fabric
(ii) bleaching and dyeing of grey fabric
(iii) fabrication of garments and
(iv) printing and finishing.

- (i) Knitting: Knitting is the first process in this industry which is organised in three major ways:

    (a) units which produce the cloth on contract basis for other employers;
    (b) units which produce cloth for their own production of goods in their factories, and
    (c) units which produce for both their own production and on contract basis.

- (ii) Bleaching and Dyeing: Bleaching the grey colour of the knit fabric is essential for dyeing. This process involves mixing bleaching powder in water through which grey fabric is made to pass. The process of dyeing and bleaching is very water intensive.

- (iii) Fabrication: This process comprises of cutting according to the pattern and then stitching. While cutting is done manually with the help of a pair of scissors or a small cutting machine, stitching is carried out on a sewing machine, manually or electrically operated. Sophisticated indigenous and imported sewing machines are used not only for high speed stitching but also for the various stitching designs.

- (iv) Printing and finishing: Printing is mainly done on the garments although it could also be done on the bleached fabric before stitching. Before the actual printing is undertaken, calendaring is done to ensure that the surface of the fabric is smooth for printing. Calendaring is the process of pressing the garment with steam. The ironing and pressing tasks are largely done by men. This is followed by the last task, 'checking', which is primarily handled by women.[12]

D. Environmental Issues

Tirupur is located in a drought prone area and has been designated as such by the Government of India. The Drought Prone Area Programme (DPAP) of the Government of India funded jointly by the Ministry of Rural Development, Government of India and the Government of Tamil Nadu has the clear objective of (i) promotion of economic development of the village community, which is directly or indirectly dependant upon the watershed through optimum utilisation of watershed, and (iii) encouraging restoration of ecological balance in watershed through sustained community action. It is being implemented in Tirupur since 1995–96.

Tirupur has a large concentration of bleaching and dyeing industries which generate between 100 to 120 MLD of effluents. These effluents have a high Biological Oxygen demand (BOD), Chemical Oxygen Demand (COD), colour and salt content. The chemicals used include wetting agents, soda ash, caustic soda, peroxides, sodium hypochlorite, bleaching powder, common salt, acids, dye stuffs, soap oil and fixing and finishing agents. The Common Effluent Treatment Plants (CETPs) and the Individual Effluent Treatment Plants (IETPs) set up by the industry do not effectively tackle the salt content or reduce the Total Dissolved Solids (TDS) in the effluents. As a result, effluent with high salt content is discharged into the River Noyyal, which then flows downstream to be stored at the Orathupalayam reservoir,[13] which in turn impacts the agricultural lands irrigated by these waters.

---

[12] Sampath Srinivas, Case of Public Interventions, Industrialisation and Urbanisation: Tirupur in Tamil Nadu, India (New Delhi: World Bank, 2000). *Source:* http://www-wds.worldbank.org/external/default/WDSContentServer/WDSP/IB/2000/11/17/000094946_00110805370683/Rendered/PDF/multi_page.pdf.

[13] The Orthupalayam dam was constructed in 1991, at the cost of Rs 16 crores with the intention of irrigating an area of 500 acres in Erode district and 9875 acres in Karur district. The dam has now been reduced to storing the industrial effluent from the textile industries.

Water is also discharged into the River Nallar. The two rivers are natural drainage courses and are seasonal, carrying water only during the monsoon period. For the rest of the year, they only carry industrial effluents that stagnate in the riverbeds and percolate into the ground water.

Another problem is the huge amount of sludge that is generated through the process of treating water. The sludge is stored in plastic bags on the ground and covered with a thick plastic sheet to prevent the wind from carrying the pollutants. However, there is every possibility of seepage through rainwater and further contamination of the soil and water. In fact, the ground water quality around the cluster of bleaching and dyeing units is polluted to such a level that it is unfit for domestic, industrial and agricultural activities. Further, pollution has adversely impacted the agricultural yields in the area; increased water-borne diseases such as diarrhoea, typhoid, malaria, jaundice and skin allergies; also adversely impacted the cattle, animals and birds; and the fish in these water bodies.[14]

The agriculturalists approached the Madras High Court against the polluting industries. The Tamil Nadu Pollution Control Board (TNPCB) and the High Court have directed the industries to implement zero discharge facilities within a stipulated period or face the threat of closure of the industries. It has also ordered the dyeing industries to pay Rs 6 crores for the reclamation of the Orthupalayam dam and Rs 140 crores as compensation to farmers downstream of the River Noyyal. Following this order, the industries have approached the Tamil Nadu Water Investment Company Limited (TWICL) to assist in developing and financing an effluent treatment plant facility.

---

[14] *See* K. Govinadrajalu, 'Industrial Effluent and Health Status – A Case Study of Noyyal River Basin', *in* Martin J. Bunch *et al.* eds., Proceedings of the Third International Conference on Environment and Health 15–17 (Chennai: Department of Geography, University of Madras and Faculty of Environmental Studies, York University, 2003).

5. Implementation and Legal Issues

A. Concession Agreement

It may be worthwhile to highlight some of the important sections of the concession agreement. The agreement provides that NTADCL shall abstract raw water from the River Cauvery up to a maximum of 250 MLD. Out of this, it shall allocate 48.70 MLD of raw water for domestic and non-domestic purposes within the TM, upto a maximum of 165 MLD for industrial units for non-domestic purposes outside the TM in the service area and up to a maximum of 36.30 MLD of raw water for domestic purposes to wayside panchayat unions alongside the main water transmission line and villages in the TLPA.

The agreement also gives NTADCL the absolute right to re-allocate the above-mentioned quantities of raw water in the event that stated quantities for domestic purpose are not offtaken or not paid for by the TM and wayside villages to other purchasers within the 'service area'. NTADCL shall be liable to supply the net quantity of potable water remaining after deduction in the quantity of transmission losses from the abstraction point to the water offtake points. NTADCL shall, at its sole discretion, supply or otherwise dispose of the potable water remaining after the offtake of the contracted quantity of potable water by the TM, wayside villages and the industrial units.

The terms of the agreement require the GoTN to issue notifications under the Tamil Nadu District Municipalities Act, 1920, and under the Tamil Nadu Water Supply and Drainage Board Act, 1970, specifying NTADCL as the entity with exclusive rights to abstract raw water, develop, finance, design, construct, own, operate and maintain the Water Treatment Facility, pipelines and waterworks in order to provide Water Treatment and Supply Services within the Service area.

The concession agreement provides for total cost recovery along with returns during the period of the concession. If there is a shortfall, the Concession shall stand extended, so as to ensure recovery of the outstanding total cost of project and return. Upon termination of the concession period, NTADCL shall transfer the facilities to the GoTN and the TM. An extension of the concession period can be granted by the GoTN if the total cost has been recovered. However, the GoTN is not obliged to extend the concession period beyond thirty-eight years from the financial closure.

Additional allocation

NTADCL shall, at any time during the concession period, have the right to make one or more requests to GoTN in writing, to increase NTADCL's water drawal rights above the existing rights of 185 MLD upto 250 MLD, if in the opinion of NTADCL, the demand for potable water by industrial units outside TM is likely to exceed the allocated 100 MLD for non-domestic purposes. NTADCL shall submit its request to GoTN and the GoTN shall, on receipt of such request supported by the above details, provide NTADCL with additional raw water abstraction rights within the abstraction area within two months from the date of the receipt of the request, provided NTADCL at the time of request is not in breach of any of its obligations under this Agreement.

Financial and project information

NTADCL is required to send to the GoTN during the construction period and the operation period, the following documents and information at the intervals described below:

(a) annual audited accounts of NTADCL delivered within hundred and eighty days of the end of each fiscal year;
(b) un-audited financial statements of NTADCL delivered within sixty days of the end of each quarter;
(c) the construction budget and operations budget for the project to be delivered within thirty days of approval by the Independent Engineer and the Independent Auditor;
(d) notification of any material adverse effect in the financial condition of NTADCL and/or the project promptly following such occurrence; and
(e) the project plan for the implementation of the project which shall comprise of the order in which NTADCL shall carry out various activities involved in the Construction of the facilities and/or the system or part thereof, repairs and refurbishment of the existing system within a period of ninety days from the construction commencement date.

Water charges

The GoTN and the TM have granted to NTADCL the right and entitlement to determine the price of potable water and the price of sewage treatment, connection fee and security deposit in accordance with the terms of the agreement. The charges shall be, more specifically, in the nature of:

(i) the water charge and
(ii) the sewage charge

NTADCL may, at its discretion, for reasons of commercial expediency, charge prices which are less than the amounts it is entitled to charge in accordance with the agreement. NTADCL may, at its sole discretion, also provide a rebate in the charges if the purchasers make prompt payments thereof prior to the due dates.

Besides the water charge, NTADCL is entitled to collect a one time connection fee, reconnection fee and security deposit. NTADCL shall be at liberty to determine by negotiations, the amount of connection/reconnection fee and security deposit. However, these charges are subject to the prior approval of the GoTN and the TM. The agreement states that the charges will be in the nature of price paid for services rendered and are not in the nature of any tax or fee. The price of potable water and the price for sewage treatment shall be determined by NTADCL and reviewed periodically.

NTADCL shall be entitled to levy a surcharge on the price of potable water supplied for non-domestic purposes to industrial units at a rate to be determined by the price review committee to set off foreign exchange rates. The price review committee would determine the amount of surcharge per kilo litre of water based on the report of the Independent Auditor.

B. Specific Issues and Concerns

1. Pricing and cost recovery

The Government Order MS No. 260 RD (WSI) dated 9 December 1998 provides for collection of water tax (user charges) by the local bodies and the O&M costs to be paid to the TWAD Board. 'As per the G.O. and in practice, this water tax is a fixed amount of Rs 360 per year collected only from those who hold individual house service connections (HSCs). The common water collection points (otherwise known as Public Fountains) are meant for the poor who cannot afford to pay for water, providing for positive discrimination in favour of the poor.

This practice is already in vogue in many of the way-side village panchayats in Tirupur. Water is supplied by the TWAD Board from a combined water supply scheme pumped from Mettupalayam (which is locally popular as the second scheme). The village panchayats are paying the O&M cost to the TWAD Board. The panchayats pay at the rate of Rs 3.50 per 1000 litres. This is very much subsidised by the Government and it would not cover the entire O&M cost to the TWAD Board. Given the polluted ground water and considering the essentiality of water service, the government is subsidising water. Presently, as per the condition given to the NTADCL, they also charge on par with TWAD Board rates only (that is, Rs 3.50 for 1,000 litres)'.[15]

NTADCL charges differing prices for water used for domestic purpose and the water used for industrial purpose. The charges are Rs 3 per kilolitre (KL) for villages, Rs 5 per KL for domestic use in the Tirupur Municipality and Rs 45 per KL for industrial and commercial consumers. The industries are required to provide a bank guarantee to NTADCL, thus ensuring offtake of water. Every year, the bank guarantee has to be renewed and it is linked to the amount of water drawn by each industry. The collection of water charges for domestic use continues to vest with the Tirupur Municipality. The municipal laws were amended to provide for the distribution of water in the Tirupur Municipality.

The water supplied to the dyeing and bleaching units initially charged Rs 45 per 1,000 litre of water supplied to the industry. This was, however, reduced to Rs 23 per 1,000 litre in July 2006, as the demand for water and offtake by the industry was very

---

[15] *See* S. Eswaramoorthi *et al.*, Zero Discharge – Treatment Options for Textile Dye Effluent: A Case Study at Manickapurampudur Common Effluent Treatment Plant, Tirupur, Tamil Nadu (Paper presented at the International Conference on Soil and Ground Water Contamination: Risk Assessment and Remedial Measure organised by Environment with People's Involvement and Co-ordination in India, Hyderabad 8–11 December 2004). *Source:* http://www.epicin.org/documents/zero_discharge zip at 4.

poor. The Project was implemented with an assessed quantum of water of 108 MLD per day but it is reported that the actual water drawn by the Tirupur industries, even after one year, is estimated at 75 MLD on normal weekdays. The water rates have again been revised in February 2007 to Rs 35 per 1,000 litre. NTADCL is offering a 10 per cent discounted rate for those industries whose monthly offtake of water is more than the agreed quantum. The discounted rate extends to the entire quantity of water drawn and not just the excess offtake of water.[16] The Project, thus, gives incentives to increased water withdrawal, marginalising optimum and sustainable utilisation of existing water resources.

The concession agreement provides for recovering, over a period of time, the capital cost and the cost of operation and maintenance. According to Palanithurai and Ramesh,

> Charging for water so as to 'recover capital cost incurred in constructing facilities' is generally viewed wrong because in a welfare state it is a state obligation. Charging the Operation and Maintenance costs is widely accepted and there are some Village Panchayats in Tamil Nadu that have the practice of collecting user charges from house service connection (HSC) holders. It is a responsibility the local body is supposed to carry out with the support of the user community. Reports in several village panchayats is that even if the user charges were collected from all HSC holders, the local bodies would not be in a position to meet the full O&M costs. One component i.e., water pumping charges (electricity bill) would eat up the entire user charge collection in one gulp.[17]

The Project does not envisage cost recovery to be effected through the water supplied to the municipality and the panchayat areas. It seeks to cross-subsidise the water supply for domestic purposes with the water charges obtained from industries. Thus,

---

[16] G. Gurumurthy, 'Dual Pricing of Industrial Water Supply in Tirupur', *Business Line*, 3 February 2007. *Source:* http://www.thehindubusinessline.com/2007/02/03/stories/2007020300050200.htm.

[17] G. Palanithurai and R. Ramesh, Public Private Partnership in Drinking Water Supply (Paper presented at the National Institute of Rural Development (NIRD) Foundation Day Seminar on Democratization of Water, Hyderabad, 10–11 November 2006).

for the Project to be viable, the industry has to offtake more and more water to effectively subsidise the domestic consumer. It must be noted, however, that the pricing does not take into account the environmental costs of pollution and treatment of the same. In the absence of standards for pricing of water in the country, it is vital that the law and policy perspectives limit and balance the cost recovery principle and profit motive with the human right needs and the ecological costs.

2. Water quality and supply

The water being supplied from NTADCL to the village panchayats is only once in fifteen days. The supply of water from NTADCL is clearly inadequate. No new investment for storage and distribution has been made for the supply of water to the domestic consumer by NTADCL. It is contended that the census figures relied on for estimating the projected water supply requirements is that of 1991 instead of 2001, according to panchayat sources interviewed at Tirupur.[18] While there is no scope for augmentation at the present head works located in Mettupalayam, no new plans, after NTADCL, are in the offing to meet the drinking and domestic water needs of Tirupur. Further, the absence of adequate sources of water and polluted ground water sources poses a challenge to future water management plans for Tirupur.

Besides, the cost of pumping water has also increased with the increase in electricity bills. It has gone up from Rs 1.60 to Rs 3.40. Nearly 60 per cent of the panchayat budget is spent on pumping water from the bore wells. Further, water to the several public taps, found in Tirupur village panchayats and the slums within the TM, are not being supplied water from the third scheme. The contractual agreement does not bind NTADCL to address the water needs of the TM and the surrounding villages, but the costs of setting up the Project adversely impacts the available financial resources with the Government to meet the water supply needs.

Residents complain that the quality of the water supplied is distinctly poorer than that supplied by the first and second Tirupur schemes. Though unconfirmed, reports from the ground indicate that the water required for the bleaching and dyeing industry is

---

[18] Interview with the elected representatives at the Chettipalayam Panchayat on 10 March 2007.

essentially unchlorinated water. The same raw water is being supplied to the TM and panchayats. On the other hand some people believe that the water supplied from the first two schemes does differ in taste from that supplied from the third scheme but it is only a matter of acclimatisation to the new water supply.

The concession agreement grants NTADCL the exclusive rights to abstract raw water, develop, finance, design, construct, own, operate and maintain the water treatment facility, pipelines and waterworks in order to provide water treatment and supply services within the service area. This and the fact that no new future projects in the service area can be conceived of and executed by the Tirupur Municipality or the TWAD Board, even when faced with a situation of dire need, brings to the fore the need for constitutionally mandated rights protecting access to available water resources in a given region.

3. Waste water recycling

Following the orders of the Madras High Court, the industries have been forced to set up Reverse Osmosis plants to stem the discharge of effluents into water bodies in Tirupur. It must be noted that the NTADCL project was conceived of at a time when the issue of pollution and management of waste water was a serious concern, with many studies indicating extensive damage to soil fertility and pollution of water bodies. It is, therefore, unconscionable that the Project was not conceived of more holistically. Supply of water ought to be viewed along with its eventual discharge as waste and management of the waste water. Hence, costs must also be determined commensurate with the environmental costs of waste management and recycling.

Experts estimate that the pricing of water would change dramatically if the cost of managing a water treatment plant and a waste water recycling plant are factored in. It is believed that the O&M cost of sewage reclamation plant may be about Rs 8/KL and hence the total water pricing of domestic water ought to be about Rs 11 to 13/KL (water supply cost of Rs 3/KL or Rs 5/KL, and reclamation cost of Rs 8/KL) and Rs 80/KL (that is, water supply cost of Rs 45/KL as per NTADCL estimation and Rs 35/KL for RO + Evaporator + Crystalliser O&M costs).[19]

---

[19] Interview with Mr. Kuttiappan, LVK Enviro Consultants, Chennai.

Not only has the Project exempted its users from environmental costs, the Government has now stepped in to rescue the industries from bearing the costs of treating the effluents, as ordered by the High Court. The Tamil Nadu Government has planned a comprehensive Rs 700 crore marine disposal scheme for the State's textile dyeing industry cluster. Textile Ecosolutions Company Limited, a Public-Private Partnership, has been incorporated to lay a 310 km long pipeline from Tirupur to Nagapattinam.

To make the NTADCL project viable, the offtake of water by the industry has to increase. This runs counter to the High Court directive requiring the industries to recycle the waste water and achieve 'zero discharge' levels. If nearly 95 per cent of the water taken is effectively recycled and reused, it would strike a huge blow to the effective functioning of NTADCL, which can only survive and make profits if there is a consistent and upward demand for the water supplied. In the absence of effective laws to prioritise effective water management in the country, the absurd scheme to transport the effluents through a 310 km pipeline to be discharged into the sea may well be carried forth, ensuring NTADCL's future and zero liability of the industries for the pollution of the environment.

4. Competing water needs

The water being supplied by NTADCL does not satisfy the needs of a large section of people in Tirupur. The water is being supplied only once in fifteen days, which lasts for about three to four days. With the booming population and increased migration, the existing projects are woefully inadequate. While the water needs of the industry are being met, the water supply demand for domestic purposes is not being met by NTADCL. The government and more particularly the TWAD Board have no new projects in the pipeline to address the water needs of the community. Given that the industrial offtake is poor, would it be viable for NTADCL to sell the excess water to the domestic consumers? The rates that industrial water fetch currently are different from those that are charged from domestic consumers. The perception on the ground, in Tirupur, also questions the prioritisation of water supply by the Project.

According to Eswaramoorthi:

The analysis of VP functionaries in Tirupur is that the scheme is primarily meant for the industries in Tirupur, secondarily for the slum dwellers most of whom are workers in the industries, and the remaining water is distributed to the wayside VPs. The reasons for the variation in the degree of importance, is obvious. The workers in the industries, most of them migrants from poorer rural areas of Tamil Nadu, should be relieved of the water crunch so that they can work for the industries. Thirdly, the wayside VPs the water passes through, if by-passed, would become a persistent headache to the TWAD Board and to Tirupur Municipality by agitations and protests.[20]

More importantly, the concession agreement provides NTADCL the absolute right to reallocate the raw water in the event the stated quantities for domestic purpose are not taken or not paid for by the TM and way-side villages, to other purchasers within the service area. Further, the agreement provides that NTADCL shall, at its sole discretion, supply or otherwise dispose of the potable water remaining after the offtake of the contracted quantity of potable water by the TM, wayside villages and the industrial units. The contract, therefore, does not lay down that failing offtake by industry, the priority shall be to supply water for domestic needs, in the local area. In other words, if the maximum of 85 MLD (out of a maximum of 250 MLD) for domestic purposes has been saturated, but there is available balance in the water allotted for industrial use, NTADCL is not obligated under the contract to supply the excess to the TM and the wayside villages.

It is interesting that the concession agreement provides for additional allocation and reallocation of water in favour of NTADCL and industrial purposes. If there is excess water after supplying to the municipalities and the wayside unions, NTADCL, under the agreement, can reallocate the water or 'otherwise dispose

---

[20] *See* S. Eswaramoorthi *et al.*, Zero Discharge – Treatment Options for Textile Dye Effluent: A Case Study at Manickapurampudur Common Effluent Treatment Plant, Tirupur, Tamil Nadu (Paper presented at the International Conference on Soil and Groundwater Contamination: Risk Assessment and Remedial Measure organised by Environment with People's Involvement and Co-ordination in India, Hyderabad 8–11 December 2004). *Source:* http://www.epicin.org/documents/zero_discharge.zip at 5.

of the potable water remaining after the offtake'. In effect, the agreement does not bind NTADCL to use the available excess water for the water-starved populations of the town and surrounding areas. The lack of clearly stated policy ensuring equitable distribution of the limited available resource and the lack of positive discrimination in favour of the poor is clearly evident in the structure of the concession agreement.

5. Human right to water

Despite three water schemes being operational, a large section of the poor in Tirupur buy water for their basic needs, from a tanker or from those who have a piped water supply. Women in the K.V.R. Nagar slum in Tirupur supplement their meager share of the public water supply with water from a well, which, due to pollution, is not of potable quality and can be used only for purposes of washing, etc. There are no set timings for the supply of water in this slum area and the women have to wait long hours. In Sukumarnagar, which is an unauthorised slum in Tirupur, there are no facilities for the supply of water. They obtain water from private tankers and the price can vary from Rs 2 per pot to Rs 4 per pot. Given this scenario and the growing demand for water for domestic use in Tirupur, the lack of positive discrimination in favour of the poor in designing the Project throws up several questions about recognition of the basic human right to water.

The right to water as a basic human right is a derivative right from several other fundamental rights entrenched in international rights instruments, such as right to life, right to an adequate standard of living, right to food and shelter, and right to physical and mental health. The concept of human right to water obligates states to ensure that populations have safe, affordable and adequate access to water through policies and strategies that create a conducive environment (be it social or economic) for such access. General Comment No. 15, which has been issued by the United Nations Economic and Social Council (ECOSOC), obligates states to realise the right to water.[21] It states that in implementing the

---

[21] Committee on Economic, Social and Cultural Rights, General Comment 15: The Right to Water (Articles 11 and 12 of the International Covenant on Economic, Social and Cultural Rights), UN Doc. E/C.12/2002/11 (2002). *Source:* http://www.ielrc.org/content/e0221.pdf.

right to drinking water, state parties have to be non-discriminatory and maintain equality. Further, states are obligated to prevent the infringement of these rights by third parties, including private companies operating water utilities. It provides that when water is distributed by the private sector, '[s]tates parties must prevent them [the private sector] from compromising equal, affordable, and physical access to sufficient, safe and acceptable water'. This obligation includes, inter alia, adopting the necessary and effective legislative and other measures to restrain, for example, third parties from denying equal access to adequate drinking water, and polluting and inequitably extracting water resources. Tested against these international norms of availability, quality, non-discriminatory accessibility and information dissemination, NTADCL clearly is in violation of the human right to water. Aside from this, the PPP abets further pollution of the environment and no measures are being taken to arrest this trend.

6. Changing structures

In the absence of a specific law and a regulatory authority governing concession agreements, which details guidelines for adequate service, the user's rights and liabilities, the tariffs policy, the bidding process, requirements of a concession contract, duties of the granting authority and concessionaire, and appropriate conditions for intervention and termination of the concession, the importance of transparency and accountability needs to be highlighted further. Changing structures and complex contractual relations emphasise the need for information in the public domain, in order to protect the rights of the citizens. Though the Planning Commission has issued some guidelines in the recent past to govern the public-private contracts, it does not specify a clear policy on public participation and accountability.

> Principles such as transparency and fairness, often associated with the state, have been brought into question by the creation of institutional linkages with private sector organisations within which the delivery of public services is now being managed. Indeed, there is growing evidence that the contractual relations of public private partnerships have led to a clear weakening of traditional notions of accountability reflecting both a shift to

new lines of accountability (private sector shareholders) and a vicious circle of monitoring and distrust between partner organisations.[22]

In practical terms, once control has been ceded to private interests, regulatory agencies face great difficulty in imposing controls on how these projects go forward. NTADCL has not proven to be an exception to this generalisation. While no information has been placed by NTADCL in the public domain, it does not even have a Public Information Officer to assist the general public. Further, it has been guarded in entertaining requests for information about the Project. In the absence of transparency, the issues of accountability, inevitably, need greater scrutiny. Thus, a more stringent and new regulatory legal framework is essential to cope with the changing structures, which impact the rights of the people.

6. Conclusion

The success of the Tirupur PPP project depends entirely on the offtake of water by the industries, thus enabling cost recovery. Implicitly, therefore, the Project discourages water recycling by the textile industries through full treatment of wastewater. The Project has been lauded as marking the beginning of a water revolution in the country. It must, however, be noted that while the water needs of the industry are being met, the water supply for domestic purposes in the surrounding areas are not being met by NTADCL. The government currently has no new projects in the pipeline to address the water needs of the community, in contradiction to the stated objective in the National Water Policy, 2002 prioritising domestic water needs above all other competing demands. Though not explored in great depth in this paper, the adverse impact on ground water caused by the pollutants from the textile industry, raise many troubling questions on the conjunctive use of water resources in the country.

---

[22] *See* Alison Mohr, Governance through 'Public Private Partnerships': Gaining Efficiency at the Cost of Public Accountability? (International Summer Academy on Technology Studies – Urban Infrastructure in Transition, 2004). *Source:* http://www.ifz.tugraz.at/index_en.php/filemanager/download/311/Mohr_SA%202004.pdf at 241–242.

Water is a precious resource and the emphasis ought to be on sustainable use and conservation. The present design of the Project, on the contrary, encourages further expansion of the water market and copious use through discounted pricing strategies. There is a need to limit consumption, fix higher responsibility on the polluters and prioritise equitable distribution of the resource. The Tirupur Water Supply and Sanitation project represents a classic case of reductionist engineering, sidelining sustainable water use planning and conservation efforts.

9

The World Bank's Influence on Water Privatisation in Argentina: The Experience of the City of Buenos Aires

ANDRÉS OLLETA

Introduction

The work of International Financial Institutions (IFIs) in developing countries has been subject to intense criticism in recent years. Detractors of the International Monetary Fund (IMF) and the development agencies forming the World Bank (WB) Group do not only deem that the contents of the policies supported by IFIs are inadequate to overcome the difficulties that member states face, but also criticise the way in which said policies are forwarded to indebted countries, which has been perceived as short of an imposition.

In a context of loss of prestige, the role of the IMF and WB Group in the contemporary wave of privatisation of public services has attracted particular attention.[1] Indeed, privatisation has been one of the central reforms sponsored by IFIs for reducing public deficit and stimulating economic growth in developing countries. Among the public services that have been transferred to private operators in recent decades, the case of water and sanitation in

---

[1] 'Privatisation' in this chapter is a term that includes selling assets to a private company, tendering a water concession to a private company, or awarding management contracts to a private company. For reasons that we will see later, in the case of water services, private companies currently prefer the latter type of contract.

urban areas has merited special study. Given their social value[2] and the magnitude of the business that they represent, the success or failure of privatisation processes of water services has attracted in-depth analysis from both supporters and critics of the work of IFIs.

This chapter will focus on the role of the WB Group in the privatisation of water services in the city of Buenos Aires.[3] It is by now undisputed that the processes of water services privatisation that IFIs encouraged in Argentina were less than successful. Popular distress provoked by deficient provision, high tariffs, mutual accusations of violations of the terms of the contracts, constant renegotiations of these, and suits filed before international arbitration organs such as the International Centre for Settlement of Investment Disputes (ICSID), have been some of the undesired consequences of the privatisation experience in Argentina. The main purpose of this chapter is to determine how the WB Group and the Argentinean authorities might have contributed to a certain extent to the failure of the Buenos Aires' water services privatisation, and whether they have appraised their own performance and learned from their mistakes. To that end, the chapter will provide the reader with a chronological survey of the events that led to the revocation of the concession and the return of water services to the state. In particular, because of the WB's importance in the water sector, it is necessary to assess whether it has changed in any way its approach to water services privatisation

---

[2] Alejo Molinari, 'Lessons from Latin America: The Potential of Local Private Operators', 1/1 *Water Utility Management International* 20, 20 (2006). The social value of water services is 'relatively high in relation to other public services. [Water and sanitation] have an effect on people's health, either directly or through their effect on environmental quality'.

[3] The chapter alludes to the concomitant role of the IMF, since both IFIs are regarded as working in unison for similar ideas and policies. Resorting to IMF work is sometimes unavoidable, given that it has traditionally been more open to the disclosure of loan negotiation documents, while the World Bank has only recently reviewed its transparency policy and given access to the public to policy and project lending documents. *See* IBRD/World Bank, The World Bank Policy on Disclosure of Information (Washington: World Bank, Policy Statement, 2002).

in other parts of the world ever since well-publicised failures such as Buenos Aires'.

The chapter will be divided in three main sections. Section I will briefly describe the WB's mission, lending mechanisms, the role of conditionality in them, and will dwell on the policies that the institution favours in water management and water services provision. Section II will review the privatisation of water services in the city of Buenos Aires, one of the first Latin American experiences of its kind, lauded as a model for many similar processes in the region. Section III will attempt to draw lessons from the previous sections and assess where the process failed, before concluding with some personal views on water services privatisation in general.

1. The World Bank and the Privatisation of Water Services

1.1 The World Bank's Lending Services and its Use of Conditionality

The WB is formed of two institutions – the International Bank for Reconstruction and Development (IBRD) and the International Development Association (IDA). Together with their affiliate agencies,[4] they form the WB Group, whose main mission is to reduce poverty and improve living standards through financial and technical assistance to developing countries. The WB provides members with low-interest loans, interest-free credits and grants in an effort to help them realise projects that are crucial for their development.[5] The importance of water and sanitation services

---

[4] The affiliate agencies of the World Bank are the International Finance Corporation (IFC), the Multilateral Investment Guarantee Agency (MIGA) and the International Center for the Settlement of Investment Disputes (ICSID).

[5] Ever since 2000, the Bank states that its work is oriented towards helping state members fulfill the Millennium Development Goals. *See* UN General Assembly Resolution 55/2, United Nations Millennium Declaration, UN Doc. A/RES/55/2 (2000). One of the goals is 'to halve the proportion of people who are unable to reach or to afford safe drinking water' (para. 19). In the 2002 Declaration on Sustainable Development, it was additionally agreed to reduce by half the population without access to basic sanitation. *See* Report of the World Summit on Sustainable Development, Johannesburg, 26 August to 4 September 2002, UN Doc. A/Conf.199/20 (2002), Plan of Implementation, Annex paras. 8 and 25. *Source:* http://www.ielrc.org/content/e0223.pdf.

determined that the WB reinforced its participation in projects for the development of water and sanitation infrastructure and in policy design related to water management and water services operation. In this sense, the WB has quickly become 'one of the most, if not the most important actor in the global water sector, be it in terms of financial aid or in terms of general policy-making in the developing countries.'[6]

IBRD and IDA credits can be classified in two different groups according to their purpose. The first category, 'project or investment lending', covers loans destined to finance infrastructure projects; in the water sector, examples would be the construction of dams, irrigation channels, sewerage plants, or the extension of the urban water services network. The second category, 'policy lending', encompasses those loans that are destined to 'help a borrower achieve sustainable reductions in poverty through a program of policy and institutional actions that promote growth and enhance the well-being and increase the incomes of poor people'.[7] This kind of loan is oriented towards the implementation of reform measures in the policy and institutional frameworks of the borrower country. In order to optimise the results of policy lending, the WB coordinates the design and execution of the loan with IMF offices, should the member state be also benefiting from IMF's financial assistance.

In policy lending programmes, the procedure for borrowing resources from the Bank —much like its counterpart in the IMF — is initiated by the state submitting a project proposal to be assessed in its economic, financial, social and environmental viability. The state declares in a document called 'Letter of Development Policy' (LDP) how the funds will be spent, and when and how the sum will be reimbursed to the Bank. The borrower also accepts that WB staff monitors the use made of the funds. However, policy loans are available to member states only after the borrower has agreed on fulfilling a set of conditions elaborated in concert with the Bank; usually, the implementation of a number of reforms that are

---

[6] M. Finger and J. Allouche, *Water Privatisation: Trans-National Corporations and the Re-Regulation of the Water Industry* 62 (London and New York: Spon Press, 2002).

[7] World Bank, Development Policy Lending, OP 8.60, August 2004, para 2. *Source:* http://go.worldbank.org/N3Y839UBH0.

considered critical for the country's social and economic recovery.⁸ This is known as 'conditionality' in both IMF and WB circles and its main function – in the Bank's own words – is to ensure that the commitments assumed by the borrowing country in its LDP are respected.⁹

There is no formal definition of 'conditionality' in the Bank's legal framework or operational policies. In fact, conditionality had not been expressly foreseen in either the IBRD's or IDA's Articles of Agreement, the constitutive instruments of both organisations. The attachment of conditions to disbursements emerged as a practice that unfolded as policy lending surfaced as a new lending service,¹⁰ although a few provisions in the Articles of Agreement have been considered to indirectly provide a legal basis for conditionality.¹¹ Currently, there is an express mention of

---

⁸ World Bank/Operations Policy and Country Services, Policy Conditions in World Bank Investment Lending: A Stocktaking, Board Report 36924, July 2006, at p 4. *Source:* http://go.worldbank.org/2A3FGDCZR1. Conditions are not usually attached to project lending programs. On the contrary, the use of conditionality in them is actually discouraged by the World Bank, mainly because 'linking the disbursement of loans to the uncertain speed of sectorwide reforms had increased the riskiness of the investments financed and affected their performance'.

⁹ World Bank/Operations Policy and Country Services, Review of World Bank Conditionality: Modalities of Conditionality (Background Paper SecM2005-0390/1, June 2005), Executive Summary, para. 2. *Source:* http://go.worldbank.org/C5UV4N0ZA0. The rationale for conditionality 'is the Bank's due diligent obligation to ensure that its resources are used effectively and responsibly by the borrowing country'.

¹⁰ Originally, only project lending was explicitly regulated in the Articles of Agreement. Article III, Section (4) (vii) of the IBRD Articles and Article V, Section (1) (b) of the IDA Articles provide that 'loans made or guaranteed by the Bank shall, except in special circumstances, be for the purpose of specific projects of reconstruction or development'. Therefore, at first policy loans were conceived as extraordinary, only allowed under 'special circumstances'. *See* IBRD Articles of Agreement, Bretton Woods, 1944. *Source:* http://go.worldbank.org/7H3J47PV51, and IDA Articles of Agreement, Washington, 1960. *Source:* http://go.worldbank.org/TSLNEK1XT0. Hereafter IBRD and IDA Articles of Agreement, respectively.

¹¹ As explained in World Bank/Legal Vice Presidency, Review of World Bank Conditionality: Legal Aspects of Conditionality in Policy-Based Lending (Background Paper SecM2005-0390/2, 2005),

(*Contd.*)

conditionality in the form of paragraph 13 of the WB's Operational Policy (OP) 8.60, which identifies three essential requirements for the Bank to make disbursements in a policy-based loan: (a) maintenance of an adequate macroeconomic policy framework; (b) implementation of an overall program in a manner satisfactory to the Bank; and (c) compliance with critical policy and institutional actions.[12]

The use of conditionality by IFIs is still controversial. Beyond its legally dubious origins, a source of debate has been the concrete degree of participation of the borrower in the design and adoption of those 'critical policies and institutional actions' that condition the grant of the WB loans. Additionally, critics of conditionality and IFI action in developing countries claim that conditionality policies are excessively intrusive in the domestic affairs of states and, most worrisome, that they benefit transnational corporations and reflect 'the neoliberal agendas and the geo-political imperatives' of G-7 governments that are in actual control of IFIs.[13]

Executive Summary, para. 4, 'the Bank's policy-based loans must be in accordance with the "purposes" identified in the Articles. Thus, where certain policy and institutional actions and measures are considered necessary for an operation to achieve the Bank's development purposes, these "conditions" may be validly justified under the Articles (IBRD and IDA Articles, Article I)'. In addition, 'the IBRD Articles recognise that the institution may provide financing for productive purposes on "suitable conditions", while under its Articles, IDA may provide financing on appropriate terms. (IBRD Articles, Article I (ii) and IDA Articles, Article V, Section 2 (b).)'

[12] Ibid., para 9. In the Bank context, therefore, conditionality can be defined as 'the set of conditions that must be satisfied for the Bank to make disbursements in a development policy operation'.

[13] S. Grusky, The IMF, the World Bank and the Global Water Companies: A Shared Agenda (Washington: International Water Working Group, Article, 2001). *Source:* http://www.citizen.org/documents/sharedagenda.pdf.

The leverage bestowed on conditionality is magnified because the non-implementation of public sector reform policies would not only entail difficulties with the creditor, but also negative word can virtually shut the country out of any other available source of financing and investment. Fund officials have been among the first to notice this phenomenon. *See* Joseph Gold, Conditionality (Washington: IMF, IMF Pamphlet Series No. 31, 1979) at p. 14, where he states that 'the Fund's endorsement, and the member's observance, of a program have become, increasingly, conditions for the entry into loan contracts by other lenders or for making resources available under contracts'.

The WB started to use conditionality in its policy lending programs in the early 1980s, mirroring the IMF's use of it in its own 'structural adjustment programmes'. WB conditions at that time addressed only short-term macroeconomic imbalances and economic distortions. By the early 1990s conditionality spanned reforms aimed at improving public sector governance; more specifically, it included policies supporting governmental efforts to strengthen public financial management and reduce public expenditures. To achieve these ends, the Bank put forth a range of distinct policies that included trade liberalisation, de-regulation, fiscal austerity and privatisation; this set of measures is comprised in IFI discourse under the general term of 'public sector reform'.

Privatisation, in consequence, has been one of the main reforms that the WB has succeeded in introducing in many countries through its lending programs.[14] Together with the IMF, the WB encourages privatisation as a polyvalent measure that simultaneously aims at regularizing fiscal accounts and reducing the role of the state in sectors where IFIs judge its participation ineffective. IFIs exhort governments to retreat from managing public services and to discard protectionist and regulatory instruments and practices that deter foreign investment. In this way, they set an ideal environment for substituting public for private operation of public services. However, the role of the WB Group in the privatisation of water services is not limited to forwarding privatisation to borrowing states through conditionality. Through its private sector arm, the IFC, the WB has also had a role in the implementation phases of privatisation processes by lending to – and even investing in – companies that become concessionaries of public services.

The following subsection will examine specifically the policies that the WB favours in water management and water services operation.

---

[14] The identification of privatisation of services – together with other distinctive liberal policies – with IFIs' action has led analysts to create the expression 'Washington Consensus' for grouping the set of free market-oriented economic reforms that IFIs have been sponsoring and spreading since the late 1980s. For a summary of these policies, *see* Multilateral Development Banks/IMF, Global Poverty Report (Washington: Report to the Okinawa G8 Summit, July 2000). *Source:* http://www.worldbank.org/html/extdr/extme/G8_poverty2000.pdf.

1.2 World Bank Policies on Water

1.2.1 Integrated Water Resources Management

The IFIs' policy of divesting the state of the operation of water services was part of the introduction of the private sector as an indispensable actor in a new scheme of water management that reflected its economic aspect. It would become known in the WB jargon as 'integrated water resource management' (IWRM). Many international declarations and reports of the early 1990s showed an 'emerging consensus that effective water resources management includes the management of water as an economic resource.'[15] The 1992 Dublin Statement on Water and Sustainable Development, for instance, called 'for fundamental new approaches to the assessment, development, and management of freshwater resources', among them the recognition of water as an economic good.[16] Agenda 21, the program for action elaborated after the 1992 Earth Summit, specified that IWRM is 'based on the perception of water as an integral part of the ecosystem, a natural resource and a social and economic good, whose quantity and quality determine the nature of its utilisation'.[17] The United Nations Development Program (UNDP) quickly followed and embraced as well this holistic new approach to water management, making it a pivotal part of its 'Safe Water 2000' decade.[18]

---

[15] John Briscoe, Water as an Economic Good: The Ideas and What It Means in Practice (Paper presented at the World Congress of the International Commission on Irrigation and Drainage, Cairo, Egypt, September 1996). *Source:* http://ln web18.worldbank.org/ESSD/ardext.nsf/18ByDocName/ WaterasanEconomic GoodTheIdeaandWhatItMeansinPracticeProceedingsofthe WorldCongressofICID1996/ $FILE/Icid16.pdf.

[16] *See* Principle 4 of the Dublin Statement on Water and Sustainable Development, issued following the International Conference on Water and the Environment, Dublin, Ireland, 31 January 1992. *Source:* http://www.ielrc.org/content/e9209.pdf.

[17] Agenda 21, Chapter 18, paragraph 18.8 *in* Report of the United Nations Conference on Environment and Development, Rio de Janeiro, UN Doc. A/ CONF.151/26/Rev.1 (Vol. 1), Annex II (1992). *Source:* http://www.ielrc.org/content/e9211.pdf.

[18] *See* United Nations Development Programme, Safe Water 2000 (New York: UNDP, 1990) at p. 3. *See also* United Nations, A Strategy for the Implementation of the Mar del Plata Plan for the 1990s (New York: UN Department of Technical Cooperation, 1991).

As a concept, IWRM was considered the solution to the two main concerns in water management globally – environmental and economic sustainability. IWRM aims at achieving a rational management of water by balancing the competing uses of it (domestic, agricultural, recreational, industrial, etc.) according to economic and social considerations.[19] To regard water as an economic good would bring important advantages – competitive market pricing and allocation would improve efficiency in water management by reducing wastage, preventing environmentally harmful uses of water and thus maximising the benefits that can be derived from the use of a finite resource.

The WB accompanied this trend with the delineation of new policies. The idea of water as an economic good actually 'constitutes the main change in the policy adopted by the main World Bank's water specialists'.[20] In the water supply and sanitation area, the Bank exhorted countries to price utility services as any other private good,[21] to implement 'full cost recovery' billing and to charge users according to their actual consumption in order to obtain the financial means for the maintenance and expansion of the service network and thus to ensure the long term viability of the provision. Given that the WB and the other IFIs were at the same time addressing the inefficiency of the state as operator of public services, calling for its withdrawal, it was logical that the private sector soon became an important actor in water services provision.

1.2.2 Decentralisation

In order to facilitate the implementation of both IWRM and privatisation at the domestic level, IFIs suggested the implementation of another reform: decentralisation.

---

[19] *See* John Briscoe, Water as an Economic Good: The Ideas and What It Means in Practice (Paper presented at the World Congress of the International Commission on Irrigation and Drainage, Cairo, Egypt, September 1996). *Source:* http://lnweb18.worldbank.org/ESSD/ardext.nsf/18ByDocName/ WaterasanEconomic GoodTheIdeaandWhatItMeansinPracticeProceedingsofthe WorldCongressofICID1996/$FILE/Icid16.pdf.

[20] *See* Finger and Allouche, note 6 above at 76.

[21] *See* World Bank, World Development Report 1994: Investing in Infrastructure 23 (New York: Oxford University Press, 1994).

Decentralisation is defined by the WB as 'the transfer of political, fiscal and administrative powers to subnational governments'.[22] Although in the beginning decentralisation was indeed a measure of reorganisation strictly within the public administration, aiming at improving the delivery and quality of the services provided by it, the definition provided by the WB is currently inaccurate. Decentralisation nowadays covers as well the divestiture of the powers of the state and their transfer to local communities, user associations and to the private sector;[23] and the decentralisation is not only 'political', but also 'administrative', 'economic' and 'financial'. In the water management context, it means that the lowest possible unit of management should be fostered.

The Bank regards decentralisation as critical to its mission of poverty alleviation and achievement of the MDG, since accomplishing such goals depends on states improving the functioning of the public sector. To the Bank and to the rest of the IFIs, an optimal public sector is one that delivers quality public services consistent with citizens' preferences and that fosters private market-led growth while managing fiscal resources prudently. Consequently, the Bank has started to support the implementation of programmes that transfer authority and responsibilities from the central government to subordinate governmental organisations, autonomous entities and the private sector. Decentralisation, as a result, is a process that is not only limited to the reorganisation of governmental structures, ascribing new and more important functions to lower levels of the administration. It is, on the contrary, an all-encompassing phenomenon that divests the state of certain powers, though not – as it is commonly affirmed – redistributing them within the governmental structure but also among non-

---

[22] World Bank, Decentralisation Home Page. *Source:* http://www1.worldbank.org/wbiep/decentralization.

[23] World Bank, Water Resources Management 15 (Washington: World Bank, Policy Paper, 1993) notes that 'the principle is that nothing should be done at a higher level of government that can be done satisfactorily at a lower level. Thus, where local or private capabilities exist and where an appropriate regulatory system can be established, the Bank will support central government efforts to decentralise responsibilities to local governments and to transfer service delivery functions to the private sector, to financially autonomous public corporations, and to community organisations such as water user associations'.

governmental entities and private companies. The link between decentralisation and privatisation is thus blatant, since the former works as a previous step to the latter by dispossessing the central government of certain functions –such as water services provision – that might be assumed by the private sector at a later stage.

It is affirmed that decentralisation has the benefit of incorporating users to a greater extent in the overall scheme of water management. The empowerment of communities and user associations in this sense has been labelled as a commendable 'democratisation' of water management and the Bank has stood behind initiatives such as Community Driven Development (CDD) that not only seek to incorporate users to decision-making levels in water management, specially in rural areas, but actually enable them to manage nearby water resources and to operate their own water services. However, as certain analysts have pointed out, the basic function of user participation, at least in the WB rationale, 'seems to be to make economic and fiscal decentralisation acceptable, in particular by (i) seeking the users' consensus on the overall project and by (ii) getting them to pay the increased fees at the local level'.[24]

As part of its promotion of decentralisation and privatisation applied to water services operation in particular, IFIs have participated in setting the proper conditions for the domestic implementation of said policies. The adaptation of the legal and regulatory framework of water services is, in this sense, indispensable. Through structural adjustment/policy lending programs, and under the supervision of IFIs' technical advisory departments,[25] indebted countries have engaged in substantial reforms in order to turn the water sector into an appealing opportunity for private investment.[26] The leverage of IFI policy advice in this sense is not to be underestimated – together with

---

[24] *See* Finger and Allouche, note 6 above at 86.

[25] These are the Private Sector Advisory Services (PSAS) and its Foreign Investment Advisory Service (FIAS), which provide governments and enterprises with advice on policy, transaction implementation, privatisation, and investment climate.

[26] World Bank Policy Paper, p. 65. Since 1993 the Bank acknowledges as one of its priorities that 'its economic and sector work, lending, technical assistance, and participation in international initiatives [aim] to promote policy and regulatory reforms' for the implementation of privatisation.

loan conditionality, policy advice forms a tandem for the execution of public sector reform in developing countries.[27]

One of such policy pieces of advice has been the suggestion to separate profitable from unprofitable areas of exploitation of water services, offering the profitable ones to private bids and keeping the unprofitable under public administration. This decision would naturally attract private investors to the former group (which usually coincides with urban areas) while relieving them from servicing rural and poorer areas. Another piece of advice has been to discard full asset sales or concessions and instead privatise in the form or management or service contracts.[28] This entails that the state keeps its ownership and responsibilities over infrastructure; however, the state usually fulfills its obligation to maintain and extend the network at the cost of falling in debt with IFIs, while private companies make easy profits from merely operating the service and taking them abroad.

1.2.3 Privatisation of Water Services

The furtherance by IFIs of privatisation of public services in general and of water services in particular is relatively recent. In fact, the WB for decades rather supported the public management of water services and emphasised the overall role of the state in it.[29] This

---

[27] Policy advice in preparation for the arrival of private operators may arise even prior to the actual loan discussions and might be used by the IFI as a precondition for opening the negotiation rounds. See Grusky, note 13 above at 3.

[28] In Argentina, private sector participation mostly took the form of concessions. The difficulties and bad results that this practice entailed in that country and others might have well influenced the decision of the Bank to recommend management contracts instead.

[29] Public Citizen/Critical Mass Energy Program, Profit Streams: The World Bank & Greedy Global Water Companies (Washington: Public Citizen Report, 2002), p. 2. Sources: http://www.citizen.org/documents/ProfitStreams-World%20Bank.pdf. From 1960 to 1990 World Bank loans to developing countries were mainly destined to building, expanding or maintaining public water utilities, especially to large projects such as dam construction. The Bank was convinced 'that investment in public utilities and other infrastructure projects would trigger the development "take off"' and that water utilities were natural monopolies 'that precluded market competition and therefore required public ownership or government regulation'. For a comprehensive review of World Bank thinking, see E. Mason and R. Asher, *The World Bank since Bretton Woods* (Washington: The Brookings Institute, 1973).

stance was not due to the belief that there were not any possible alternatives to public management of said resources, but mainly to the fact that the Bank was involved in project rather than policy lending, granting loans for the construction and development of infrastructure that would be operated and managed by the state, and that private companies themselves did not regard the provision of water services as profitable ventures. Notwithstanding, once the scarcity of water resources came to the forefront of public attention, corporations realised how lucrative the water sector could become. Coincidentally, this happened as the debt crises that hit developing countries in the early 1980s made it clear that 'state-owned enterprises in core sectors of their economies, like infrastructure, were suffering from severe performance problems' and were 'constraining economic growth and undermining international competitiveness.'[30]

The first institution to propel privatisation as a policy conditioning access to the institution's resources was the IMF. Through their structural adjustment programs, the IMF succeeded in incorporating privatisation to the domestic agenda of borrowing countries, including Argentina. In the late 1980s and early 1990s the WB would join those efforts in the water sector specifically, both downplaying the performance of the state operating the provision of water services and underlining the active role that the private sector could have in it.[31] Specifically, private sector involvement was deemed essential to increase investment in infrastructure and expand the network coverage[32] – two of the

---

[30] Ioannis N. Kessides, 'The Challenges of Infrastructure Privatisation', 1/2005 *DICE Report* 19, 20 (2005).

[31] *See* World Bank, note 26 above at 67. The Bank admitted to have learnt lessons from the review that its own Operations Evaluation Department (OED) had conducted of World Bank-funded water endeavours. This review, while acknowledging mistakes in the planning phases of past projects, also concluded that management in public hands had been less than stellar. *See* World Bank/ OED, Water Supply and Sanitation Projects: The Bank's Experience 1967– 1989 (Washington: World Bank, Report 10789, June 1992) at para. 36.

[32] *See* World Bank, *World Development Report 1994: Investing in Infrastructure* (New York: Oxford University Press, 1994) and World Bank, *World Development Report 1997: The State in a Changing World* (Washington: World Bank, 1997). World Bank, note 23 above at 27, claimed that states lacking resources were incapable of extending the service network so as to reach the poorest sectors of the population or of modernising it to meet current health and environmental standards.

main goals that the concession in the city of Buenos Aires had – , while the public sector was to be in charge through regulation of ensuring that social and other policy goals were preserved during the performance of the private company. The next section will assess whether the parties involved achieved those ends.

2. The Privatisation of Water Services in Buenos Aires

2.1 The Concession Contract and Early Problems

The modifications to the legal and regulatory framework that made possible the privatisation of water services in the city of Buenos Aires can be traced back to 1989. One of the very first initiatives of the recently-elected president Carlos Menem became Law 23.696 – also known as State Reform Law –,[33] a legal instrument that was the result of the dialogue and negotiations that the country had initiated with IFIs while seeking an exit to the major economic crisis hitting the country in the late 1980s. Law 23.986 reflected the policies that the WB was sponsoring at the time and heralded a comprehensive privatisation process of public enterprises and enterprises in which the state participated, whether or not they provided public services. The first of President Menem's years in office saw the WB participating intensively in the reform process through non-lending services, in particular through informal Economic and Sector Work and policy dialogue.[34] The Bank readied

---

[33] Approved on 17 August 1989, promulgated the next day and reglamented by National Decree 1105. Published in Anales de Legislación Argentina No. XLIX-C, 1989, 23 August 1989, pp. 2444–2457. Menem had taken office on 8 July 1989. World Bank/OED, Argentina: Country Assistance Review (Washington: World Bank, Report 15844, 1996), Executive Summary, para. 5 notes that Argentines were at that moment suffering 'the trauma of extended recession and hyperinflation', which had created a 'fertile ground for reform'.

[34] *See* World Bank/OED, note 32 above at paras. 8 and 9. As a consequence of the 1989 Argentinean crisis, 'the Bank reduced its lending program in Argentina but continued, or intensified, its policy dialogue, and the respective roles of the Bank and the Fund were clarified through a concordat whereby the Fund agreed to take the lead in short-term macroeconomic programming and monitoring, while the Bank agreed to deal with the institutional underpinnings of macroeconomic policy' (para. 8).

loans that became essential to the initial success of the public sector reform. In 1991 it granted Argentina a Public Enterprise Adjustment Loan and a related Technical Assistance Loan which supported the comprehensive privatisation program.[35]

In the water services sector, a single national utility, Obras Sanitarias de la Nación (OSN), used to cover the entire Argentine Republic. However, under the pressure of the WB, water services were decentralised in 1980,[36] have been under provincial jurisdiction ever since, and were managed by provincial companies or local/community entities such as cooperatives until water services were targeted for privatisation.[37] Because of the previous decentralisation, at the time of introducing privatisation, the federal government did not have the authority to implement it beyond its own jurisdiction. The National Executive could just take the initiative to liquidate and privatise the utilities that remained under federal jurisdiction – among them OSN, which after the decentralisation only covered the services of the capital of the Republic and some of its neighbouring suburbs located on territory of the province of Buenos Aires.[38] The central government then started to pressure the provinces to do their share in their

---

[35] Ibid., para. 24.

[36] OSN had endured budget cuts under the pressure of IFIs' structural adjustment programs until the decision to decentralise was imminent. Law 18.586 of 1970 instrumentalised the decentralisation. However, the decision to implement it was not taken until 1979 and became effective only in 1980 through National Decree 258/80.

[37] The World Bank aid offered to the country since the late 1980s, in particular to extend water services coverage, already insisted on the privatisation of the company. World Bank/OED, note 33 above at para. 4 argues that the downfall of the economic stabilisation programs designed by Argentine authorities 'was the failing of measures to rein in the public sector deficit, particularly that generated by public enterprises'. The underwhelming performance of OSN during the 1970s and 1980s cannot be contested: low investment, lack of control, high indebtedness of users, absolute absence of public participation, etc. made the company an early candidate for reform.

[38] The water services provided to these suburbs form an indivisible unity with the services provided to the city of Buenos Aires.

jurisdictions in order to also bring a balance to their public deficits.[39]

The privatisation of OSN was undoubtedly the biggest operation in the water sector, both in terms of infrastructure and number of users involved. The water services of the city and of thirteen of its neighbouring suburbs,[40] home to some nine million inhabitants, were only privatised in 1993 even if they had already been appointed for privatisation three years earlier.[41] The delay was due to the work that was necessary on OSN prior to its transfer to private management, and to the strong opposition of labour groups. Workers at OSN expected personnel cuts as one of the first measures to be taken to improve the efficiency records of the enterprise. The opposition was eventually weakened and the ultimate adhesion of workers was won over by using the denominated 'programs of shared ownership' introduced under Law 23.696.[42] These programs allowed working forces to purchase

---

[39] The federal government pressured the provinces through the distribution of the GDP and the refusal to underwrite their credits with international institutions. Provinces such as Santa Fe, Tucumán, Córdoba and Buenos Aires also privatised their water services, with negative results. Many of them had to cancel the contract early and the Argentine government has therefore been sued before the ICSID by the private operators. *See* the pending ICSID cases No. ARB/97/3, No. ARB/03/17, No. ARB/03/19, No. ARB/04/4, No. ARB/03/30, No. ARB/07/17, No. ARB/07/26. The restatisation of water services in the province of Buenos Aires motivated another suit (No. ARB/01/12) that saw Argentina condemned to pay reparation. Argentina initiated annulment proceedings against that decision that are still pending. The case of the concession in the province of Córdoba (No. ARB/03/18) was settled through an agreement between the parties.

[40] The district of Quilmes would later become the fourteenth suburb by voluntarily joining this privatisation. Later, the suburbs became seventeen due to the subdivision of the Moron district.

[41] National Decree 2074/90 (10 October 1990). The announcement calling for the presentation of bids was made a year after through Resolution of the Secretary of Public Works and Services No. 178 (13 December 1991).

[42] Law 23.696, Chapter III (Articles 21–40), *see* note 33 above.

a portion or the totality of the privatised entities' shares,[43] and the profits expected from the operation gradually convinced several labour groups to yield. While workers got a ten per cent of the shares of the company, in the months following the operation half of the 7,200 jobs at OSN would be cut.[44]

In the end, the Argentinian government ceded management of the services to a consortium called Aguas Argentinas,[45] formed by French, British and Spanish capital, together with local partners.[46] The contract, which took the form of a concession of the services of drinking water and sewerage, was granted to this conglomerate not on the basis of the biggest investment commitment but upon presentation of the largest tariff rate reduction bid.[47] Prior to the privatisation operation, the government had however raised tariffs

---

[43] This strategy is criticised by privatisation detractors as representing a sophisticated, institutionalised form of bribery. Its use as a common practice in Argentinean privatisations for 'buying' the consent of workers has been acknowledged in papers elaborated for the Inter-American Development Bank (IADB). See D. Artana, F. Navajas, and S. Urbiztondo, Regulation and Contractual Adaptation in Public Utilities: The Case of Argentina 21 (Washington: IADB, Paper IFM-115, 1998). Source: http://www.iadb.org/sds/publication/publication_267_e.htm.

[44] Public Citizen Web Page, Report on Argentina. Source: http://www.citizen.org/cmep/Water/cmep_Water/reports/argentina. It is estimated that between 1990 and 1994 280.000 Argentineans lost their jobs in the public sector throughout the whole privatisation process, the majority of which through official programs of early retirement. Only 40 per cent of that number was absorbed by the new private operators. See 'El Estado tiene deudas por 7000 millones tras las privatizaciones', Argentinean Journal *La Nación* [hereafter '*La Nación*'], 16 September 1996.

[45] The operation was approved by National Decree 787/93 on 22 April 1993.

[46] The partners were, respectively, Lyonnaise des Eaux (SUEZ Group), Compagnie Génerale des Eaux S.A (Veolia Group), Anglian Water PLC, Aguas de Barcelona S.A and the local partners Meller S.A., Sociedad Comercial del Plata S.A., and Banco de Galicia y Buenos Aires S.A. Lyonnaise des Eaux (SUEZ) was named main operator. See Official Bulletin of the Argentine Republic, 24 March 1993.

[47] Aguas Argentinas offered to reduce the rate in 26.9 per cent, slightly above the bid of the runner-up (26.1 per cent). Only three bidders reached the final stage of the bidding process.

by 62 per cent and additionally increased them by 18 per cent through the creation of a new tax. Many failed to realise that this was a strategy of the Argentinean government for convincing public opinion that privatisation was the sole alternative to deficient and expensive services handled by the public sector. In a context where memories of the penuries endured during the hyperinflationist crisis of the late 1980s were still fresh, this tactic was successful.[48] Aguas Argentinas indeed lowered rates as promised, but the disproportionate rise that the Argentinean government had put into effect before the transfer of the utility to the company, had given the latter margin for doing so and still making considerable profits.[49]

The privatisation of water services was paired with the creation of a regulatory entity in charge of the supervision of the performance of the private operator and its compliance with the terms of the contract, the Ente Tripartito de Obras y Servicios Sanitarios (ETOSS).[50] ETOSS was financed through a percentage of Aguas Argentinas' billing, a fact that has shed doubts over its capacity to objectively intervene on tariff-related issues. Moreover, it later became clear that the entity, with its tripartite composition, was victim of constant politicisation. The composition of its directory made it difficult to find agreement among the representatives of the three jurisdictions involved, since all six directors (two per jurisdiction) had the same power and attributions, thus preventing the entity from acting quickly and effectively whenever disagreement arose. However, it was particularly remarkable how its work and role would be progressively neglected

---

[48] It was applied in the same way in relation to the privatisation of ENTEL, the national telecommunications company.

[49] As it will be seen below, the private operators preferred to present an opportunistic bid in order to become concessionaires, and then engage in multiple renegotiations to accommodate the contract to their interests.

[50] Created by Agreement among the Central Government, the Government of the Province of Buenos Aires and the Mayor of the City of Buenos Aires on February 10, 1992. The 'Tripartito' refers precisely to its composition of representatives from the three jurisdictions involved in the provision of water services by OSN in Buenos Aires: the city, the province and the central government.

by higher governmental levels, to the point of being virtually deprived of authority or support as circumstances called for.[51]

The story of the privatisation of water services in the city of Buenos Aires was one of constant revisions and multiple breaches of the contract. Only one year into operation, Aguas Argentinas pressured ETOSS to allow for a rate increase even though the company had agreed not to raise prices for ten years.[52] ETOSS agreed[53] and an increase of 13.5 per cent for consumption, disconnection and reconnection of the service, and of 42 per cent for new connections, was adopted in exchange for accelerating the connection of slum communities to the service network.[54] The entity disregarded the fact that this first resolution that had informally modified the terms of the privatisation contract could have serious precedent-setting effects.[55] In reality, it became the first of many modifications, thereby creating a climate of legal and regulatory insecurity for all actors involved, including users of the service.

---

[51] *See* L. Alcazar, M.A. Abdala, and M.M. Shirley, The Buenos Aires Water Concession (Washington: World Bank Policy Research Working Paper No. 2311, 2000). *Source:* http://papers.ssrn.com/sol3/papers.cfm?abstract_id= 630683, where they criticise the politicisation of ETOSS, the inexperience of its directors and correctly underlined that the weakness of the regulatory institutions was an obstacle to the success of the concession.

[52] This compromise emerges after an attentive reading of the original contract. It must also be said that, from what is inferred from Article 12.7 of National Decree 787/93, there was no room for modifications in the expansion plans for at least five years since the beginning of the concession.

[53] ETOSS Resolution 81/94, 30 June 1994.

[54] In practice, the company raised tariffs over 16 per cent for consumption, disconnection and reconnection. The 42 per cent raise for new connections was not noticed by users and did not cause the upheaval it did until the corresponding bills were sent two years later. *See* Artana, note 43 above at 20–21.

[55] A. Loftus and D.A. McDonald, 'Of Liquid Dreams: A Political Ecology of Water Privatisation in Buenos Aires', 13/2 *Environment & Urbanization* 179, 191 (2001) note that all price increases that were effectuated in the first five years of the concession 'implied that the contract was negotiable and that the company could push for tariff increases whenever it wished to, particularly if they could show that new demands were extra-contractual and had to be paid for by the consumer'.

Aguas Argentinas succeeded in establishing a connection fee that ranged from 600 pesos/dollars for drinking water and from 1,000 pesos/dollars for sewerage. The WB, using its common rhetoric of full cost recovery and placing the expansion of the service as the main priority, publicly defended Aguas Argentinas' posture and not only granted to it new loans through the IFC, but it even became its partner through the purchase of a share of the company.[56]

Despite tariff increases, in 1996 the company still boasted that it had lowered tariffs by 17 per cent with regard to the numbers that users were paying to OSN in 1993 and that the process was considered successful.[57] This statement, however, as we mentioned earlier, was taking into account the tariffs artificially inflated by the government in order to introduce the privatisation of services. In reality, the 26.9 per cent reduction offered in the winning bid by Aguas Argentinas had little chance of materialising, since tariffs started to mount as soon as 1994. However, most of the profits the company was making[58] were not the result of tariff rises but of not fulfilling its engagements in infrastructure investment. Aguas Argentinas called for a renegotiation of the original contract, arguing that most new users were unable to pay the connection charges introduced in 1994. In reality, the private company was interested in obtaining new delays for the construction of sanitation facilities and the expansion of the service network.

---

[56] The share was acquired by the IFC and it was actually an exchange for the debts Aguas Argentinas had with it. As Loftus and McDonald, note 55 above at 185 affirm, 'not only does this testify to the instant profitability of the firm (the IFC wanted a share in these profits), it raises questions about the objectivity of World Bank research into the privatisation initiative. It also makes the Bank's aggressive promotion of the Argentinean model abroad problematic'. The IFC also participated in Correos Argentinos, the postal company, which was also among the first public services to have their privatisation contract revoked.

[57] World Bank/OED, note 33 above at 9 was proud of having contributed 'finance and advice for the most far-reaching public enterprise privatisation program ever carried out by a developing country'.

[58] According to Artana, the publication 'The Economist' considered the Argentinean concession the most profitable in the sector worldwide, with rates of return close to 40 per cent. *See* Artana, note 43 above at 21.

2.2 The 1997 Renegotiation of the Contract

In April 1996, Aguas Argentinas approached ETOSS concerning the need to renegotiate the contract on the grounds that, if tariffs were not raised as a compensation for the unpaid new connections, it would be impossible for them to meet contractual obligations related to the maintenance and expansion of the network. The sums owed to Aguas Argentinas escalated to 30 million by October of the same year, prompting ETOSS to ask the company to suspend thousands of suits filed by the latter against users for their delay in payment and to accept to open a dialogue for the modification of the concession contract.[59] To the Argentinean public opinion, this revision was presented as indispensable on the grounds of 'new exigencies of public order that had not been foreseen in the original contract and had emerged ever since'.[60] The representatives of Aguas Argentinas concurred on the dialogue rounds not being about renegotiating the original contract, but rather about 'adapting' it to an unforeseen situation that the company was then forced to face.[61]

When ETOSS expressed its disagreement with the company's arguments, the Ministry nominated the Secretary of State of Environment as principal negotiator, unauthorising the regulator.[62] The state representatives in the renegotiations of 1997 had an ample

---

[59] By the time the renegotiations started, the company was arguing that its operative deficit had reached 60 million dollars/pesos.

[60] National Decree 149/97, 14 February 1997, Article 1.

[61] Interview with Guy Canavy, General Manager of Aguas Argentinas. *See* Javier Blanco, 'Obras y Negocios para Evitar Aumentos', *La Nación*, 22 February 1997. Jerome Monod, head of the Lyonnaise des Eaux Group, visiting Argentina in March 1997 as part of French President Jacques Chirac's retinue in the latter's official visit to the country, equally insisted on the rounds being more of an 'update' of the original contract rather than a renegotiation. *See* Interview by Tomás Farchi, 'Cuando el Futuro Impulsa los Negocios del Presente', *La Nación*, 26 March 1997.

[62] ETOSS was bypassed in favour of the Public Works Secretariat and the Natural Resources and Sustainable Development Secretariat on the basis that the renegotiations would address environmental issues as well, such as the recuperation of the Matanza river, one of the most polluted streams in the world.

margin for bargaining, being entitled by the Executive to bargain on basically every aspect of the contract in order to reach an agreement with Aguas Argentinas.[63] It was the first time that a concession contract with a private operator would be modified in Argentina. For this reason, the development and outcome of the renegotiation rounds were closely watched by IFIs and private companies operating in Argentina; it could become a precedent for eventual changes to similar contracts in force. The WB even decided to send one of its senior water management authorities as a consultant for Aguas Argentinas. It is to be underlined that the WB Group by then had spent millions of dollars in loans to the company, had already invested more directly in it by the acquisition of a share and had put forth this particular contract as a success story in privatisation.

Aguas Argentinas agreed to reduce charges for new connections in exchange for a postponement of the infrastructure investments of the original project and the extension of the privatisation contract; a deal that had already been foreseen by the specialised media.[64] The reduced connection charge (CIF) was paired with a new surcharge called SUMA to be perceived by Aguas Argentinas from 1998 on.[65] It was argued, once again, that the surcharge would guarantee the expansion of the network in a context of alleged diminution of the company's profits due to the acute recession that Argentina had started to endure.[66]

---

[63] *See* National Decree 149/97, note 60 above.

[64] Ernesto de Paola, 'Bajarán un 5 Por Ciento las Tarifas de Electricidad', *La Nación*, 26 October 1996.

[65] Apparently, the state and the company had agreed to delay the perception of this surcharge until the elections of 26 October 1997. Thus, the official instrument that finally approved all the modifications to the contract was signed shortly after (National Decree 1167/97, 7 November 1997) and SUMA would appear on water bills from March 1998 onwards.

[66] The product of both CIF and SUMA literally meant that Aguas Argentinas' investment in infrastructure would mainly be afforded by the users of the service (which had not been given the opportunity to participate in the renegotiation process) and had to be paid prior to the execution of any works, a measure that eliminated any risks for the company. *See* Loftus and McDonald, note 55 above at 192–193.

In addition to this, the renegotiation condoned all previous blatant breaches of the contract by Aguas Argentinas.[67] Moreover, according to a report made by both the ETOSS and a panel of technicians from the Danish Hydraulic Institute, among other institutions, the company had put on the table for discussion a proposal that omitted or modified several investment projects whose realisation it had previously assumed and which did not even count with an environmental impact assessment.[68] The most notable of these was the Berazategui Plant for treating sewerage waters. The report stated that with each year that its construction was delayed, Aguas Argentinas was increasing its current value of future profits in 35 million dollars.[69] The suggestion of replacing primary and secondary treatment by pre-treatment is especially puzzling in a renegotiation supposedly summoned for addressing environmental concerns since, according to the report under comment, this option would increase the rate of polluting substances spilled in the Rio de la Plata River.

Equally perplexing is the introduction in the new version of the contract of a provision that violated the Convertibility Law – the annual indexation of the tariffs, which were issued in pesos, according to the inflation rate of the United States of America.[70] This provision further relieved the company from a major financial risk in its operations in the country, granting intangibility to its profits. Conversely, the impact of the said norm would be a major source of concern for users, given that devaluation could hit their

---

[67] While user associations were denouncing unreached investment goals worth 400 million dollars, the government in the renegotiation rounds had only addressed works worth 201 million that were included in a new schedule. The eight million dollars in fines that the ETOSS had sanctioned the company with were not even discussed.

[68] In fact, it is highly possible that the ETOSS was excluded from the renegotiation rounds due to having solicited said report, given that the document in question was extremely critical of the company's plan and performance. *See* ETOSS, Final Report of the Expert Commission for the Analysis of the Aguas Argentinas S.A. Proposal for the Modification of the Plan of Improvement and Extension Service, 29 March 1996, in particular at 4–9. Available on file with the author.

[69] Ibid., 4.

[70] Such measure was forbidden by Article 10 of the 1991 Law 23.928 (Law of Convertibility).

household income while tariffs for an essential service would remain untouched. In 1998 Aguas Argentinas profited for the first time of this provision and asked for a further 11.7 per cent rise in tariffs. This raised protests by users, which had constantly been left out of negotiations, denied information and anyway been compelled to pay for the investment projects of Aguas Argentinas.[71]

In summary, the 1997 renegotiation of the contract between the government and the company clearly proved that the combination of a monopolistic private provider of water services and of the state neglecting its role of regulator can only work in prejudice of the users.[72] The process revealed that the company would not hesitate to paralyse connection works as a way of exerting pressure for new tariff rises, and that the company counted with a complacent attitude on behalf of the Secretary of State of Environment, which prevented the ETOSS from participating in the renegotiation rounds, trivialised its work and inexplicably supported a company with disappointing pollution records. The Secretary disregarded the institutional mechanisms that had been created for guaranteeing the preservation of users' rights and their ultimate wellbeing. In a context where there was no political will to dispute Aguas Argentinas' pretensions, the frail regulatory framework did not have an actual chance of working in defense of users' rights.

---

[71] The whole renegotiation operation was impugnated by the National Ombudsman and users' associations in February 1998. *See* Anonymous, 'La justicia definirá los aumentos', *La Nación*, 21 February 1998. In March 1998, the Justice endorsed the Ombudsman's impugnation and as a precautionary measure ordered the state and the company to suspend the application of the 'SU' part of 'SUMA' (that is, the surcharge that all users were paying for the extension of the service network to low-income areas in the outskirts of Buenos Aires). This setback prompted Aguas Argentinas to warn that it would be impossible to continue with the expansion projects in order to connect the three million inhabitants that still did not count with basic services. The use of this argument forced the mayors of several districts to ask for the revision of the decision on the grounds that the works for the expansion of the sewerage network were completely paralysed. The Executive joined the appeal and finally the Tribunal revoked the suspension ordered in the first instance.

[72] As a result of the renegotiation, tariffs increased by 36 per cent throughout 1998.

2.3 The Service after the 2001 Economic Collapse

The administration that took office in 1999, after President Menem's second mandate came to a close, was unable to provide a remedy for the worsening state of affairs that affected the privatisation of water services in Buenos Aires. On the one hand, when the government called for renegotiations to adapt all privatisation contracts to the economic crisis situation, the foreign companies that were operating public services in Argentina complained about the lack of legal security. On the other hand, the antagonism against foreign economic actors quickly deepened in a country in recession, since they were seen as exploiters of users that constantly failed to satisfy their own investment commitments.[73] In the case of Aguas Argentinas, the strategy of making dependent the expansion of the service network and other investment projects on new tariff rises still proved successful for the company and new rises of 3.9 per cent were agreed upon in January 2001. Nevertheless, the courts started to accept complaints by users and nullified certain tariff rises that had been approved in violation of due procedure, while the ETOSS applied heavy fines to the company for its delay in the construction of new infrastructure.

The beginning of the end for the privatisation contract of water services in Buenos Aires came with the economic, political, institutional and social collapse of Argentina in December 2001.[74] As a result of the ensuing devaluation, the huge profits made by Aguas Argentinas for almost a decade were threatened as inflation increased its mounting debts. In 2002, Emergency Law 25.561 took

---

[73] One of the early suits was filed by authorities of the Berazategui District on the grounds that Aguas Argentinas was spilling untreated sewerage waters to the Rio de la Plata River. The company argued that the construction of the plant for the district had been postponed in the 1997 contract renegotiations, and thus the government itself had chosen to reschedule the works and overlook the pollution.

[74] For an analysis of the end of the privatisation of Aguas Argentinas and its nationalisation, see D. Aspiazu and N. Bonofiglio, Nuevos y Viejos Actores en los Servicios Públicos: Transferencia de Capital en los Sectores de Agua Potable y Saneamiento y en Distribución de Energía Eléctrica en la Post-Convertibilidad (Buenos Aires: FLACSO-Argentina, Documento de Trabajo No. 16, 16–19, 2007).

the first step towards the redefinition of the contractual relation with all privatised firms, calling for a revision of the contracts in force.[75] The company publicly warned that a minimal profit would have to be guaranteed to Aguas Argentinas or else it would not hesitate to rescind the contract and sue the government of the country. It started an aggressive campaign in preparation of this event and deployed its usual means of lobbying, which included diplomatic meetings of French government representatives with national authorities.[76] It also conditioned the fulfillment of its contractual obligations on a series of demands that were unrealistic in the context of the Argentine economic crisis.[77] Among them, an

---

[75] *See* Law 25.561 of 6 January 2002 and National Decree 293/02 of 14 February 2002, which entrusted the Ministry of Economy with the mission to conduct the renegotiations. Article 8 of said instrument eliminated the privileges Aguas Argentinas was holding, such as the 'dolarisation' of prices and their indexation according to variations in the United States' price figures. Article 4 confirmed on the other hand what had already been stated in judicial decisions, namely that this practice violated the Convertibility Law. This could open the door to the review of all of the water tariffs increases imposed under the contract. At the same time Law 25.561 froze tariffs for six months.

[76] The involvement of French government representatives in the many renegotiations of the contract was blatant. Francis Mer, Minister of Economy of France, warned Argentina that France would defend the interests of Suez and other French companies operating in the country, and complained that Nestor Kirchner's presidency lacked a genuine will to adjust public services tariffs. He did not hesitate in reminding Kirchner about the importance of France's support to Argentina in reaching an agreement within the IMF after the default. *See* Anonymous, 'Francia Promete Defender a sus Compañías', *La Nación*, 24 January 2004. The linking of tariff adjustment and support before the IMF and World Bank was also mentioned by Dominique de Villepin (French Minister of Foreign Affairs) during his visit to Argentina in February 2004. *See* Florencia Carbone, 'De Villepin Mezcló Reclamos con Promesas ante Kirchner', *La Nación*, 4 February 2004.

[77] Note 35.050/02. These demands included the unilateral suspension of all investment projects, the extension of the contract to compensate for losses and the suspension of the fines imposed by ETOSS. They were grounded in the argument that the company had taken an important external credit (before the World Bank Group and the IDB) in order to finance investment projects and that it needed the state to sell to it dollars at the usual parity for servicing such debt.

exchange rate insurance that would have implied that the state (in fact the society as a whole) should bear more than half the firm's external debt, contracted with national and international banks and with multilateral organisms such as the IDB and the WB.[78] The Argentine government refused to offer such exchange rate insurance and Aguas Argentinas defaulted on 10 April 2002, with an avowed debt of 700 million dollars.

The renegotiations proved unsuccessful in solving the company's problems caused by the devaluation of the currency. As the period banning tariff increases expired, Aguas Argentinas prepared an estimation of losses that amounted to 500 million Euro and notified the government of its intention of making use of the recourse foreseen in the bilateral treaty for the protection of investment signed between Argentina and France.[79] The company eventually filed a suit before ICSID against the state.[80]

In view of the mounting hostility between the company, the government and users – who had successfully prevented new tariff

---

[78] D. Aspiazu and K. Forcinito, Privatisation of the Water and Sanitation Systems in the Buenos Aires Metropolitan Area: Regulatory Discontinuity, Corporate Non-Performance, Extraordinary Profits and Distributional Inequality (Paper presented to the First Project Workshop entitled 'Private Sector Participation in Water and Sanitation: Institutional, Socio-Political and Cultural Dimensions' organised by the School of Geography and the Environment, University of Oxford, 22–23 April 2002).

[79] Signed on 3 July 1991; in force since 3 March 1993, a month before Aguas Argentinas assumed the provision of water services to the city of Buenos Aires.

[80] The Ministry of Economy replied by publicising an official report revealing that the company was not enduring an operational deficit that allowed it to ask for further tariff raises. This document also surveyed the non-performance of its infrastructure investment plans and how tariffs had risen between 54 and 65 per cent since 1993. In this light, users' associations demanded that the state ended the contract. *See* Comisión de Renegociación de Contratos de Obras y Servicios Públicos, Informe de Cierre Fase II, Buenos Aires: Ministerio de Economía (2002). *Source:* http://www.mecon.gov.ar/crc/aguas_final_fase_ii.pdf. Aguas Argentinas' suit was dropped by the company in February 2006 in order to facilitate the transfer of the management of the service to potential replacements in the concession. Suez and Aguas de Barcelona, on the other hand, did not withdraw their individual suits as shareholders of the company.

increases through the courts – the WB and IMF sent representatives to the country in order to mediate among all actors and add a layer of pressure on the government for resolving the conflict. Both IFIs confirmed via a note to the Argentine Ministry of Economy that the mission did not seek 'to make recommendations for specific changes to the contracts or tariffs, but to get acquainted with the general situation of the renegotiation and to assess the framework in which it is being carried out'.[81] However, the team would in the end make specific recommendations for the reform of regulatory frameworks and concession contracts of privatised public services in Argentina,[82] and despite denying any involvement in the 'adjustment' of tariffs, it was not a secret that after the peso devaluation IFIs were pushing for new tariffs that reflected the operation costs of providing public services in this new context.[83] The IMF/World Bank mission also expressed to local functionaries its concern over the 'excessive judicialisation' of the contract and tariff renegotiation processes, thus upsetting user associations and ombudsmen who had legitimately resorted to courts to stop the tariff rises.[84]

---

[81] Text of the note quoted in Anonymous, 'Servicios: Empieza el Análisis del FMI y el BM', *La Nación*, 11 February 2003. The note was co-signed by the World Bank and IMF directors for the region.

[82] *See* Josefina Giglio, 'Informe Final sobre las Privatizadas', *La Nación*, 24 February 2003.

[83] Anne Krueger, then head of the IMF, had said that tariffs should be increased between 30 and 50 per cent. *See* EFE, 'Llegan Mañana los Auditores del FMI y del BM', *La Nación*, 16 February 2003. The tariffs of course were a prominent topic of discussion. The IMF/World Bank team met with executives from the foreign companies operating public services in Argentina in order to discuss the restructuring of their defaulted debts with the WB Group, and the executives, in turn, manifested to the IFIs' representatives their need for an emergency tariff adjustment in order to keep providing the services, the setting of an indexation mechanism for tariffs and that the regulatory framework was redesigned to establish more clearly the concessionaries' obligations. *See* Anonymous, 'El Banco Mundial Recibió a las Privatizadas', *La Nación*, 20 February 2003.

[84] *See* Josefina Giglio, 'Un Nuevo Fallo Contra los Aumentos', *La Nación*, 5 March 2003.

By September 2003, with a new government in charge of the renegotiations, the attitude towards the companies operating public services became more stringent. Every privatisation contract in force was re-examined and the nationalisation of services therefore became a concrete option in official discourse.[85] Aguas Argentinas' failures to meet its commitments made it an early candidate for contract revocation. Simultaneously, the IFIs increased their pressure by linking the rescheduling of their credits with Argentina to the success of the renegotiations with all firms operating privatisations, accusing the country of delaying the talks on purpose and of not really having the will to negotiate. The team of technical assistants of the IMF and WB returned in 2004, with the renegotiations still pending, urging the country to conclude them once and for all. A transitory act was signed with Aguas Argentinas,[86] but the final contract and the new regulatory framework to be sent to Congress was still in the works. In late June 2004, when the WB had to discuss the granting of two credits to Argentina amounting 700 million dollars, the Board of Directors was divided: G7 countries that had nationals affected by Argentina's private debt default opted to abstain or vote against Argentina, and insisted on a final solution to the renegotiation of contracts and tariffs. Analysts deemed that the approval or denial of these credits by the WB would in turn influence the decision that the IMF would take on the restructuring of its own credits and the disbursement of new ones to Argentina.[87] It became evident that complying with the

---

[85] *See* Anonymous, 'Kirchner Afirmó que No Aceptará Presiones de Estados Unidos', *La Nación*, 15 September 2003. An ultimatum to invest under sanction of revoking the privatisation was released in November 2003 and accepted by the company. The state also wanted to assume a more active role, divesting the company of some of its competences. For example, it started to collect the sums that were destined to infrastructure projects and directed the works itself, deciding where and when to construct, only leaving to the company the maintenance and management of the service.

[86] The agreement included the promise not to raise tariffs until December 2005, the suspension of the payments of the debts accumulated by the company from ETOSS's fines, the implementation of a 242 million pesos investment program and the suspension of the suit filed before the ICSID.

[87] *See* Jorge Rosales, 'El Banco Mundial Define Dos Créditos con el G-7 Dividido', *La Nación*, 29 June 2004.

demands of the private companies operating private services in Argentina would be a precondition to mend the relationship with IFIs after the collapse of the local economy. Once the country committed itself to end the renegotiations processes, the WB Board of Executive Directors unanimously consented to the new loans, which were in part aimed at re-establishing an investment climate in Argentina.[88] The support of the WB Group to the company was strengthened when the IFC accepted to restructure Aguas Argentinas's private debt with them, reducing it by 35 per cent.

Suez threatened to end the contract, warning about the effects that the nationalisation could have on the investment climate of the country and on the pending suits before ICSID, which amounted to more than 20,000 million Pesos.[89] The decision to end their operations in Buenos Aires was finally taken by Suez together with the other major share-holder, Aguas de Barcelona, in September 2005, arguing two years of fruitless negotiations for new tariffs. The Argentine state threatened to sue the company if they did not ensure the provision for the following year, while the transition to a new operator was implemented. In March 2006, the state notified the company that it had instead decided to rescind the contract, arguing Aguas Argentinas' many breaches of it, and to put water services back in public hands.[90] Following the announcement, the IFC and the WB Group in general opted to remain silent.

---

[88] *See* Jorge Rosales, 'Apoyo Condicionado del Banco Mundial', *La Nación*, 30 June 2004.

[89] Suez, who was also handling the water services of fifteen important cities in the province of Santa Fe, announced shortly after, in May 2005, that it would end its operations under the name Aguas de Santa Fe and concentrate in pursuing the renegotiations to save its contract in force for the city of Buenos Aires. It still operated Aguas Cordobesas, in the province of Córdoba, though it would announce in April 2006 its intention to step aside from that concession as well.

[90] National Decrees 303/2006 and 304/2006, 21 March 2006. The provision would be in charge of a new public enterprise called AYSA (Aguas y Saneamiento Argentinos Sociedad Anónima), with 90 per cent of its share in the hands of the state and ten per cent in the hands of the employees of Aguas Argentinas through the programs of shared ownership. Currently, the water services of the provinces of Buenos Aires, Santa Fe, Catamarca and Tucumán are public as well after their own failed experiences with privatisation.

3. Lessons to Learn from the Privatisation of Water Services in Buenos Aires

The failure of the privatisation of water services in the city of Buenos Aires is due to the confluence of two main factors – the lack of an adequate legal and regulatory framework, and the fixation of the WB with spreading the privatisation model around the world.

3.1 The Importance of a Sound Legal and Regulatory Framework

The privatisation of public services was introduced in Argentina by the Bank through its policy lending programs and technical advice services as a neoliberal measure that would become a first step in remedying the grave crisis hitting the country in 1989.[91] Consequently, the Argentinean popular opinion and statesmen embraced it without previously undertaking the design of a solid legal and regulatory framework or the creation of those institutions that are indispensable for protecting the interests of users. Many of the problems related to the privatisation of OSN that would eventually surface were the result of a rushed and poorly executed privatisation process and had roots in its unprofessional implementation. This is well illustrated by the regulatory entity, ETOSS. Its efficiency was put in question on several occasions, having been a victim of politicisation and bypassed many times by higher organs within the national administration that were suspiciously more willing to comply with the terms that the company put forward in the renegotiations of the contract. In addition, the inexperience and lack of training of many of the directors of ETOSS hindered its role of supervisor of the performance of Aguas Argentinas.

Two of the most significant deficiencies of the legal and regulatory framework included, first, the inoperativeness of the regulatory agency's work and, secondly, the absence of user

---

[91] World Bank/OED, note 33 above at 27 notes that '[t]he earlier ESW studies supplied data that provided functional input for the [Argentine] Government's privatisation program, and provided justification for the Bank to begin lending later in the period to help finance the effort'.

participation in the oversight of the company's performance. Indeed, ETOSS collected the relevant information to assess the compliance of the company with the terms of the contract, but failed to react opportunely due to its delays in processing the data, and to the political pressures within the directory, which prevented it from adopting prompt sanctionatory measures against the company. Moreover, during the first years of the concession, some WB technicians supported the idea of 'light-handed regulation', according to which the regulator should not interfere with the work of the private operator. For many years too, ETOSS did not publicly disclose available information and thus prevented user associations from participating in the monitoring of Aguas Argentinas via available channels within the administration or before the courts.[92] The government equally minimised the importance of public participation by not convoking users to expose their grievances in the major renegotiations that took place, including those which took place in 1997. The summoning of public audiences by the regulatory entity to inform users and allow them to be heard did not happen until later in the process, and often coincided with electoral periods when local politicians sought to ingratiate themselves with the population by addressing their concerns. These hurdles to stakeholder participation in the monitoring of water services provided by a private company even represented a direct violation of constitutional provisions[93] and added to the legal uncertainty that reigned over the concession in Buenos Aires. It must also be said that water users, in turn, failed to organise themselves appropriately, and this prevented them from effectively defending their interests for much of the concession term. User activism became prominent only after the effects of the 1997 renegotiation were fully impacting on the income of households against the background of a grave economic recession.

---

[92] Users associations denounced on numerous occasions that ETOSS was putting a major obstacle to their optimal participation in the surveillance of the company by not sharing information. *See* for example Laura Barbuto, 'El Alza del Agua Es Todavía un Misterio', *La Nación*, 20 February 1998. This situation was remedied when the ETOSS created a 'Users Committee' involving local NGOs, which had access to all the information that ETOSS collected and was entitles to make recommendations to the directory.

[93] *See* Article 42 of the Constitution of Argentina, last reformed in 1994.

Perhaps the main lesson to learn from the forgettable experience of the city of Buenos Aires with privatisation of water services has to do with the importance of establishing clear rules by which both sides of the privatisation contract must abide. These rules must ensure the existence of effective monitoring of private companies by the state, safeguarding of water users' interests, transparency and public participation. Only in the presence of legal certainty and of effective institutions can the state monitor the performance of multinational water companies operating in its territory. This ideal scenario includes the use of the mechanisms and alternatives that democratic institutions offer. For instance, it is essential that the decision to privatise essential public services such as water and sanitation is debated in Congress and that independent parliamentary commissions oversee the execution of the contract together with the regulatory entities, drawing the attention of the latter to problems and irregularities. In the case of Argentina, the regulatory role was left early on in the hands of the Executive. While this option can accelerate decision-making, it can also mean that decisions are rushed and are the result of pressures from local and international lobbies.

The revocation of the contract in 2006 offered the Argentinean authorities an opportunity to design a new regulatory framework, this time for the provision of water services by a state company. The new rules would, in principle, address the points above and correct the deficiencies that the privatisation experience had uncovered in the previous framework. Nevertheless, such opportunity seems to have been wasted. The current regulatory framework – this time, enacted by law[94] – can indeed be criticised, mainly on the institutional design and public participation fronts. On the one hand, the institutional organisation is lacklustre – it concentrates both the operation and regulation of the services in the hands of the Executive, in a manner not dissimilar to the scheme of proved inefficiency that led to the privatisation in the first place. The service is operated by a state company, Agua y Saneamientos Argentinos (AySA), but there are two regulatory agencies under the control of the Executive. The ETOSS – now known as ERAS (Ente Regulador de Agua y Saneamiento) – is in charge of

---

[94] Law 26.221 of 13 February 2007, published on 2 March 2007.

monitoring the quality of the service and receiving the complaints of the users. The Agency of Planning, conversely, is in charge of designing the investment plans for the maintenance and expansion of the network.[95] The study of the tariffs and subsidies is not the task of either of them but falls within the competence of the Undersecretary of Hydric Resources, who is under the authority of the national Executive. This functional division is less than optimal – the concentration of operation and regulation in the hands of the national Executive dampens private investors from participating in any capacity and favours corruption in a country of the unfortunate institutional immaturity of Argentina.

On the other hand, the current regulatory framework is also lacking in what is related to access to information and participation. In a note elevated to Argentinean authorities by the Centre on Housing Rights and Evictions (COHRE), the ONG deplored that the draft bill 'had not undergone public scrutiny' and that it did not 'provide for the participation of users in decisions that will affect their right to water and sanitation.'[96] Moreover, even if the financial support of IFIs is still essential for the improvement of the services sector in developing countries, the projects supported by their loans are not discussed with the public to the extent that IFIs claim them to be. The WB and the IDB are currently considering the approval of a loan to AySA that would finance the investment plans of the company, yet said plan is a copy of the strategy that AA had put forward in its day. It foresees pouring untreated sewage waters to the Rio de la Plata river, a practice that would increase pollution in the area. Needless to say, once again, the plan has not been disclosed by Argentinean authorities to the public, in order to enable any input from stakeholders. It has neither been the object of a consultation, nor has it been the subject of any public audience in order to discuss it with users.

---

[95] Ibid., Articles 4 and 5.
[96] COHRE, Recommendations on Revisions of the Buenos Aires Water and Sanitation Regulatory Bill in order to fully Integrate International Human Rights Standards (Geneva: Centre on Housing Rights and Evictions, Letter to Dr. Néstor Kirchner, President of Argentina, 9 February 2007). *Source:* http://www.cohre.org/store/attachments/COHREWaterRegulatoryReformletter9Feb07.pdf.

3.2 Increasing the Accountability of the International Institutions Involved

The second contributing factor to the failure of the privatisation of water services in Buenos Aires goes beyond the responsibilities of the state and can therefore be seen as an 'external' factor. The degree of accountability of the IFIs involved in the process, in particular the WB, needs to be addressed. In effect, even though WB staff were aware from early on of the negative effects that a deficient legal and regulatory framework could have in the mid and long term on the outcomes of a privatisation process,[97] the Bank failed to react to the warnings of its own research teams by continuing to support the Buenos Aires water privatisation operation and granting loans to Aguas Argentinas through the IFC despite the poor performance of the company.

Much like the IMF, the WB had put itself in a difficult position by holding Argentina as an example of the benefits that could derive from reform. With its reputation and the adequacy of its trademark policies at stake, the Bank could not but try to save the privatisation in Buenos Aires for as long as possible. The frequent visits by WB/IMF teams to the country, their meetings with governmental and company representatives in order to intercede during the toughest phases of contract renegotiations, and the disbursements and debt reductions granted by the IFC to Aguas Argentinas are thereby explained.

However, the WB's support of Aguas Argentinas is still remarkable. By the premature end of the concession, 'the World Bank together with the Inter American Development Bank [(IDB)] and local Argentine banks [had] provided all but 30 million of the 1 billion dollar needed investment for infrastructure to Suez when it took over the operations ... '[98] And yet all these disbursements could not prevent that to this day 1.5 million households in Buenos Aires still lack access to drinking water and 3 million are not connected to the sewerage network. They also could not prevent the company's criminal prosecution for the high levels of nitrates in tap water or the legal actions related to the pollution of the Rio

---

[97] *See* World Bank/OED, note 33 above at 6.
[98] IBON Databank and Research Center, *Water Privatization: Corporate Control versus People's Control* 90 (Manila: IBON Books, 2005).

de la Plata River initiated before national courts – both suits originating in constant delays in the maintenance and construction of new infrastructure that had been foreseen in the original version of the contract.

A review of the WB's operations in Argentina by its own accountability organs is essential in ensuring the proper application of Bank policies and procedures.[99] In particular, since the main involvement of the WB Group in the privatisation of water services in Buenos Aires took the form of financial support to Aguas Argentinas by the IFC, the role of the Office of the Compliance Advisor Ombudsman (CAO) can be raised. The aim of the CAO performing its ombudsman function is to handle a complaint in order 'to identify problems, recommend practical remedial actions and address systemic issues that have contributed to the problems, rather than to find fault'.[100] The CAO in its advisory function has provided valuable guidelines for the IFC to supervise more efficiently the projects it is involved in and address adverse environmental and social consequences. In this sense, the CAO states that the 'IFC should seek to increase and exercise its leverage. Environmental and social issues should be included in legal covenants. Similar to the World Bank and private banks, IFC should consider suspending loans or withdrawing from projects whose environmental and social performance present unacceptable risks to IFC'.[101] This explicit recommendation came too late for the case of the city of Buenos Aires, though its strict application by the IFC is yet to be proved.

In summary, the Buenos Aires experience with water privatisation illustrates the deficiencies of a privatisation process dominated by improvisation and disregard of fundamental social

---

[99] *See* the role of the OED and the Inspection Panel. The latter, however, covers mainly IBRD and IDA infrastructure projects (project lending).

[100] CAO, Information about the CAO Ombudsman Process. *Source:* http://www.cao-ombudsman.org/html-english/ombudsman.htm.

[101] CAO, Review of IFC's Safeguard Policies 53 (Washington: CAO/IFC/MIGA Report, January 2003). It also affirmed that 'during supervision IFC should track appropriate indicators to monitor whether its project-specific development objectives are being met, particularly whether its intermediary operations are financing sustainable and environmentally sound private enterprises. It should take action if they are not' (Ibid., 58).

and environmental factors by the government, breaches of established rules and practices by the private operator, and unconditional support from an IFI that fails to address its own mistakes or shortcomings.

The failed privatisation experience in Buenos Aires brings to the forefront important issues that the WB needs to address, should it continue to support the privatisation of water services. Let us recall in the following final paragraphs what the Buenos Aires case study teaches us about privatising water services in general, and about bringing privatisation initiatives to countries that lack the institutional capacity and maturity to implement them.

4. Concluding Remarks

It is evident that the Bank still regards the involvement of the private sector in water resources management and service provision as an essential requisite for granting access to everyone to safe drinking water and adequate sanitation. The Bank remains convinced that the private sector is key in reducing the cost of services and increasing the accountability of utilities.[102] This insistence towards the privatisation of water services is both remarkable and disconcerting; above all, because there are no in-depth studies by the Bank that have assessed the existence of alternatives to privatisation of the water sector. If the reason argued for promoting privatisation is the need to expand the service network and to improve the efficiency of water services, making water available to all groups regardless of their economic capabilities, then the privatisation solution should be reconsidered. Primarily, because private companies – as the privatisation in Buenos Aires demonstrates – privilege serving those areas where profit is bigger and risk-free. This means that there is a fundamental contradiction in resorting to privatisation as the solution for connecting the poorest sectors of the population to the service. Should the behaviour of Aguas Argentinas be any indication of the general *modus operandi* of private companies, the provision of water services by private companies will strictly be ruled by a

---

[102] World Bank/IBRD, Water Resources Sector Strategy: Strategic Directions for World Bank Engagement 19 (Washington: World Bank, 2004).

profit-making mentality. The provision of drinking water and sanitation is precisely an example of a public service where not every opportunity for making profit should be seized, but where social considerations should prevail.

The WB believes that private companies are interested in extending the network coverage in order to increase their profits by incorporating new users. However, as Aguas Argentinas proved, sometimes it can be more profitable to service areas already covered by the network and delay every investment commitment related to its expansion; especially the construction of sanitation infrastructure, which is particularly costly.[103] Private companies, in the absence of a solid legal and regulatory framework that defines concrete, enforceable obligations for the concessionaire, will be reminded of the opportunities that it has at reach for maximising their profits. The story of the privatisation in Buenos Aires can be read, in this sense, as an attempt by Aguas Argentinas to denaturalise through renegotiations intrinsic features of water services provision.[104] In effect, the operation of water services is by definition capital-intensive and demands substantial investment at an early stage. Only in the long term is the initial investment capable of generating return profits. Because investing in the long term in emerging economies as in Argentina is highly risky, Aguas Argentinas tried to change the concession contract from a long term profit venture to a short term one.

The WB's purpose is confessedly to eradicate poverty, but the privatisation of public services, if feebly implemented, may contradict its core mission. When a solid legal framework is not properly in place, and the institutional capacity to privatise such a sensitive public service is still lacking, privatisation should not be attempted at the expense of the population. In Buenos Aires, it was the sector with the lowest income that suffered the most with the constant modifications of tariffs and the delays in investment

---

[103] In this sense, the failure of Aguas Argentinas to invest is still perplexing given the huge numbers that it made during the 90's, before it could argue that the macroeconomic policies that Argentina took for trying to palliate the recession had affected the financial balance of the concession contract and prevented them to follow the investment plans as agreed.

[104] *See* Molinari, note 2 at 21.

that the company pursued in order to ensure profitability and reduce risks. As Karina Forcinito exposed at the Third World Water Forum in March 2003, Aguas Argentinas succeeded in changing the nature of the original contract in only nine years, transferring investment risks to users, introducing new fixed charges to the bill and raising tariffs by 88.2 per cent. By 2003, the poorest families in the metropolitan area of Buenos Aires were spending 9 per cent of their income on drinking water and sewerage.[105]

If anything, the Buenos Aires experience should be taken as a case study that demystifies privatisation as an instant solution to deficient water services operation. The Bank still trusts privatisation with improving water services, though it now seems open to partnerships with the public sector in water infrastructure, and stresses the obligation of the public sector to establish a sound legal and regulatory framework as an essential requisite for privatising public services.[106] In this sense, recent Bank studies show that the organisation is aware of what is needed for privatisation to succeed and which mistakes should be avoided. Taking into account the experience with privatisation of water services in Ghana in 2003,[107] it appears the Bank cannot apply the lessons learnt from past failures to new privatisation endeavours. What is more discouraging is the fact that this incapacity is linked to its preoccupation for saving face by not acknowledging and remedying its own errors.

---

[105] D. Aspiazu and K. Forcinito, Privatización del Agua y Saneamiento en Buenos Aires. Historia de un Fracaso (Paper presented at the III World Water Forum, Kyoto/Shiga/Osaka, March 2003). On file with the author.

[106] World Bank, note 102 above at 12: 'While private investment and management are playing, and must play, a growing role, this must take place within a publicly established long-term development and legal and regulatory framework'.

[107] *See* R.N. Amenga-Etego and S. Grusky, 'The New Face of Conditionalities: The World Bank and Water Privatization in Ghana', *in* D. McDonald and G. Ruiters eds., *The Age of Commodity: Water Privatization in Southern Africa* 275–290 (London: Earthscan, 2005).

10

The Linkages between Access to Water and Water Scarcity with International Investment Law and the WTO Regime

FRANCESCO COSTAMAGNA AND FRANCESCO SINDICO

1. Introduction

Water cannot be replaced and is crucial for human and animal life on earth. These two simple statements clearly highlight the importance of water for the international community and its unique nature among other natural resources. The fact that more than a billion people lack adequate access to water should give a picture of the gravity of the problem we are facing.[1] The problem has two main dimensions – the individual access to water services and the overall availability of water as a physical resource.[2]

Against this background the paper seeks to address both these dimensions, by first considering whether foreign direct investments and international trade can be seen as instruments for coping with

---

[1] World Health Organisation, The Right to Water (Geneva: WHO, 2003). *See also* United Nations, *Water – A Shared Responsibility* (Paris: UNESCO, 2006). *Source:* http://unesdoc.unesco.org/images/0014/001454/145405E.pdf.

[2] *See* Paul J.I.M. de Waart, 'Securing Access to Safe Drinking Water through Trade and International Migration', *in* Edward H.P. Brans *et al.* eds., *The Scarcity of Water: Emerging Legal and Policy Responses* 101 (London: Kluwer Law International, 1997).

inadequate access to water for the individuals (2.1) and water scarcity (2.2) respectively.[3] Then the paper explores the impact of both the international investment system (3) and the World Trade Organisation (WTO) regime (4) on States' capacity to deal with these problems. The main aim is to evaluate whether these two international regimes are capable of reconciling the conflicting needs in relation to access to water services and water scarcity.

2. Coping with Access to Water and Water Scarcity through Foreign Investments and International Trade

Ensuring universal access to water services and the fight against water scarcity are daunting challenges for several States. Financial constraints, infrastructural inadequacies and adverse natural events affecting the availability of the resource are just some of the factors that may hamper the pursuit of such objectives. The recourse to international market forces has often been presented as a solution for allegedly inadequate domestic capabilities. As for the first dimension, foreign investments represent a good case in point – the influx of foreign capital and operators is widely regarded as a factor that may help to enhance access to water services by making greater financial resources available, transferring technology and improving the management of these services. This chapter aims to analyse the main legal issues arising out of the privatisation[4] of water services and, in particular, the extent to which the international investment protection system can affect States' regulatory autonomy in this field. Secondly, if the country lacks water resources, or if these have been hindered by serious environmental problems, then water could be brought to the

---

[3] In our paper we are approaching the access to water and water scarcity problems by taking into account those options whose goal is to bring drinking water to the people. There is a third possible solution, which focuses on bringing the people to the drinking water. This option has been explored by de Waart, note 2 above at 109–114.

[4] The term is used to refer to a wide array of different phenomena. In this paper, it will indicate processes that increase the participation of formal private enterprises in water and sanitation provision but do not necessarily involve the transfer of assets to the private operator.

country through, *inter alia*,[5] international trade.[6] Here the main question is to assess whether the WTO regime has a role to play in the regulation of the trade of water, should the latter be taken as a tradable good.

2.1 Coping with Access to Water through Foreign Direct Investments

For years, privatisation of water services and infrastructures has been advocated as the main, and sometimes the only, viable solution for the failures of public authorities in ensuring universal access to water services. According to its advocates,[7] the entry of private capital and operators would have helped to overcome States' budget constraints to improve infrastructures, and it would have

---

[5] 'Water exchanges' among States could be another option.

[6] It must be recognised that today this scenario can apply to a relatively small number of situations, since in general the main problem lies with the management of water resources rather than with physical unavailability of water. However, the situation is likely to change in a near future, as the total quantity of water is not expected to change significantly over the next years and the amount of clean fresh water will not be as abundant as it used to be. This, linked with the tremendous increase in the world population, is likely to cause serious *water scarcity* problems, further exacerbated by current climate change trends that are leading to severe droughts and reduced rainfall worldwide. In fact, by 2025 48% of the projected population will suffer from serious water shortages and by the same year more than thirty countries will have to deal with severe water scarcity problems. *See* Centre for International Environmental Law (CIEL), Going with the Flow: How International Trade, Finance and Investment Regimes Affect the Provision of Water to the Poor (CIEL Law Issue Brief, July 2003). *Source:* http://www.ciel.org/Publications/Waterbrief_3Sept03.pdf1. *See* also Katsumi Matsuoka, Tradable Water in GATT/WTO Law: Need for New Legal Frameworks? 1 (Paper presented at the AWRA/IWLRI – University of Dundee International Specialty Conference, Dundee, 6–8 August 2001). *Source:* http://www.awra.org/proceedings/dundee01/Documents/Matsuoka.pdf.

[7] The international financial institutions, namely the World Bank and the International Monetary Fund, played a big part in pushing forward this approach. On their role in the water industry, *see* Sara Grusky, The IMF, the World Bank and the Global Water Companies: A Shared Agenda (International Water Working Group, 2001), available at www.citizen.org/documents/sharedagenda.pdf.

brought about a more efficient management of services, reducing the waste of resources. Results have not always lived up to these expectations,[8] with limited improvements in terms of both infrastructures and access.[9] In certain cases these two targets were even found to contradict each other. In Buenos Aires, for instance, post-privatisation connection fees remained high, and in fact unaffordable to poor households, because of the 'infrastructure charge' that was added to finance the expansion of secondary water distribution and sewer networks.[10]

Private investments in water utilities pose unique challenges for both the State and the investor, as water is not a normal commodity, its value going well beyond the economic dimension. Access to an adequate amount of drinking water is crucial to maintain basic health and the fulfilment of other rights, while urban water, drainage and sanitation provide important public benefits, such as protection from infectious diseases, and they are widely considered as impure public goods.[11] States thus have the duty,[12] and not just the right, to adopt all the necessary measures to prevent

---

[8] Today some 1.1 billion people in developing countries have inadequate access to water and 2.6 lack basic sanitation. For further data, *see* United Nations Development Programme, *Human Development Report 2006 – Beyond Scarcity: Power, Poverty and the Global Water Crisis* (New York: UNDP, 2006), Chapter 1.

[9] Naren Prasad, 'Privatisation Results: Private Sector Participation in Water Services after 15 Years', 24 *Development Policy Review* 669 (2006).

[10] For a detailed analysis of the experience of water privatisation in Buenos Aires, *see* Olleta in this volume.

[11] Committee on Economic, Social and Cultural Rights, General Comment 15: The Right to Water (Articles 11 and 12 of the International Covenant on Economic, Social and Cultural Rights), UN Doc. E/C.12/2002/11 (2002). *Source:* http://www.ielrc.org/content/e0221.pdf, 1–6. 'Impure public goods' are goods that provide important public benefits but, unlike public goods, they are either rival or exclusive in consumption.

[12] Although far from settled, there is a growing consensus in the international community that access to water can be considered as a human right. See, *e.g.*, Pierre-Marie Dupuy, Le droit à l'eau, un droit international? (Florence: EUI Working Paper No. 2006/06, 2006). On the relationship between international investment law and human rights, *see* Lahra Liberti, 'Investissements et droits de l'homme', *in* Philippe Kahn and Thomas Wälde eds, *Les aspects nouveaux du droit des investissements internationaux/New Aspects of International Investment Law* 791 (Leiden: Martinus Nijhoff Publishers, 2007).

the privatisation process from restricting access to water services. Public infrastructure investments entail high risks also for private investors. These projects are characterised by massive sunk costs that require long amortisation periods, thus making private operators particularly vulnerable to eventual changes of attitude by host governments.[13]

States and foreign investors seek to prevent reduction in social welfare as well as to guarantee a sufficient return for the investment by agreeing to terms and conditions under which the service has to be provided. Although very detailed, these legal devices are often inadequate to solve the political problems arising between the parties. The case of tariffs represents an illuminating example in this regard, as their regime is normally regulated by complex contractual arrangements that seek to ensure both the affordability of the service and the profitability of the investment. This notwithstanding, the item represents by far the main cause for conflict in water privatisation projects. The reason is that a short-term consequence of the privatisation process is an increase in water rates if compared with those charged under public management. These dynamics are likely to create resentment among the population[14] and, in turn, to put further pressure on the relationship between the investor and the State. Under these conditions, contract's renegotiation is often beyond reach, as the private party refuses to yield the favourable conditions contained

---

[13] Thomas Wälde and Stephen Dow, 'Treaties and Regulatory Risk in Infrastructure Investment: The Effectiveness of International Law Disciplines versus Sanctions by Global Markets in Reducing the Political and Regulatory Risk for Private Infrastructure Investment', 34 *J. World Trade* 1 (2000).

[14] The 'Water War' that erupted in Cochabamba (Bolivia) in 1999 represents a good case in point. For a detailed analysis of the case, *see* Maria McFarland Sánchez-Moreno and Tracy Higgins, 'No Recourse: Transnational Corporations and the Protection of Economic, Social, and Cultural Rights in Bolivia', 27 *Fordham Int'l L.J.* 1663, 1747–1789 (2004) and Eric J. Woodhouse, 'The 'Guerra del Agua' and the Cochabamba Concession: Social Risk and Foreign Direct Investment in Public Infrastructure', 39 *Stanford J. Int'l L.* 295 (2003). Public protests have also erupted in other countries such as Ecuador, Paraguay, Thailand, Indonesia, Pakistan, India, South Africa, Poland, and Hungary. *See* Violeta Petrova, 'At the Frontiers of the Rush for the Blue Gold: Water Privatisation and the Human Right to Water', 31 *Brook. J. Int'l L.* 577, 577–580 (2006).

in the original terms of the agreement, while host governments cannot soften their negotiating stance for fear of losing political support.[15] Considering such a troublesome context, it is hardly surprising that several water related projects have recently ended in failure.

2.2 Coping with Water Scarcity through Trade

Water export is rising and 'bulk transfers of water occur between different kinds of actors and in different modes'.[16] Almost any kind of water export has been foreseen, from private companies shipping water by tankers from one continent to the other,[17] to iceberg trading.[18]

Now, two questions must be posed – is water a tradeable good, thus falling under the realm of the WTO legal regime? And, are all water related international transfers covered by the multilateral trading system?

Without going into the discussion of water as the common heritage of mankind,[19] as a human right,[20] or as an instrumental

---

[15] These dynamics are often explained through the 'obsolescing bargain' model.

[16] Edith Brown Weiss, 'Water Transfers and International Trade', *in* Edith Brown Weiss, Laurence Boisson de Chazournes and Nathalie Bernasconi-Osterwalder eds., *Fresh Water and International Economic Law* 61–89 (Oxford: Oxford University Press, 2006). The author highlights government-to-government transfers by treaty (Lesotho – South Africa), government-to-government contractual transfers (Turkey – Israel), transfers between government and foreign private party (Bolivia), and transfers between private parties in different countries (Canada – US).

[17] Ibid., at 76.

[18] Antoinette Hildering, *International Law, Sustainable Development and Water Management* 111 (Delft: Eburon Academic Publishers, 2004).

[19] Dannielle Morely ed., *NGOs and Water – Perspectives on Freshwater, Issues and Recommendations of NGOs* 1 (London: United Nations Environmental and Development Forum, 2000).

[20] *See* M.A. Salman, The Human Right to Water: Legal and Policy Dimensions (Washington: World Bank, 2004); Enrico Fantini, 'The Human Rights to Water: Recent Positive Steps and the Way Ahead', 2 *Pace Diritti Umani* 123 (2005); Stephen C. MacCaffrey, 'The Human Right to Water', *in* Edith Brown Weiss, Laurence Boisson de Chazournes and Nathalie Bernasconi-Osterwalder, note 16 above at 93.

right for the fulfilment of other human rights, such as the right to life, one cannot deny that water in itself cannot be considered like any other economic good.[21] Water is essential for life; while other goods are not.[22]

What does the multilateral trading system have to say about water? The WTO deals with trade in goods through the General Agreement on Tariffs and Trade (GATT), but the latter does not include any definition of what a good is. The presence of water in the World Customs Union's Harmonising Commodity Description and Coding System, which is used in the GATT for classification purposes, has been seen by some authors as a first element in favour of considering water as a good under trade law.[23] However, the latter just tells us where water would be classified, should it be considered a good,[24] but it does not convert its content into a tradable good *per se*.[25] Furthermore, in the WTO

---

[21] There has been considerable discussion on whether water can be considered an economic good at all. Traditionally water has been excluded from the economic sphere, being considered as an exclusively social good pertaining to the society as a whole. This perception seems to have changed after the various industrial, agricultural and demographic revolutions have increased the pressure on water resources to such an extent that it begun to be perceived as a scarce resource. This has led water to be considered as an economic good. This is not to say that water is an exclusively economic good, its value going well beyond the economic dimension. On this point, *see* Hildering, note 18 above at 95–122.

[22] Some authors stress that water is like oil, *see* William M. Turner, The Commoditisation and Marketing of Water 3 (Paper presented at the Annual Meeting of the Council of Canadians, Vancouver, Canada, 5–8 July 2001). *Source:* http://www.waterbank.com/Newsletters/nws35.html. However, it is self-evident that human and animal life can continue without it. Oil is a fungible natural resource that can, and should, be replaced in modern economies.

[23] *See* World Customs Organisation (WCO), *Harmonised Commodity Description and Coding System: Explanatory Notes* 186 (Brussels: WCO, third edition 2002).

[24] Steven Shrybman, Water Export Controls and Canadian International Trade Obligations, Legal Opinion Commissioned by the Council of the Canadians, 1999. *Source:* http://www.canadians.org/water/publications/Trade_Shrybman.html. *See also* Katsumi Matsuoka, note 6 above at 2–3.

[25] *See* Edith Brown Weiss, note 16 above.

there is no tariff binding on water and other regional trade agreements go as far as to exclude water in its natural state from their scope of application.[26]

This last consideration deserves further attention. It seems to imply that water can take different forms and it may well be that these deserve different legal treatment. In this paper two main kinds of water transfers are identified,[27] each calling for different treatment under the WTO regime. The first kind of water transfers takes place through the diversion of the river flow from one country to another. In these cases two arguments play against any application of trade law regulations – the water has not been captured, it is still in its natural state and, therefore, it is not apt for commerce.[28] Furthermore, such water transfers are normally regulated by bilateral international treaties, which are the only instruments that should deal with any issue arising from the use of the water in the river.[29]

The second kind of water transfers includes bulk water transfers through complex systems of dams and pipelines (Lesotho-South Africa treaty) or via tankers bringing water to the thirsty country (Turkey-Israel agreement). With regard to these commercial transactions, international trade law can only play a subsidiary role. We agree with E. Brown Weiss' conclusion that 'the precautionary approach in international law as developed and applied to fresh water makes it important to exclude them [bulk water transfers] from the reach of trade law, at least for now until

---

[26] 1993 Statement by the Governments of Canada, Mexico and the United States: 'Unless water, in any form, has entered into commerce and becomes a good or product, it is not covered by the provisions of any trade agreement, including the NAFTA. And nothing in the NAFTA would oblige any NAFTA Party to either exploit its water for commercial use, or to begin exporting water in any form. Water in its natural state in lakes, rivers, reservoirs, aquifers, water basins and the like is not a good or product, is not traded, and therefore is not and never has been subject to the terms of any trade agreement'. This statement does not appear to have a formal name or number.

[27] The two kinds of water transfers are not the only ones that can be stipulated between countries. However, for the purposes of this paper they highlight cases in which the WTO cannot, or should not, play a role, and cases in which the WTO can play a subsidiary role.

[28] We agree on this point with Katsumi Matsuoka, note 6 above at 3.

[29] See Edith Brown Weiss, note 16 above at 70.

more experience is gained with them'.[30] However, trade law may step in if the export of water does not fall under a bilateral international treaty or contracts. In this case, the WTO Dispute Settlement Body (DSB) should take into account the particular nature of the good that is being traded, water.[31]

3. Access to Water and Investment Law

3.1 The International Protection System and State's Right to Regulate: Overview of Some Key Issues

The international legal framework for the protection of foreign investments currently consists of more than 2500 bilateral treaties, few multilateral treaties with regional or sector-specific coverage, some recent free trade agreements with investment protection provisions,[32] but no multilateral treaties having universal scope. Notwithstanding such a scattered background, there is an undeniable convergence between all these instruments, with regard to both substantive and procedural provisions. Bilateral Investment Agreements (BITs), regional agreements and Free Trade Agreements (FTAs) are indeed concluded on the basis of 'model agreements' elaborated by both Organisation for Economic Co-operation and Development (OECD) and emerging countries,[33] so that it is possible to speak about an international system for the protection of foreign investments.

A key feature in this regard is the overall objective of the system that is enhancing the stability of the legal framework to better

---

[30] Ibid., 67.
[31] Furthermore, once the water has entered into the importing country, the access to the water by its population may fall under the realm of the GATS. This is a very important issue, but it falls out of the scope of this article. For details, see Francesco Costamagna, 'L'impatto del GATS sull'autonomia regolamentare degli stati membri nei servizi idrici ed energetici', 19 *Diritto del commercio internazionale* 501 (2005).
[32] *See* the UNCTAD Database. *Source:* http://www.unctadxi.org/templates/DocSearch_779.aspx.
[33] On the approaches taken by these different groups of States, *see* Konrad Von Moltke, An International Investment Regime? Issues of Sustainability (Winnipeg: International Institute for Sustainable Development, IISD Paper 16, 2000). *Source:* www.iisd.org/pdf/investment.pdf.

protect foreign investments.³⁴ This investor-friendly purpose sets the tone of the whole protection system, as it informs the interpretation of treaty clauses,³⁵ which, when in doubt, 'should be interpreted *in favorem* investors'.³⁶ Arbitrators follow the rule by adopting expansive readings of some key provisions,³⁷ downplaying the negative effects of such interpretations on States' regulatory autonomy or other competing interests.³⁸ This could deter States, especially less powerful ones, from adopting regulatory measures for fear of violating treaty clauses and being, thus, obliged to pay compensation for that. According to this view, the system has the potential to freeze States' willingness to intervene, despite compelling welfare objectives which would require them to do so. Conversely, other scholars³⁹ categorically reject the chilling effect hypothesis, on the basis that international agreements are flexible enough to leave national public authorities free to choose their course of action.⁴⁰

The need to strike a balance between legislative stability and protection of the right of access to privatised water services calls

---

³⁴ *See* Rudolph Dolzer, 'The Impact of International Investment Treaties on Domestic Administrative Law', 27 *NY Univ. Journal of Int'l Law & Pol.* 953 (2005); Thomas Wälde and Abba Kolo, 'Environmental Regulation, Investment Protection and 'Regulatory Taking' in International Law', 50 *Int'l & Comp. Law Quarterly* 811, 825 (2001).

³⁵ Art. 31 of the Convention on the Law of Treaties, Vienna, 23 May 1969, 1155 UNTS 331, 340.

³⁶ Rudolph Dolzer, 'Indirect Expropriations: New Developments?', 11 *NY Univ. Env. L. J.* 64 (2003).

³⁷ *CMS Gas Transmission Company v. The Argentine Republic*, ICSID Case No ARB/01/8, Final Award of 12 May 2005, 44 *Int'l Leg. Mat.* 205 (2005).

³⁸ Stuart G. Gross, 'Note, Inordinate Chill: Bits, Non-NAFTA MITS, and Host-State Regulatory Freedom—An Indonesian Case Study', 24 *Mich. J. Int'l L.* 893, 899 (2003) and Howard Mann, Private Rights, Public Problems: A Guide to NAFTA's Controversial Chapter on Investor Rights 17 (Winnipeg: International Institute for Sustainable Development, 2001).

³⁹ Joe J. Coe and Noah Rubins, 'Regulatory Expropriation and the *Tecmed* case: Context and Contributions', *in* Todd G. Weiler ed, *International Investment Law and Arbitration: Leading Cases from the ICSID. NAFTA, Bilateral Treaties and Customary International Law* 599 (London: Cameron May, 2005).

⁴⁰ Jan Paulsson, 'Indirect Expropriation: Is the Right to Regulate at Risk?', 3 *Transn. Dispute Management* 1, 12 (April 2006).

for careful consideration of two elements that may impinge on State's regulatory autonomy – the dispute settlement mechanism and the uncertain definition of basic substantive provisions.

3.2 Access to Water and Public Interest Concerns in Investor-State Arbitration

Most investment treaties provide foreign investors with a unique mechanism for dispute resolution, allowing them to challenge regulatory measures of the host states directly in front of an international arbitral tribunal.[41] The mechanism departs from traditional international law principles, which did not give individuals a direct cause of action against a State for violations of international law that affected their rights. Instead, they were dependent upon the willingness of their home country to espouse their claims, or they could decide to initiate litigation before host State's national courts. Granting direct standing to private investors in front of an international arbitration forum greatly enhances the effectiveness of the protection system, as it allows the private investor to bypass the allegedly partisan domestic courts[42] and to avoid his litigation being restrained by foreign relation considerations.[43]

International arbitration is meant to bring both parties on a level playing field, providing private investors with an effective mean to balance State's sovereign powers.[44] Therefore, it is not

---

[41] The most relevant one is that administered by the International Centre for the Settlement of Investment Disputes, established by the *Convention on the Settlement of Investment Disputes between States and Nationals of Other States*, Washington, 18 March 1965, 575 *UNTS* 159; 4 *Int'l Leg. Mat.* 532 (1965).

[42] Giorgio Sacerdoti, 'Bilateral Treaties and Multilateral Instruments on Investment Protection', 269 *Recueil des Cours* 251, 413–414 (1997).

[43] Susan D. Franck, 'The Legitimacy Crisis in Investment Treaty Arbitration: Privatising Public International Law Through Inconsistent Decisions', 73 *Fordham J. Int'l L.* 1521, 1538 (2005).

[44] Nathalie Bernasconi-Osterwalder, 'Who Wins and Who Loses in Investment Arbitration? Are Investors and Host States on a Level Playing Field? The Lauder/Czech Republic Legacy', 6 *Journ. World Investm. & Trade* 69, 69–71 (2005).

surprising that it has an inherently asymmetric character,[45] starting from the fact that only investors can sue Governments. Several other features help to strengthen the position of the investor, so much that the mechanism has been depicted as a powerful 'weapon' that can be used to restrain State's willingness to intervene for the protection of essential welfare objectives[46]. Although the claim seems to overstate the impact of direct arbitration, there is still the need to examine it, by singling out certain aspects that have emerged in recent water-related disputes.

Investment arbitration has a hybrid nature, being composed of elements drawn from private and public law contexts. As for the first dimension, it must be observed that most of its procedural features derive from the commercial arbitration model. As the latter had been created to solve private disputes, it is debatable whether such a transplant is suitable in the context of non-commercial disputes that entail significant public policy considerations.[47] Privacy and confidentiality of the process are fitting examples to this regard – although legitimate in a purely commercial context, they look much less acceptable when matters of basic public interest are at stake.[48] The need to find a new balance between the integrity of the arbitral process and the protection of the public interest[49] has emerged in two cases arising out of disputed water privatisations.

---

[45] Thomas Wälde, 'The Specific Nature of Investment Arbitration', in Philippe Kahn and Thomas Wälde eds., *Les aspects nouveaux du droit des investissements internationaux/New Aspects of International Investment Law* 43, 54 (Leiden: Martinus Nijhoff Publishers, 2007).

[46] Howard Mann and Konrad Von Moltke, NAFTA's Chapter 11 and the Environment (Winnipeg: International Institute for Sustainable Development, IISD Paper, 1999).

[47] Guus Van Harten and Martin Loughlin, 'Investment Treaty Arbitration as a Species of Global Administrative Law', 17 *Europ. J. Int'l L.* 121, 121–150 (2006). *See* Mann and Von Moltke, note 46 above.

[48] Loukas A. Mistelis, 'Confidentiality and Third Party Participation: *UPS v. Canada* and *Methanex Corp. v. United States*', in Todd G. Weiler ed., *Investment Law and Arbitration: Past Issues, Current Practice, Future Prospects* 169, 169–199 (Ardsley: Transnational Publishers, 2004).

[49] Jeffery Atik, 'Legitimacy, Transparency and NGO Participation in the NAFTA Chapter 11 Process', in Todd G. Weiler ed., *Investment Law and Arbitration: Past Issues, Current Practice, Future Prospects* 169, 169–199 (Ardsley: Transnational Publishers, 2004).

In the *Aguas Argentinas* case,[50] a dispute arising from the privatisation of the water distribution and sewerage system of the city of Buenos Aires, the ICSID arbitral panel held that it has the power to entertain participation of non-disputant parties as *amici curiae* because of 'the particular public interest' of a dispute that can potentially affect the operation of systems 'provid[ing] basic public services to millions of people'.[51] Some months later, in the *Biwater Gauff* case, the water firm asked the tribunal to forbid Tanzania from disclosing documents related to the dispute. Although conceding the 'need for greater transparency in this field',[52] the panel held that controls and restrictions on the disclosure of documents are warranted to protect the integrity of the procedure and avoid the aggravation of the dispute. Unlike in the *Aguas Argentinas* case, the eminent public character of the interests at stake was not deemed strong enough for displacing the traditional secretive character of arbitration proceedings. The conclusion is far from convincing, as any decision affecting the water distribution system of a large metropolitan area, such as that of Dar Es Salaam, is a matter that cannot be discussed behind closed doors, but calls for a high degree of transparency and accountability.

The investor position is further strengthened by other features that go beyond the commercial arbitration model. The issue of consent represents a good case in point: unlike in the commercial context, investment arbitration is no longer premised on a specific agreement between the parties to submit the dispute to arbitration. The host State's consent is often given in the investment treaty concluded with the investor's home State, while the investor can provide its consent by simply accepting this standing offer at a later stage, once the dispute has actually arisen. In fact, this evolution provides the private party with a right to set up unilaterally the arbitration process. On the one hand, the so-called 'arbitration without privity' model represents an evolution that

---

[50] *Aguas Argentinas, S.A., Suez, Sociedad General de Agua de Barcelona, S.A. and Vivendi Universal, S.A. v The Argentine Republic*, ICSID Case No. ARB/03/19, Order in response to a petition for transparency and participation as *amicus curiae* of 19 May 2005.

[51] Ibid., 19.

[52] *Biwater Gauff (Tanzania) Ltd. v. United Republic of Tanzania*, ICSID Case No. ARB/05/22, Procedural order No. 3 of 19 September 2006, 133.

allows the system to better perform its function, by bringing it closer to those judicial review mechanisms such as the European Court of Human Rights.[53] On the other hand, this evolution seems to go beyond what had been originally envisaged, further unbalancing the relationship between the parties, as it strengthens the position of foreign investors. For instance, there have been cases where investors resorted to their increased bargaining power[54] to put pressure on host governments.[55] In *Azurix*, a case concerning the privatisation of the Buenos Aires Province's water distribution system, the US water firm threatened to resort to the mechanism when discussions to find a new agreement with Argentina were still underway, just to gain a leverage in the negotiations.[56]

The deterrent effect of the threat is strengthened by the use by international arbitrators of damages as main form of remedy. The feature may possibly deter States, especially poorest ones, from adopting regulatory measures that could negatively affect the investment, for fear of being forced to pay compensation. This is particularly the case for disputes concerning privatised water services, because of the large amounts of money involved. Suffice to say that in the *Azurix* case the US water firm sought US$ 600 million in compensation and the Tribunal awarded it US$ 165 million, although it rejected most of the investor's claims.

3.3 Access to Water and the Uncertain Definition of Key Substantive Provisions

Nowadays the international discipline on foreign investment protection develops mainly out of cases.[57] Treaty provisions are often drafted in rather general terms, thus leaving international arbitrators wide discretion in interpreting them. So far their activity

---

[53] *See* Wälde, note 45 above at 60.

[54] Charles H. Brower II, 'Investor-State Disputes under NAFTA: The Empire Strikes Back', 40 *Columbia J. Transn. L.* 43, 73 (2001).

[55] Thomas Wälde observes that '[t]he impact of the arbitration clause is less in its actual use, as its implicit threat to both parties'. *See* Wälde, 'Law, Contract and Reputation in International Business: What Works?', 3 *CEPMLP Internet Journal* (1998).

[56] *Azurix Corp. v. The Argentine Republic*, ICSID CASE No. ARB/01/12, Award of 14 July 2006. It was the first case in which an international arbitral tribunal decided on the merit a disputed water privatisation.

[57] *See* Wälde, note 45 above at 46.

has not yet coalesced in a coherent case law and the definition of some key investment protection provisions is still less than certain. Unclear rules 'obscure the boundaries of appropriate conduct'[58] and, hence, they may affect the balance between ensuring adequate protection to foreign investments and protecting basic social interests. In the water sector, such a condition is apparent especially with regard to the notions of regulatory expropriation and the fair and equitable treatment standard.

3.3.1. Regulatory expropriation

Regulatory expropriation is a form of indirect expropriation, taking place when a State's regulatory action infringes upon the economic value of the investment without any formal transfer of the property's title. This said, it is still to be determined whether any regulatory measure affecting the investment is to be compensated or if some measures can be exempted. The arbitral practice has dealt with the issue in a highly casuistic fashion, but, although a case-by-case evaluation seems somehow inevitable,[59] its potential impact on States' regulatory autonomy would call for the adoption of a more principled approach. Concerns are strong especially in those sectors, such as privatised water services,[60] where regulation now represents the main, and sometimes even the only, instrument at States' disposal to satisfy basic social needs. An excessively broad interpretation of regulatory expropriation could indeed jeopardise universal access to water by unduly tying the hands of national authorities.[61] On the other side, there is the need to avoid that 'a

---

[58] *See* Franck, note 43 above at 1585. *See also* Vaughan Lowe, 'Regulation or Expropriation?', 55 *Curr. Leg. Prob.* 447, 453 (2002).

[59] Steven R. Ratner, 'Regulatory Takings in Institutional Context: Beyond the Fear of Fragmented International Law' 102 (3) AJIL 475 (2008). Andrew Newcombe, 'The Boundaries of Regulatory Expropriation in International Law', 20 *ICSID Review – FILJ* 1, 6 (2005). For a sharp critique of 'hadocism', *see* Susan Rose-Ackermann and Jim Rossi, 'Disentangling Regulatory Takings', 86 *Virginia L. Rev.* 1435, 1444–1448 (2000).

[60] *See* Wälde and Kolo, note 34 above at 813.

[61] David A. Gantz, 'The Evolution on FTA Investment Provisions: From NAFTA to the United States – Chile Free Trade Agreement', 19 *Am. Univ. Int'l L. Rev.* 679, 684 (2004) and Howard Mann, 'The Right of States to Regulate and International Investment Law', *in* UNCTAD, *The Development Dimensions of FDI: Policy and Rule-Making Perspectives* 211–223 (Geneva-New York: UNCTAD, 2003).

blanket exception for regulatory measures [could] create a gaping loophole in international protection against expropriation'.[62]

Even a cursory look at the international arbitral case-law clearly indicates that the decisive criterion to define regulatory expropriation is the measure's impact on the investment,[63] while the purpose of State's action is less important[64] or even utterly irrelevant.[65] As for the water sector, this approach has been recently adopted in the *Vivendi II* case, where the Tribunal recalled that 'the *effect* of the measure on the investor, not the State's intent, is the critical factor'.[66] However, an exclusively effect-oriented approach does not allow for adequate consideration of States' duty, and not just right, to regulate. The problem has been expressly recognised in Azurix, where the Tribunal observed that: '[i]n the exercise of their public policy function, governments take all sorts of measures that may affect the economic value of the investments without such measures giving rise to a duty to compensate'.[67]

Problems arise when it comes to identifying measures not entailing the duty to compensate. Some international lawyers have relied upon the police powers' exception[68] to this end, but the

---

[62] *Pope & Talbot Inc. v The Government of Canada*, Interim Award of 26 June 2000, UNCITRAL, 40 *Int'l Leg. Mat.* 258 (2001), 99.

[63] For a detailed overview of the international case-law, *see* Christoph H. Schreuer, 'The Concept of Expropriation under the ECT and other Investment Protection Treaties', 2 *Trans. Disp. Manag.* 1, 28–39 (November 2005).

[64] *Tippetts, Abbett, McCarthy, Stratton v. TAMS-AFFA Consulting Engineers of Iran*, 6 Iran-US C.T.R. 219.

[65] *Metalclad Corporation v United Mexican States*, ICSID Case no. ARB(AF)/97/1, Award of 30 August 2000, 40 *Int'l Leg. Mat.* 36 (2000), 111.

[66] *Compañia de Aguas del Aconquija S.A. and Vivendi Universal S.A. v. Argentine Republic*, ICSID Case No. ARB/97/3, Award of 20 August 2007, 7.5.20. See also Biwatez Gauff (Tenzania) Limited v. United Republic of Tanzania, ICSID Case No. ARB/05/22, Auszd of 24 July 2008, 463–464.

[67] *See Azurix Corp. v. The Argentine Republic*, ICSID CASE No. ARB/01/12, Award of 14 July 2006, 310.

[68] The concept, drawn from the US jurisprudence, can be understood as encompassing the basic power vested in the state/government to regulate, restrict or limit the private rights in interest of the public welfare, law and order and security.

doctrine is a rather controversial tool,[69] as its scope is yet to be clearly defined in international law.[70] Moreover, the relationship between the exception and international customary law, which requires compensation also for measures enacted in the public interest,[71] is far from clear.

To sort the matter out, the *Azurix* Tribunal proposed to complement the traditional approach by looking at the proportionality between the regulatory measure and the aim to be achieved.[72] The adoption of a balancing test may represent a promising development in the definition of regulatory expropriation, ushering a more nuanced approach than the classical *effect doctrine/police powers exception* dichotomy.[73] If properly applied, such a tool does not indeed put into question 'the due deference owing to the State when defining the issues that affect its public policy or the interests of society as a whole

---

[69] This notwithstanding Mann and Von Moltke consider it as a principle of customary international law. Howard Mann and Konrad Von Moltke, NAFTA's Chapter 11 and the Environment (Winnipeg: International Institute for Sustainable Development, IISD Paper, 1999) at 18.

[70] Newcombe, maintains that the exception 'allows the state to protect essential public interests from certain types of harms' (*See* Newcombe, note 59 above at 26), while Banks, takes the concept as covering not just public health, safety, morals or welfare, but also anti-trust, consumer protection, securities, environmental protection and land planning [*see* Banks, 'NAFTA's Article 1110 – Can Regulation Be Expropriation?', 5 *NAFTA L. Bus. Rev. Am.* 499, 510 (1999)].

[71] To be lawful any expropriation must be in the public interest, non-discriminatory, consistent with due process and against the payment of full compensation.

[72] *See* Azurix Award, note 56 above, at 312. The proportionality test is widely used by the European Court on Human Rights also to deal with right of property cases, but *Azurix* represented only the second investment case where the test has been applied, the first being *Tecnicas Medioambientales Tecmed S.A. v The United Mexican States*, ICSID Case No. ARB(AF)/00/2, Award of 29 May 2003, 43 *Int'l Leg. Mat.* 133 (2004).

[73] Another element considered in *Azurix* was whether the Argentine measures frustrated the investor's legitimate expectations. The criterion is playing an increasingly important role not just in the context of regulatory expropriation, but also with regard to the fair and equitable treatment and, hence, it will be dealt with in the next section of the paper.

[…]',[74] while allowing for adequate protection of investor's rights. Similar balancing tests have been recently incorporated in international investment treaties to provide further guidance to prospective arbitrators.[75] The shift towards more flexible approaches to define regulatory expropriation can help to find a better balance between States' regulatory discretion for the defence of fundamental social interests, such as access to water, and foreign investment protection.

3.3.2. Fair and equitable treatment

The definition of fair and equitable treatment (FET) is 'somewhat vague',[76] as its precise nature, scope and meaning continue to remain beyond reach.[77] Fairness and equity are inherently flexible concepts that are bound to continuously evolve and cannot 'be frozen in time'.[78] The lack of a clear definition has not prevented it from becoming 'the most important standard in investment disputes',[79] being used to articulate a variety of rules necessary to

---

[74] *See* Tecnicas Medioambientales Tecmed S.A., note 72 above, at 122.

[75] *See*, for example, 2004 Canadian Model BIT, Annex B 13(1) on the clarification of indirect expropriation. *Source:* http://www.naftaclaims.com/files/Canada_Model_BIT.pdf and 2004 US Model BIT, Annex B available on http://www.naftaclaims.com/files/US_Model_BIT.pdf. *See also* OECD, 'Indirect Expropriation and the Right to Regulate in International Investment Law', Working paper on international investment no. 2004/4, 10 September 2004.

[76] *See CMS Award*, note 37 above, at 274. For a comprehensive analysis, *see* Ioana Tudor, *The Fair and Equitable Treatment Standard in the International Law of Foreign Investment* (Oxford: Oxford University Press, 2008).

[77] Barnali Choudhury, 'Evolution or Devolution? Defining Fair and Equitable Treatment in International Investment Law', 6 *Journ. World Investm. & Trade* 297, 298 (April 2005).

[78] *ADF Group Inc v United States*, ICSID Case No. ARB/(AF)/00/1, Award of January 2003, 18 *ICSID Review – FILJ* 195 (2003), paras 179–181. *See also* Azurix Award, note 56 above, para. 361 and Peter Muchlinski, 'Caveat Investor'? The Relevance of the Conduct of the Investor Under the Fair and Equitable Standard', 5 *Int'l & Comp. L. Quarterly* 527, 532–533 (July 2006).

[79] Christoph Schreuer, 'Fair and Equitable Treatment in Arbitral Practice', 6 *Journ. World Investm. & Trade* 357 (June 2005). Rudolph Dolzer observes that '[…] hardly any lawsuit […] is filed these days without invocation of the relevant treaty clause requiring fair ad equitable treatment'; *see* R. Dolzer, 'Fair and Equitable Treatment: A Key Standard in Investment Treaties', 39 *The International Lawyer* 87 (2005).

achieve the treaty object and purpose.[80] The standard has indeed become a sort of catch-all formula,[81] covering any regulatory measure that upsets the stability of the investment environment.[82]

Although dictated by international customary norms[83], such a purposive reading has greatly broadened the scope of the standard, giving it a 'potentially very considerable impact on the freedom of a government to regulate its economy'.[84] The risk that an expansive interpretation may tilt the balance between the respect for State's sovereignty and the protection of investor's rights too in favour of the latter has already prompted some governments to adopt normative and interpretative acts that aim at limiting the scope of the obligation.[85]

The trend can be better appreciated by looking at one of the FET components that is highly relevant for investments in the water sector – the protection of investor's legitimate expectations.[86] The principle is recognised in most domestic administrative systems

---

[80] *See* Brower II, note 54 above at 56.

[81] *See* Dolzer, note 79 above at 88.

[82] *See*, in regard to the water sector, *Azurix Corp. v. The Argentine Republic*, ICSID CASE No. ARB/01/12, Award of 14 July 2006, at 372.

[83] Codified in arts. 31–32 of the Convention on the Law of the Treaties, Vienna, 23 May 1969, 1155 UNTS 331.

[84] *See* Lowe, note 58 above at 455. Dolzer, observes that the standard has 'wide-ranging repercussions for the sovereignty of the host state' (*see* Dolzer, note 79 above at 964).

[85] *See*, for instance, the Notes of Interpretation of Certain Chapter 11 Provisions, NAFTA Free Trade Commission, 31 July 2001, available at http://www.international.gc.ca/trade-agreements-accords-commerciaux/disp-diff/nafta-interpr.aspx?lang=en which state that the fair and equitable treatment does not require treatment beyond that required by customary international law. Similar interpretative provisions have been included in other US FTAs with Singapore, Chile, Australia, Dominican Republic-Central America, Morocco. Arguing for a narrower interpretation of FET Graham Mayeda, 'Playing Fair: The Meaning of Fair and Equitable Treatment in Bilateral Investment Treaties', 41 *Journ. World Trade* 273, 287–288 (2007).

[86] On the principle, *see International Thunderbird Gaming v The United Mexican States*, UNCITRAL (NAFTA), Separate Opinion (Professor T. Wälde) of 26 January 2006. *See also* Francisco Orrego Vicuña, 'Regulatory Authority and Legitimate Expectations: Balancing the Rights of the State and the Individual under International Law in a Global Society', 5/3 *International Law FORUM du droit international* 188, 193–195 (2003).

as well as in the European Union legal order,[87] where it is considered a key component of the principle of legal certainty.[88] In investment law, it requires States not to unfairly upset basic conditions that have been taken into account by the investor to make the investment. It has become 'the dominant element'[89] of the FET and even 'an independent basis for a claim' under this heading.[90] Furthermore, in privatised water services investments, like in most public infrastructure undertakings, expectations of the private operator usually arise from legal and contractual provisions. The formality of the source determines the legitimacy of the expectations, i.e., their strength.[91] Accordingly, reliance upon a detailed regulatory framework reinforces investor's claims and, hence, it could narrow down States' regulatory space. Should the rule be interpreted as a sort of stabilisation clause, this could prevent host governments from imposing universal access obligations upon the private operator, if this was not expressly provided for in the original agreement. However, it is widely accepted that the duty not to alter the investment's regulatory framework cannot be taken as an absolute principle requiring the State to freeze its legal system for the investor's benefit.[92]

Interestingly enough, the existence of a detailed regulatory framework may also help States to avoid liability for changes that

---

[87] John Temple Lang, 'Legal Certainty and Legitimate Expectations as General Principles of Law', *in* Ulf Bernitz and Joakim Nergelius eds., *General Principles in European Community Law* 163 (The Hague: Kluwer Law International, 2000).

[88] Being recognised by a multiplicity of States pertaining to different legal traditions, it can be said that the legitimate expectation principle is a general principle of law. *See* Elizabeth Snodgrass, 'Protecting Investors' Legitimate Expectations: Recognising and Delimiting a General Principle', 21 *ICSID Rev. – FILJ* 1 (2006).

[89] *Saluka Investment BV (The Netherlands) v The Czech Republic*, UNCITRAL, Partial Award on Jurisdiction and Liability of 17 March 2006, at 301.

[90] *International Thunderbird Gaming v The United Mexican States*, UNCITRAL (NAFTA), Separate Opinion (Professor T. Wälde) of 26 January 2006, at 37.

[91] Ibid., at 31.

[92] *See* Christoph Schreuer, 'Fair and Equitable Treatment in Arbitral Practice', 6 *Journ. World Investm. & Trade* 357 (June 2005) at 374 and R. Dolzer, 'Fair and Equitable Treatment: A Key Standard in Investment Treaties', 39 *The International Lawyer* 87 (2005) at 105.

affected foreign investments. This flows from the emerging awareness that the FET imposes duties also upon the investor, 'given the inherent balancing process that lies at his heart'.[93] Among them, the duty to conduct the business in a reasonable manner requires any operator to be aware of the legal environment which he is entering in. Accordingly, the decision to invest in highly regulated sectors, such as water public utilities, would entail the implicit acceptance of the risk that public authorities will further intervene to adapt the rules to the evolving needs.[94] In these circumstances, the modification of the regulatory environment could not constitute a valid basis for a claim under the FET, as investor's expectations for a fixed legislative framework are not legitimate. Any other solution would turn investment treaties into insurance policies covering all sorts of risks, an outcome that goes beyond their scope.

4. Water Scarcity and the WTO

In this paper we have focused on two different kinds of water transfers – trade in water through the diversion of the river flow from one country to another and bulk water transfers. We have argued that just the latter may be dealt with through WTO law, and even in this case only if the water transfer is not covered by any other specific agreement, or if the latter is not strong enough. The goal of this section is to determine whether a water importing country can bring a claim to a WTO Panel against restrictive trade measures adopted by a water exporting country in bulk water transfers operations.[95] The objective here is to see whether Articles

---

[93] *See* Peter Muchlinski, 'Caveat Investor'? The Relevance of the Conduct of the Investor Under the Fair and Equitable Standard', 5 *Int'l & Comp. L. Quarterly* 527, 532–533 (July 2006) at 542.

[94] The principle was made explicit, albeit with regard to expropriation, in *Methanex Corp. v United States*, UNCITRAL, Final Award on Jurisdiction and Merits of 3 August 2003, Part IV – Ch. D – Page 5, 9–10.

[95] *See* the situation in Canada where in 1999 the federal government announced a strategy to protect Canadian water. This strategy was based on an accord for the prohibition of bulk water removal from Drainage Basins that clearly provided for the possibility to restrain water exports in order to protect the environment. *Accord for the Prohibition of Bulk Water Removal from Drainage Basins*, available at http://www.scics.gc.ca/pdf/accord.pdf.

XI and XX can help bridge the gap between these competing positions.

4.1 The Compatibility of Water Export Bans with WTO Law

We have already argued that it is very difficult to conceive water as a good under the WTO. However, it is wise to explore this possibility due to the strength of the WTO DSB and of the organisation itself.[96] Should that be the case, water transfers may fall under the realm of the WTO. A country that suffers an export ban on water may bring a dispute before the DSB and it may argue that the country applying the trade measure has violated Article XI GATT, which provides for the elimination of quantitative restrictions.

Against this background, could a State ban water exports, and how could this be balanced with the needs of the importing country?

According to Article XI, a country can temporarily prohibit exports in order to 'prevent or relieve critical shortages of foodstuffs or other products essential to the exporting contracting party'.[97] Does water fall under the category of essential product? Can the lack of water cause a critical shortage of food? It is evident that this is the case with water and that, therefore, a water export ban taken by a country suffering a serious water crisis would be WTO compatible.

However, the scenario provided for in Article XI does not seem to reflect the characteristics of those countries that most likely may consider water export bans as policy options. Now, has the multilateral trading system taken into account the fact that trade in water may also take place between countries that are in desperate

---

[96] This is a common feature in the trade environment interface. One example comes from the debate between climate and trade in which, despite emission allowances being considered neither goods nor services by many authors, their trade has been subject to a study of their WTO compatibility. *See inter alios* Christina Voigt, 'WTO Law and International Emissions Trading: Is there Potential for Conflict?', 2 *Carbon and Climate Law Review* 54, 55–56 (2008).

[97] *See* Article XI 2 (a) of General Agreement on Tariffs and Trade (GATT), Marrakesh, April 1994.

need of water? Usually water crisis that lead to critical shortages of food occur in water importing countries. Does the WTO acknowledge the consequences of export bans on importing countries? The Agreement on Agriculture (AoA) contains a provision, which is explicitly linked to Article XI.2.a):

> Where any Member institutes any new export prohibition or restriction on *foodstuffs* in accordance with paragraph 2(a) of Article XI of GATT 1994, the Member shall observe the following provisions: the Member instituting the export prohibition or restriction shall give *due consideration to the effects of such prohibition or restriction on importing Members' food security*;[98]

We consider this to be a crucial provision, as it may help to balance the interests of exporting countries that are applying the trade measure and those of importing countries that may be suffering the consequences thereof.

In order for this provision to be applicable, it must first be determined whether water qualifies as foodstuff. We consider that, once again, this is self-evident. Furthermore, there is growing literature on virtual water trade that argues that water is essential in the production of most food products.[99] If water stands for foodstuff, a country adopting a water export ban must take into account the effects of the measure on the water importing countries' food security.

In sum, Article XI must be read together with Article XII AoA in order to strike a balance between water exporting developed countries and water importing developing countries. The linkage between these two provisions may provide a pathway to deal in a sustainable manner with water export bans. However, more work

---

[98] *See* Article 12 of Agreement on Agriculture (AoA), Marrakesh, April 1994. *Source:* http://www.wto.org/english/docs_e/legal_e/14-ag.doc. Emphasis added.

[99] Trade in virtual water can be considered as a further kind of water transfer. Its coverage under the WTO realm and further issues exceed from the scope of this paper. For more information, *see* Ashok K. Chapagain, Arjen Y. Hoekstra and Huub H.G. Savenije, Saving Water through Global Trade, UNESCO Value of Water Research Report Series No. 17 (Delft: UNESCO-IHE, 2005). *Source:* http://www.waterfootprint.org/Reports/Report17.pdf. For trade aspects of virtual water, *see, e.g.*, Alix Gowlland-Gualtieri, Legal Implications of Trade in 'Real' and 'Virtual' Water Resources (Geneva: IELRC, Working Paper 2008-01, 2008). *Source:* www.ielrc.org/content/w0801.pdf.

is needed in order to clarify the requirements of this sustainable approach.[100]

4.2 Can Water Export Bans be Justified under Article XX GATT?

Should water export bans be considered a violation of Article XI, Article XX could be invoked to justify the measure, as it entitles WTO Members to deviate from the Agreement's rules for legitimate non-commercial goals, such as the conservation of natural resources. The general exception entails a double layered analysis that deals with the *content* and the *application* of the measure.[101]

In relation to the former, the first point is to see whether water export bans are 'necessary to protect human, animal or plant life or health'.[102] On the one hand, water export bans fulfil this condition because water is crucial for both human life and for the conservation of biodiversity. On the other hand, one must also see if the water export ban is actually *necessary* to secure domestic health. According to current jurisprudence the importance of the non-commercial goal pursued by the trade restrictive measure, the effectiveness of the measure in fulfilling it and the restrictive effects of the measure on international trade are the parameters taken into consideration to evaluate the necessity of the measure.[103] A clear-cut water export ban would be necessary only having regard to the first of these parameters.

The second issue in relation to the content of a water export ban is whether it is 'related to the conservation of exhaustible natural resources [and] if the measure is made effective in

---

[100] A first condition has already been written down and it obliges the country that intends to adopt an export ban in foodstuff to notify affected Parties in advance (*See* AoA, note 98 above, Article 12.1.b). Further conditions, such as monetary compensation, enhanced international cooperation, may be developed and discussed within the WTO Committee on Agriculture.

[101] The analysis must start from the paragraphs of Article XX; *see United States – Import Prohibition of Certain Shrimp and Shrimp Products,* WTO Appellate Body Report, WT/DS58/AB/R, 1998.

[102] General Agreement on Tariffs and Trade (GATT), note 102 above, Article XX (b).

[103] *Korea - Measures Affecting Imports of Fresh, Chilled and Frozen Beef,* WTO Appellate Body Report, WT/DS161/AB/R, 2000.

conjunction with restrictions on domestic production or consumption'.[104] Three points must be addressed. First, in some countries water can be a scarce resource and it may be deemed as an exhaustible natural resource. Second, the measure must be reasonably related to the conservation purpose,[105] a requirement that is easier to meet than the necessity test.[106] Third, the export restriction must be taken together with similar restrictions at a domestic level and such criterion is likely to be crucial.

The next step is to determine if the *application* of the water trade related measure meets the requirements provided for in the *chapeau* of Article XX. The first one is that the measure must not constitute an arbitrary and unjustifiable discrimination. This requirement will be met if the adopting country demonstrates sufficient flexibility and prior negotiation efforts. The latter entails that a State must take action internationally before adopting a water export ban. A unilateral trade restrictive measure should be the last policy option, once all international efforts have failed.[107] However, a State does not have to wait until the international negotiations succeed, but it can adopt unilateral actions in the meanwhile.[108]

The last requirement provided for in the *chapeau* of Article XX is that the measure must not be a *disguised* restriction of international trade. The latter will be revealed by the structure of the measure, which must be thus carefully analysed.[109]

In sum, current trade and environment case law on Article XX does not seem to fully protect a State that wishes to include a

---

[104] General Agreement on Tariffs and Trade (GATT), note 102 above, Article XX (g).

[105] *United States – Import Prohibition of Certain Shrimp and Shrimp Products*, note 101 above.

[106] The difference between *necessary* and *related to* has been underlined already in the old GATT dispute settlement system. See *Canada – Measures Affecting Exports of Unprocessed Herring and Salmon*, GATT Panel Report, Doc. L/6268 – 35S/98, 22 March 1988.

[107] *United States – Import Prohibition of Certain Shrimp and Shrimp Products – Recourse to Article 21.5 of the DSU by Malaysia*, WTO Appellate Body Report, WT/DS58/AB/RW, 2001.

[108] Ibid.

[109] *European Communities – Measures Affecting Asbestos and Asbestos – Containing Products*, Report of the Panel, Doc. WT/DS135/R, 2000.

water export ban in its domestic water conservation policy. Despite the fact that so far no dispute has arisen on a similar situation, the adopting State may face problems related both to the content and to the application of the measure.

4.3 Bridging the Gap between the WTO and International Water Law

A new *institutional* and a new *normative* setting are needed in order to bridge the gap between international trade and water law. Water scarcity related problems can be dealt with through trade, but not through the current WTO legal regime. From an institutional point of view three options have been proposed – to strengthen international water agreements, to reform the multilateral trading system or a *status quo* situation.

The first option is to strengthen international water governance. Usually water transfers are dealt with by bilateral international treaties (Lesotho – South Africa) or other kind of international bilateral agreements (Turkey – Israel). These must contain a mechanism that the parties therein may resort to in case a dispute arises. Furthermore, a global international water agreement dealing with trade in water could deal comprehensively with water trade related disputes.[110] Again, the presence of a strong dispute settlement mechanism is crucial. Global water governance must be enhanced in order to counterbalance the power of the WTO and, in particular, of the DSB. A consequence, and some would argue a risk, of strengthening of international water agreements, be they bilateral or multilateral, is the possibility of forum shopping between trade and water courts, which could lead to a conflict of jurisdictions.[111] In order to prevent it, bilateral and international water agreements dealing with trade should include an explicit prevailing clause over the WTO. The DSB should then refrain itself from accepting water disputes that can be solved through international water agreements.

The second option is to reform the WTO. A wide array of solutions in relation to the overall trade and environment

---

[110] *See* Steven Shrybman, note 24 above at 15.
[111] *See* Edith Brown Weiss, note 16 above at 82–83.

relationships has already been proposed.¹¹² Three possibilities seem to enjoy higher consideration. First, a waiver could be proposed to exempt water export bans from certain specific WTO provisions.¹¹³ Second, an amendment of the GATT could be sought in order to either exclude water export bans from the WTO discipline or include it as a specific exception in the framework of Article XX.¹¹⁴ The last option would be an authoritative interpretation either by the WTO Ministerial Conference or by the General Council that clarifies if and how the WTO regime should deal with trade in water and serves as a basis on which the DSB would decide any dispute arising from a water trade restrictive measure.¹¹⁵ The 1993 Statement regarding the application of the North American Free Trade Agreement (NAFTA) to water could serve as a precedent.¹¹⁶

The third option is doing nothing. A *status quo* option would imply that the current multilateral trading system already balances environmental and trade interests in the best possible way and that there is no reason to modify the WTO or to strengthen the international water governance system.

From a normative point of view, what is needed is a new paradigm in the relationship between water scarcity and international trade. This new approach must be linked to any of the above mentioned institutional settings. The balance between water exporting states' interests and water importing states' interests must be put at the centre of the debate, as Article XII AoA is able to do with Article XI. There must be a bridge between the environmental concerns of the water exporting country that wants to retain its water for conservation purposes and the development

---

¹¹² *See* for example Frank Biermann, 'The Rising Tide of Green Unilateralism in World Trade Law Options for Reconciling the Emerging North–South Conflict', 35 *Journal of World Trade* 421 (2001).

¹¹³ Waivers must be decided by consensus according to Article IX.3 of the *Marrakech Agreement Establishing the WTO* (*Marrakech Agreement*), Marrakesh, April 1994. *Source:* http://www.wto.org/english/docs_e/legal_e/04-wto.doc.

¹¹⁴ *See* Edith Brown Weiss, note 16 above at 86. The procedures to amend the WTO are provided for in Article X of the *Marrakech Agreement*.

¹¹⁵ Brown Weiss considers it the best option if States decide to reform the WTO. *See* Edith Brown Weiss, note 16 above at 87.

¹¹⁶ *See* note 26 above.

concerns of importing countries whose need of water may vary depending on sociological, geographical and political factors.[117] Our approach wishes to combine the much needed anticipatory caution suggested by E. Brown Weiss with the developing needs of a water scarce country. The key to establish this linkage is *Comment 15 to the United Nations Committee on Economic, Social, and Cultural Rights* (*Comment 15*) that enshrines access to water as a right;[118] or even a human right.[119] This means that in those cases in which a water importing country is facing serious water scarcity, water is not just a commodity but it becomes a right. According to *Comment 15* 'international cooperation and assistance [must] take joint and separate action to achieve the full realisation of the right to water'. A strict interpretation of this provision may lead water exporting countries to have obligations towards water importing countries that suffer water scarcity rather than just commercial rights. This position is to be supported since it forces the trade regime to pay due consideration to the fundamental social interests at stake and, thus, it embodies the evolution that has taken place in the international order to this regard.[120]

---

[117] A different balance is required if we are dealing with a water export ban between Canada and the US or a ban imposed by Canada on a developing country that is suffering a water crisis.

[118] Committee on Economic, Social and Cultural Rights, General Comment 15: note 11 above.

[119] *See* Amy Hardberger, 'Whose Job Is It Anyway?: Governmental Obligations Created by the Human Right to Water', 41 *Texas International Law Journal* 553, 541–542 and 545 (2006).

[120] It is important to highlight that despite the fact that most studies, including ours, focus mostly on possible conflicts between trade and environment, there is scope for mutual supportiveness between international environmental law and the international trade regime. One can find this in several Multilateral Environmental Agreements such as in the preamble of the Cartagena Biosafety Protocol, or in Article 2.3 of the Kyoto Protocol to the United Nations Framework Convention on Climate Change, Kyoto, 11 December 1997. *Source:* http://unfccc.int/resource/docs/convkp/kpeng.pdf. Furthermore, especially from the trade side there seems to be an interest to promote mutual supportiveness arguing that the enforcement of trade rules could actually benefit the environment, as the Director General of the WTO has postulated in relation to climate change. *See,* on this point, Pascal Lamy, 'Preface', 2 *Carbon and Climate Law Review* 1 (2008).

In sum, if the goal pursued by the water exporting state through the water export ban is an environmental objective, and the goal undermined by the measure in the importing country is related with the life and health of its citizens, a sustainable development issue arises. Despite the difficulties of translating the politics of sustainable development into legally binding principles, this is the realm where the solution is to be searched,[121] since, as rightly emphasised by A. Hildering, 'an economic approach to water... is not necessarily compatible with sustainable development'.[122]

5. Concluding Remarks

Water access and water scarcity have been dealt with by international investment law and the WTO just recently, and still in a rather limited fashion. However, international investment and WTO law are posed to play a greater role in regulating water-related issues in the near future, although in these contexts water is not considered as a natural resource, but only as an economic good that may be provided as a service or traded. Their compliance mechanisms have proven to be highly effective and, hence, they are likely to further displace other tools for global water governance. Against this framework, it becomes even more compelling to stress that water is not just an economic good, as its value goes well beyond the economic dimension.

The inextricable relationship between all these dimensions is to be fully considered when international investment law and WTO provisions are applied in this field. Both legal frameworks considered here would already allow for such a special treatment, provided that great care is exercised in the use of enforcement mechanisms as well as in the interpretation of key substantive provisions. Flexibility is key to finding a fair balance between the economic and social interests at stake. States must be free to choose their course of action in guaranteeing universal access to water services or fighting against water scarcity. Consequently, inter-

---

[121] Francesco Sindico, Unravelling the Trade and Environment Debate through Sustainable Development Law Principles, ESIL Inaugural Conference Agora Paper 2005. *Source:* http://www.esil-sedi.org/english/pdf/Sindico.PDF.

[122] Antoinette Hildering, note 18 above at 122.

national investment rules and the WTO are not to be applied in a way that could hamper such efforts, but only to avoid these issues being abused by domestic authorities to pursue protectionist ends. More intrusive approaches would not just undermine the protection of fundamental social needs, but they could also damage the credibility of both international legal regimes.

11

More Drops for Hyderabad City, Less Crops for Farmers: Water Institutions and Reallocation in Andhra Pradesh

MATTIA CELIO

As water is a scarce resource there must be some rule over precedence in its use. Paramount is the right to quench thirst. Even appropriated water is overruled by the necessity to provide water for man and beasts where no other suitable supply is available.[1]

Introduction

Global urban population growth, particularly in developing countries, is happening at an unprecedented rate. The world population rose from 750 million in 1950 to 2.9 billion in 2000, and the number of people living in urban areas has equalled the rural population in 2007, and is on the way to reaching sixty per cent by 2030.[2] The sustainability of such a vibrant growth is contingent upon the availability of sufficient water for covering agricultural,

---

[1] John C. Wilkinson, 'Muslim Land and Water Law', 1 *Journal of Islamic Studies* 54 (1990).

[2] United Nations Secretariat, World Urbanisation Prospects – The 2001 Revision, Report of the Department of Economic and Social Affairs – Population Division, UN Doc. ESA/P/WP.173 (2002). *Source:* http://www.un.org/esa/population/publications/wup2001/wup2001dh.pdf.

domestic, commercial, industrial, environmental as well as other minor demands. If urban demand for water is growing, the availability of the resource has shrunk over the last decades due to massive diversions for agricultural needs. As hydrologists like to put it, many river basins around the world are reaching the stage of closure, which occurs when all available water in a basin is utilised. Reallocating water then becomes necessary, for instance when a particular user such as a city wants to increase its withdrawals. Under these conditions, water conflicts are likely to develop, and appropriate rules, policies, and organisations responsible for transferring water between users need to be in place.

The case of Hyderabad, capital of the state of Andhra Pradesh in south India, is exemplar of the challenge of sustaining a rapid urban growth in a water-scarce environment. The city has grown on average at a rate of 2.4 per cent from 1991 to 2001, rising from 2.5 to 5.5 million,[3] and is expected to reach 13.6 million by 2021.[4] Over the last three decades, water has been brought to the city from the Musi River initially and further from the Manjira and Krishna Rivers. In the latter two cases, the government of Andhra Pradesh has administratively reallocated water from the agricultural sector.

Institutions for water reallocation between sectors is the main focus of this article, which builds upon the existing literature and empirical findings of a research conducted by the author between 2005 and 2007 at the International Water Management Institute (IWMI) in Hyderabad. This chapter considers water institutions (meant as water policies, law, and administration) in India and specifically in the state of Andhra Pradesh, and discusses their implications for water transfers from agriculture to Hyderabad.

Following this introduction, the main theoretical tenets of intersectoral water reallocation are illustrated, and the general principles of the institutional approach to water management reviewed. Then, after having outlined the history of Hyderabad water supply, the main focus of this chapter is addressed, i.e.,

---

[3] K.C. Sivaramakrishnan, A. Kundu and B.N. Singh, *Handbook of Urbanization in India* (New Delhi: Oxford University Press, 2005).

[4] Hyderabad Urban Development Authority, Hyderabad 2020: Draft Master Plan for Hyderabad Metropolitan Area (Hyderabad: 2003).

water institutions and intersectoral reallocation. Water policy, law, and administration in India and Andhra Pradesh are presented to the extent to which they have been influential for water transfers in Andhra Pradesh from agriculture to Hyderabad. The article concludes by discussing the main lessons drawn from the study of Hyderabad water supply and reallocation from irrigated agriculture.

Theoretical Tenets: Water Scarcity, Intersectoral Water Competition, and Institutional Reforms

The development of river basins can be represented by a tri-phased model.[5] In the beginning, water supplies largely outstrip the demand; dams are constructed at the most convenient locations; and the quantities of water used for domestic purposes are relatively modest when compared against the utilisation in other sectors, notably irrigated agriculture. In the second phase, successive increases in the consumptive use of water lead to sporadic resource shortages, particularly during dry seasons, and developing new resources requires conspicuous financial investments since the most convenient locations have already been exploited. Eventually, water becomes chronically scarce and the basin, once it reaches its most advanced phase of development, is said to become 'closed' since all the existing water resources are used by established water users. At this stage, reallocation is necessary for meeting the demand for water of additional users or for the priority-based increasing of supply to existing ones. Efforts are then usually directed at reallocations towards the most economically valuable uses; and new institutions are needed to address intersectoral competition and manage river basin resources in an integrated manner.[6] Though

---

[5] Jack Keller, Andrew A. Keller and Grant Davids, 'River Basin Development Phases and Implications for Closure', 33 *Journal of Applied Irrigation Science* 145 (1998) and David Molden, R. Sakthivadivel and Samad Madar, 'Accounting for Changes in Water Use and the Need for Institutional Adaptation', *in* Charles L. Abernethy ed., *Intersectoral Management of River Basins* 73 (Colombo and Feldafing: International Water Management Institute and German Foundation for International Development, 2001).

[6] *See* François Molle, Development Trajectories of River Basins – A Conceptual Framework (Colombo: International Water Management Institute, Research Report No. 72, 2003).

simplified and only capturing the outlines of the complex and multifaceted development of a river basin, the model presented above clearly depicts the changing relationship between rivers and their water users.

Over the last decades, water withdrawals for agricultural production have brought about a significant decrease in water availability in many river basins around the world. Growing cities are today competing for a share of this water, and intersectoral reallocations from the agricultural to the urban sector are increasingly taking place.[7] The extent of this phenomenon and its complexity has provided fertile ground for the emergence of an area of research specifically dealing with water reallocation mechanisms. The reasons explaining why water is transferred out of irrigated agriculture are that the agricultural sector has traditionally received the lion's share of all the water diverted from rivers (around seventy per cent worldwide),[8] and it is blamed of wasting water since only a small fraction of the water diverted is actually used by the crops for their growth.[9] Moreover, it is reckoned that even small increases in the efficiency of irrigated agriculture might free enough water for entirely covering present and future urban needs.[10] These aspects have generally been put on the table as justifications for transferring water out of the agricultural sector. Furthermore, in order to reduce the costs of urban water supplies, cities often exploit or even appropriate existing irrigation water reservoirs instead of going for new ones, thus curtailing the supply that was formerly available for crop production.

---

[7] David Molden ed., *Water for Food, Water for Life: A Comprehensive Assessment of Water Management in Agriculture* (London: Earthscan, 2007).

[8] United Nations, *Water for People – Water for Life* (Paris: UNESCO, 2003). *Source:* http://www.unesco.org/water/wwap/wwdr/wwdr1/table_contents/index.shtml.

[9] World Water Council, A Water Secure World: Vision for Water, Life, and the Environment (Hague: World Water Council, Commission Report, 2000). *Source:*http://www.worldwatercouncil.org/fileadmin/wwc/Library/Publications_and_reports/Visions/CommissionReport.pdf.

[10] James Winpenny, *Managing Water as an Economic Resource* (London: Routledge, 1994).

Water is generally reallocated between the agricultural and the urban sectors through competitive markets or through administrative procedures.[11] The mechanism in vogue depends essentially upon the way water rights are defined. Basically, in market systems users hold property rights over water, so that they can decide to permanently or temporarily transfer their rights according to competitive market forces; whereas in the administrative allocation water rights are vested within the state, and users are only entitled to use the resource they are provided with. Both mechanisms present advantages as well as shortcomings. Administrative allocation is notably supposed to better taking into account equity aspects, as resource supply to the poorest section of the population, and to exercise a tighter control over reallocation third-party effects as modifications of return flows (that is the water not actually utilised after diversion that returns in the natural system). On the other hand, it is well known that administrative allocation provides little or no incentives for water conservation,[12] that the decision-making process is often highly politicised, and that water is reallocated without proper compensations for those that are forcibly deprived.

Because of these and other shortcomings, some scholars have advocated the reform of water institutions so as to enable the shift from administrative to market-based water allocation, supporting their arguments by successful (or at least claimed as such) cases as reforms undertaken in Chile, Mexico or Australia.[13] Though such reforms appear necessary for regulating the reallocation of water

---

[11] Ariel Dinar, Mark Rosegrant and Ruth Meinzen-Dick, Water Allocation Mechanisms. Principles and Examples (Washington, DC: World Bank Policy Research Working Paper 1779, 1997). *Source:* http://www.worldbank.org/html/dec/Publications/Workpapers/WPS1700series/wps1779/wps1779.pdf.

[12] Ruth S. Meinzen-Dick and M.S. Mendoza, 'Alternative Water Allocation Mechanisms: Indian and International Experiences', 31 *Economic and Political Weekly* 25 (1996).

[13] Charles W. Howe, Dennis R. Schurmeier and Douglas W. Shaw Jr., 'Innovative Approaches to Water Allocation: The Potential for Water Markets', 22 *Water Resources Research* 439 (1986) and Mark Rosegrant and Renato Gazmuri Schleyer, Tradable Water Rights: Experiences in Reforming Water Allocation Policy (Washington: Applied Study Prepared for the Bureau for Asia and the Near East of the U.S. Agency for International Development, 1994).

among users in situations of scarcity, they entail high transaction costs notably for the setting up and enforcement of new legal and institutional frameworks. Moreover, because of the high strategic value of water and the uncertainties in predicting the outcomes of reforming the water law, many governments are generally adverse to give up their monopoly over the resource.[14]

Hyderabad Water Supply

Since its creation in 1591,[15] Hyderabad used to rely on water impounded in tanks as well as on ground water tapped through shallow dug wells. In the beginning of the twentieth century, the seventh Nizam of Hyderabad H.E.H Osman Ali Khan commissioned the construction of two reservoirs approximately 8 km upstream of the city, namely Osmansagar on the Musi River, and the Himayatsagar on the Esi (*see* Figure 1).[16] These two water works provided for protection against the recurring floods that used to hit Hyderabad, and for a supply of 205,000 m^3 of water per day. The quantity of water conveyed to the city was further increased in 1965 and again in 1982, by bringing water from the Manjira Barrage across the Manjira River.

In 1972, the government constituted an expert committee in charge of recommending options for substantially increasing Hyderabad water supply. The committee recommended the transfer of water from the Krishna River.[17] Further on, in 1975 and 1978,

---

[14] Paul Holden and Mateen Thobani, Tradable Water Rights: A Property Rights Approach to Resolving Water Shortages and Promoting Investment (Washington: World Bank Policy Research Working Paper 1627, 1997). *Source:* http://econ. worldbank.org/external/default/main?pagePK=64165259&theSitePK=469372&piPK=64165421&menuPK=64166093&entityID=000009265_3961214131318.

[15] M. A. Nayeem, *The Splendour of Hyderabad: The Last Phase of an Oriental Culture (1591–1948 A.D.)* (Hyderabad: Hyderabad Publishers, 2002).

[16] Hyderabad Metropolitan Water Supply and Sanitation Board, Feasibility Report on Augmenting Water Supply to the Twin Cities of Hyderabad and Secunderabad, Manjira Water Supply Scheme Phase-III (Singoor Project) (Hyderabad: HMWSSB, 1983).

[17] Government of Andhra Pradesh, Report of the Committee on Drawing Additional Water to Twin Cities from Srisailam or Nagarjunasagar or Other Projects (Hyderabad: Government of Andhra Pradesh, 1973).

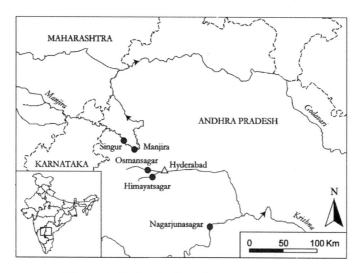

Figure 1: Hyderabad surface water sources.

Maharashtra and Karnataka signed two separate agreements with Andhra Pradesh, allowing it to draw 113 Mm^3 of water per year from the Manjira River for Hyderabad through the construction of a new reservoir. The option of taking water from the Krishna River was put aside, and according to the interstate agreements Andhra Pradesh built the Singur reservoir on the Manjira River, and started transferring water from there to Hyderabad in 1991. Nevertheless, even before completion of the Singur project it became apparent that the Manjira River transfer project would not be sufficient to keep pace with the demand of Hyderabad's expanding urban population and industrial sector. In 1986 the government appointed a second expert committee which recommended the transfer of 467 Mm^3 per year from the foreshore of Nagarjunasagar reservoir on the Krishna River. Carrying water from the Godavari, the other major river in Andhra Pradesh, was not deemed cost-effective since the river was located at a greater distance as compared to the Krishna.

Endorsing the recommendations of the expert committee, the Telugu Desam Party (TDP) that was then heading the government issued a Government Order in 1988, administratively sanctioning the increase of Hyderabad water supply from Nagarjunasagar. This decision to convey water to the city from the Krishna, the first ever taken by a government, was vehemently opposed by Members of the Legislative Assembly (MLAs) belonging to the Indian National Congress (INC) that was in opposition at that time. Protests were conducted in the Legislative Assembly, and the MLAs ended up staging a protracted hunger strike.[18] One year later in 1989, the TDP lost the state general elections and the project of drawing water from the Krishna was initially withdrawn. However, in subsequent years it was again harshly debated, and eventually started in 2003. Factors explaining the opposition to the project are related to the regional set-up of Andhra Pradesh, as well as to the general context of confrontation so characteristic of political debates in the state.[19]

For historical reasons, Andhra Pradesh is divided into three regions namely Telangana in the North, Rayalseema in the South and Coastal Andhra. The political discourse very often proceeds along regional fault lines, to the extent that a strong political party in the state, the Telangana Rashtra Samiti, has as its main political objective the creation of a separate Telangana state out of the Telangana region. The reallocation of water from the Krishna to Hyderabad has been invariably opposed by farmers and politicians from the drought-prone Rayalseema region on the basis of regional considerations. Notably, MLAs from Rayalseema have argued that before bringing water to Hyderabad which is located in Telangana, projects that had been previously sanctioned for diverting water from the Krishna River to Rayalseema, but never implemented, had to be taken up first.[20] The political sensitivity of supplying Hyderabad from the Krishna stalled the decision-making process till 2003, when a major drought brought Hyderabad to the brink of a major water crisis and the government eventually took and implemented the decision to convey water from the Krishna River through the Krishna Drinking Water Project. The first phase carrying around 409,000 m^3/day to Hyderabad was completed in 2005.

---

[18] Anonymous, 'MLAs' Fast Continues', *Deccan Chronicle*, 20 August 1988 and Anonymous, 'R'seema MLAs Stage Walk-Out', *Deccan Chronicle*, 28 July 1988.

[19] T. Bhaktavatsalam, Politics of Opposition in Andhra Pradesh: 1983–88 (Guntur: Nagarjuna University, M.Phil. Dissertation, 1991).

[20] Debates at the Legislative Assembly of Andhra Pradesh, 18 August 1988.

Because of opposition against water transfer from agriculture to Hyderabad and the delays in increasing urban water supply, water conveyance to the city has never met the demand, as shown in Figure 2.[21]

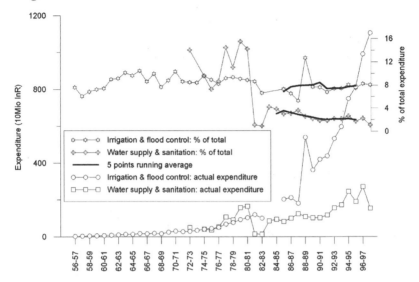

Figure 2: Andhra Pradesh state government's annual expenditures under the heading of 'irrigation and flood control' and 'water supply and sanitation'; and percentages to the total state annual outflow.

Water Policies

In this section water policies in India and Andhra Pradesh are discussed, stressing those aspects relevant for water allocation. In particular, attention is paid to (i) the National Water Policy issued in 2002 by the Indian Central Government; (ii) central funding schemes related to drinking water supply; (iii) trends in Andhra Pradesh state expenditures; and (iv) urban water policies in Hyderabad. The section is organised according to the scale of analysis – the national level is addressed first, then Andhra Pradesh, and then water policies in Hyderabad are discussed last.

---

[21] Population data drawn from HUDA, Hyderabad Urban Development Authority, *Hyderabad 2020: Draft Master Plan for Hyderabad Metropolitan Area* (Hyderabad: 2003).

Indian National Context

Indian National Water Policy

The Indian government issued its first exclusive National Water Policy (NWP) in 1987, whose main concerns were the promotion of conjunctive use of surface and ground water; supplemental irrigation; water-conserving; cropping patterns; and irrigation and production technologies.[22] Later on in 2002, the Central Government has issued a reviewed NWP that heavily draws from its predecessor of 1987.

Drinking water is held in high consideration in the NWP presently in force, which has a crucial target to provide adequate safe drinking water facilities for the entire Indian population in urban and rural areas. The paramount importance of water for drinking purposes is also addressed in the NWP when it comes to prioritise water uses. Notably, water allocation priorities should be as follows – drinking water should come first, then irrigation, hydro-power, ecology, agro-industries and non-agricultural industries, and at last navigation and other uses.[23] Notwithstanding this declared goal to put drinking water high in policy and action, prioritisation is somehow watered down, since '... the priorities could be modified or added if warranted by the area/region specific considerations'.[24] Unfortunately, prioritising drinking water over other uses is a weak instrument for transferring water to cities, since water conveyed to urban centres is used for a multitude of uses extending well beyond drinking. Furthermore, transferring water out of irrigation projects doesn't mean that only agriculture will be affected, since in many cases irrigation canals also feed surface reservoirs that are utilised for drinking purposes. This, for instance, is the case in the Nizamsagar irrigation project along the Manjira River, where a share of the water stored in the reservoir is reserved for Nizamabad city and Bodhan town in the case of protracted drought that can put at risk urban water supply.

---

[22] Maria Saleth, Strategic Analysis of Water Institutions in India: Application of a New Research Paradigm (Colombo: International Water Management Institute, Research Report No. 79, 2004). *Source:* http://www.iwmi.cgiar.org/Publications/IWMI_Research_Reports/PDF/pub079/Report 79.pdf.

[23] Government of India, National Water Policy, 2002. *Source:* http://www.ielrc.org/content/e0210.pdf.

[24] Ibid., Section 5.

Irrigation and drinking uses are moreover intimately intertwined because of the physical interconnectedness between surface and ground water, which impinges upon the effectiveness of water use prioritisation policies. For instance, strictly implementing a policy that aims at reducing water supply for irrigation in favour of a transfer to a city for domestic purposes would have consequences on ground water recharge and return flows in irrigation schemes. This would very likely affect drinking water supply in rural areas, which generally depends to a large extent on ground water availability. This linkage between water use at the river basin level, named 'cascading reuse systems',[25] significantly reduces the scope of prioritising water use as a means of reallocating water to high value sectors.

The 2002 NWP also advocates for planning and setting-up institutional mechanisms capable of dealing with intersectoral water use and competition: 'Water resources development and management will have to be planned ... multi-sectorally ...'.[26] Moreover, the importance of decentralising the administration of water resources and handing over responsibilities to water users is clearly spelled out: 'With a view to give effect to the planning, development and management of water resources on a hydrological unit basis, along with a multi-sectoral, multi-disciplinary and participatory approach ..., the existing institutions [meant as "organisations"] at various levels under the water resources sector will have to be appropriately reoriented/reorganised and even created, wherever necessary'.[27] Further, the NWP states that 'Special multi-disciplinary units should be set up to prepare comprehensive plans taking into account not only the needs of irrigation but also harmonising various other water use ...'.[28]

Central Funding Schemes for Drinking Water Supply

Central schemes for rural and urban water supply are dealt with in the following section of this chapter. Funding schemes provide for evidence on public authorities' water policies and priorities.

---

[25] David Molden and M.G. Bos, 'Improving Basin Water Use in Linked Agricultural, Ecological, and Urban Systems', 51 *Water Science and Technology* 147 (2005).
[26] *See* National Water Policy, note 24 above, Section 3(3).
[27] Ibid., Section 4(1).
[28] Ibid., Section 4(2).

Drinking water supply was first recommended priority in India within the works of the Bhore Committee in 1946 and the Environmental Committee in 1949.[29] In 1954, the Central Government provided assistance to the states for establishing special investigation divisions in the fourth five-year plan, so as to identify villages with particularly difficult access to protected water sources. The Accelerated Rural Water Supply Programme was introduced in 1972–73 to speed up the pace of water access in villages without a safe water source within a distance of 1.6 km; where water is available at a depth of more than 15 meters; or if ground water quality was not suitable for drinking use. In 1986, a Technology Mission (TM) was set up to assist the states in the provision of drinking water supply. The TM was further renamed 'Rajiv Gandhi National Drinking Water Mission' whose main objective was to provide sufficient and safe drinking water to all villages in the country. The Central Government also introduced in 1993–94 the Accelerated Urban Water Supply Programme (AUWSP) to cover towns having a population of less than 20,000 as per the 1991 Indian Census, and allow the states to receive 50 per cent matching funds from the centre.[30]

Water supply in mega cities is sponsored by the Central Government through the Infrastructure Development of Mega Cities (IDMC) scheme initiated in 1993–94 and the Jawaharlal Nehru National Urban Renewal Mission (JNNURM) in 2005. The IDMC primary objective is to '... undertake infrastructure development projects of city-wide/regional significance covering a wide range of components like water supply and sewerage, roads and bridges, city transport, ...'.[31] Cities that can benefit from this scheme are Mumbai, Kolkata, Chennai, Bangalore, and Hyderabad. The overall

---

[29] Meena Panickar, State Responsibility in the Drinking Water Sector. An Overview of the Indian Scenario (New Delhi: International Environmental Law Research Centre, Working Paper 2006–07. *Source:* http://www.ielrc.org/content/w0706.pdf.

[30] Government of India, Ministry of Urban Development, Accelerated Urban Water Supply Programme, available at http://urbanindia.nic.in/moud/programme/uwss/auwsp.htm.

[31] Government of India, Ministry of Urban Development, Infrastructure Development of Mega Cities. *Source:* http://urbanindia.nic.in/moud/programme/ud/infrastructuredevel.htm.

goal of the JNNURM, the more recent scheme, is to '... encourage reforms and fast track planned development of identified cities'. The JNNURM comprises of two sub-entities – the Sub-Mission for Urban Infrastructure and Governance, whose main thrust encompasses water supply and sanitation infrastructure projects, and the Sub-Mission for Basic Services to the Urban Poor.[32]

Andhra Pradesh State's Expenditures: Inferring Irrigation and Urban Water Supply Policies

Andhra Pradesh has still not endorsed an exclusive water policy, though a draft is currently being worked upon. In absentia of a formal water policy, the government attitude and priorities towards water allocation can be inferred from funds allocation. The plot of Figure 3 builds upon annual financial data spanning from 1956–57 to 1996–97.[33] Actual expenditures classified under the headings of 'water supply and sanitation' and 'irrigation and flood control' are reported, as well as the percentage they represent against the total state outflow.

Data represented in Figure 3 does not specifically refer to financial outflows in urban water supply, since sanitation is also included. Moreover, expenditures are for rural and urban water supply and sanitation altogether. Notwithstanding this, the plot shows important trends in state funds allocations over 50 years under consideration. In absolute terms, between 1956–57 and 1996–97 expenditures for irrigation and flood control have been steadily increasing: till around 1982–83 at moderate pace, further on more vigorously. Expenditures for water supply and sanitation tend to show a comparatively more attenuate increase, and have been lower in absolute terms when compared to irrigation and flood control. This data suggests that irrigation has received more attention by

---

[32] Government of India, Ministry of Urban Development/Ministry of Urban Employment and Poverty Alleviation, Jawaharlal Nehru National Urban Renewal Mission, Overview. *Source:* http://urbanindia.nic.in/moud/programme/ud/jnnurm/Overview.pdf.

[33] *See* V. Hanumantharao, N. K. Acharya and M. C. Swaminathan eds, *Andhra Pradesh at 50: A Data-Based Analysis* (Hyderabad: Data News Features, 1996).

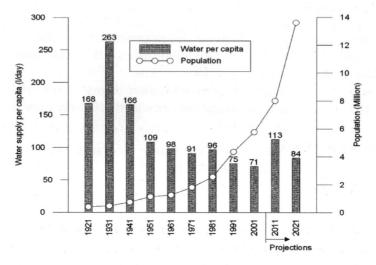

Figure 3: Water supply per capita and population growth in Hyderabad. Data-source: water supply per capita has been reckoned based on the conveying capacity of Hyderabad water sources, and accounting for an estimate of 40% water losses because of leakages and water tampering in the supply system.

policy-makers, fact explained by the high dependency of the state population on the agricultural sector and concerns related to food security. Priority on the irrigation sector is also highlighted by the percentage of expenditure for irrigation and flood control, respectively water supply and sanitation, as compared against the states' total annual financial outflow. A detailed interpretation of the trends observed in Figure 2 would require a thorough analysis of history and events in Andhra Pradesh, which would go far beyond the purpose of this article. It is nevertheless worth noting that from 1984–85 to 1994–96 the expenditures for irrigation and flood control have represented a relatively constant percentage of the total state outflow; whereas the percentage of spending for water supply and sanitation had decreased. This observation tends to reassert the importance that irrigation is given in Andhra Pradesh when compared to water supply and sanitation.

A recent and illustrative example of the paramount importance of the irrigation sector in Andhra Pradesh is brought by the Jalayagnam program promoted and carried out by the present government headed by the Chief Minister Dr. Y.S. Rajasekhara

Reddy.³⁴ The program aims at completing 30 major and eighteen medium irrigation projects in five years, at a cost of INR 460 Billion.³⁵

Hyderabad Urban Water Policies

Since the constitution of the Hyderabad Metropolitan Water Supply and Sewerage Board (HMWSSB) in 1989, a number of policies aimed at improving services to water users and promote water conservation have been introduced. Water tariff policies have been among those. In an effort to cover the operation and maintenance (O&M) costs of Hyderabad water supply and promote water conservation at the end-user level, water fees in the city have been increased. Water fees and connection charges were raised by 17 per cent in 1993, by another 25 per cent in 1997, and again by a sizable 64 per cent in 2002. It is estimated that if all the water users in Hyderabad paid their water bills, the present tariff structure would cover the O&M costs of the urban water supply system. In reality, only around 50 per cent of metered users pay for the water they use, and most of those who receive water from the HMWSSB pay flat rates still not having installed a meter on their connection.³⁶

Another water conservation policy in Hyderabad has consisted in tackling water transmission and distribution losses and water tampering (Unaccounted For Water – UFW). The UFW problem was dealt with in 1996 by setting-up a new division within the HMWSSB in charge of conducting area-wise campaigns to identify and replace damaged pipes and consumer service lines.³⁷ Notwithstanding this, present estimates indicate that UFW still ranges between 40 to 55 per cent of the water withdrawn at the sources.³⁸

---

³⁴ Information available at http://jalayagnam.org/index1.php.
³⁵ Around U$S 11.6 Billion at 3 October 2007 exchange rate.
³⁶ Ibid.
³⁷ Hyderabad Metropolitan Water Supply and Sewerage Board, 8th Annual Report 1996–1997 (1997).
³⁸ Jennifer Davis and Sunil Tanka, The Hyderabad Metropolitan Water Supply and Sewerage Board (Massachusets: Harvard University, Paper Prepared for the Kennedy School of Government, undated). *Source:* http://ocw.mit.edu/NR/rdonlyres/Urban-Studies-and-Planning/11-479Spring-2005/3AFFDD6C-7096-4CC8-85ED-89839F8BAC15/0/pm2background.pdf.

Another initiative of the Government of Andhra Pradesh for promoting water conservation and coping with ground water depletion has consisted in establishing in 1997 a Rain Water Harvesting Cell within the HMWSSB.[39]

The last and certainly most ambitious plan of the HMWSSB and of the Government has been to reuse treated wastewater generated by Hyderabad for non-drinking purposes. According to the feasibility study commissioned by the government, there is a scope for reusing 700,000 m^3 of water per day by 2031.[40] Depending upon its quality after treatment, the reclaimed wastewater would be reused by industries, for irrigated agriculture, for ground water recharge, for commercial establishments, and replenishment of existing drinking water reservoirs.

Water Law

Indian Water Law

Water law is supposed to give effect to the course of action indicated in existing water policies, or in any case to provide for regulations that respond to existing water challenges and opportunities. The following of this section is structured in two distinct parts, the first looking at surface and ground water law as specified in national legislation and constitutional provisions; the second discussing water law reforms in the state of Andhra Pradesh. For the national as well as the state-level analysis, the emphasis is put on those legal provisions having implications on water allocation between water sectors.

Surface Waters
Constitutional powers

The Constitution of India provides for the separation of powers between Central Government and states in regard to water. Notably, the right to legislate over 'Water, that is to say, water supplies, irrigation and canals, drainage and embankments, water

---

[39] Hyderabad Metropolitan Water Supply and Sewerage Board, 9th Annual Report 1997–1998 (1998).

[40] PA Consulting Group, Water Reuse Preliminary Concept and Feasibility Study (Hyderabad: PA Consulting Group, 2004).

storage and water power ...' is given to the states through the seventh Schedule of the Constitution, State List Entry 17.[41] Notwithstanding this, the states' power over water resources can be curtailed in case of disputed interstate rivers. With Entry 56 of the Union List of the seventh Schedule of the Constitution, the Central Government is notably conferred with powers to regulate and develop interstate rivers, to the extent that parliament declares it by law to be expedient in the public interest. Therefore, as most of the important rivers in India are interstate, water is potentially as much a central subject as a state subject.[42] Powers of the Central Parliament in the matter of interstate watercourses are further specified through Article 262, which rules that the parliament may by law provide for the constitution of a tribunal for the adjudication of any dispute or complaint with respect to the use, distribution or control of interstate river waters. The legislative competence of a state under Entry 17 must thus be exercised in such a manner as not to prejudice the interests of other states and create a dispute in the meaning of Article 262. Under Article 262 the Parliament further enacted the Inter State Water Disputes Act, 1956 and the River Boards Act, 1956.

Since 1994, there is a third tier in the constitutional structure established by the enactment of the 73rd and 74th constitutional amendments, that is local bodies of governance at the village and city level – the village panchayats (Art. 243G) and the city *nagarpalikas* (Art. 243W). The Eleventh and Twelfth Schedules of the Constitution lay down the lists of subjects to be devoted to the *panchayats* and *nagarpalikas*. The panchayats list includes, inter alia, minor projects, drinking water, water management, watershed development and sanitation.

Statutory provisions

In India, there is no exclusive national water law. Water-related legal provisions are dispersed across various irrigation acts, central

---

[41] Schedule VII of the Constitution of India specifies subject-matters of law made by Indian Parliament (List I, or 'Union List'), by the Legislatures of States (List II, or 'State List') and by the concurrence of both (List III or 'Concurrent List').

[42] Ramaswamy R. Iyer, *Water: Perspectives, Issues, Concerns* (New Delhi: Sage, 2003).

and state laws, constitutional provisions, court decisions, customary laws, and various penal and criminal procedure codes.

British legislation in India during 1859–1877 recognised the customary water rights of individuals and groups.[43] But a change occurred with the Easement Act of 1882, which is still today a determinant national statute defining powers over surface and ground water. The Easement[44] Act was passed under the pressure of reallocating powers over water in the late nineteenth century, which was necessitated by changes in political structures and boosted by the Industrial Revolution which made possible new technologies for water harvesting.[45] Section 2 of the Easement Act declares that states' rights over surface water are not affected by easement and customary rights and that: 'Nothing herein contained shall be deemed to affect any law not hereby expressly repealed; or to derogate from … any right of the government to regulate the collection, retention and distribution of the water of rivers and streams flowing in natural channels, and of natural lakes and ponds, or of the water flowing, collected, retained or distributed in or by any channel or other work constructed at the public expense for irrigation'. In addition to the Indian Easement Act, state governments' rights over surface water are generally further reasserted and reinforced in states' laws, notably those dealing with irrigation.[46]

Ground water

Ground water rights in India are appurtenant to land, that is contingent upon land ownership. Section 7(g) of the Indian Easement Act declares that every landowner has the right to '… collect and dispose within his own limits of all water under the

---

[43] Maria Saleth, Strategic Analysis of Water Institutions in India: Application of a New Research Paradigm (Colombo: International Water Management Institute, Research Report No. 79, 2004). *Source:* http://www.iwmi.cgiar.org/Publications/IWMI_Research_Reports/PDF/pub079/Report 79.pdf.

[44] Section 4, Easement Act, 1882 defines an easement a 'right which the owner or occupier of certain land possesses, as such, for the beneficial enjoyment of that land, to do and continue to do something'.

[45] Chhatrapati Singh, 'Water Rights in India', *in* Chhatrapati Singh ed., *Water Law in India* p. 8 (New Delhi: Indian Law Institute, 1992).

[46] Ibid.

land which does not pass in a defined channel ...'. Hence, by this Act, the owner of a piece of land does not own the ground water under the land or surface water on the land; he only has the right to collect and use the water. However, it is customarily accepted across India that a well on a piece of land belongs to the owner of that land, and others have no right to extract water from the well or restrict the landowner's rights to use the water. This belief and practice is indirectly supported by various laws such as land and irrigation Acts that list all things on which the government has a right, and ground water is not mentioned. Additionally, interpretations of the Transfer of Property Act of 1882 and the Land Acquisition Act of 1894 also support the position that a landowner has proprietary rights to ground water; that it is connected to the dominant heritage (the land), and cannot be transferred apart from it.[47]

Indian Water Law and Rights: Implications for Water Allocation between Sectors

Indian water law is determinant in shaping the manner in which water is allocated, and a fundamental difference exists between surface water on the one hand, and ground water on the other.

Surface waters are allocated by the Indian states, but only in the case of disputed inter-state watercourses. States overarching powers upon surface waters have notably permitted to massively develop water resources after India gained independence in 1947, during the so-called 'green revolution'. Holding powers over surface waters, states are also the main actors in reallocating water from agriculture to cities.

The situation for ground water is by far different, notably because of the link between land ownership and ground water rights. The most striking feature of ground water utilisation in

---

[47] Poorest Areas Civil Society, Water Law and Rights in India (New Delhi: PACS Backgrounder, Undated). *Source:* http://www.empowerpoor.com/backgrounder. asp?report=616. *See also* Maria Saleth, Strategic Analysis of Water Institutions in India: Application of a New Research Paradigm (Colombo: International Water Management Institute, Research Report No. 79, 2004). *Source:* http://www.iwmi.cgiar.org/Publications/IWMI_Research_Reports/PDF/pub079/Report 79.pdf; Chhatrapati Singh, 'Water Rights in India', *in* Chhatrapati Singh ed., *Water Law in India* (New Delhi: Indian Law Institute, 1992).

India and other Asian countries is certainly the massive spread of borewells and prosperous ground water markets. In India, the total number of mechanised wells and tube wells grew from a small fraction of a million in 1960 to some 19 million in 2000.[48] Ground water markets are particularly active in the states of Gujarat, Tamil Nadu, West Bengal, Orissa, Bihar, and Uttar Pradesh, where it is held that they provide for wider access to water as compared to situations in which water is not shared and is utilised only by the owner of the ground water structure.[49]

Though generally available in lesser proportions when compared to surface water and not tightly controlled by a central authority (namely the state), ground water can also play an important role for urban water supply. In general, ground water is supplied by tanks privately operated or temporarily or permanently contracted by urban water boards. The role of these private providers can be particularly determinant in case of protracted droughts, during which the availability of surface water dwindles. Ground water supply to Hyderabad in Andhra Pradesh during the 2003 drought is an exemplar of the crucial service that this resource can provide to cities.[50] Another case of urban dependency upon ground water is Chennai in Tamil Nadu, where farmers would sell water to private tankers that further convey it to the city.[51]

Water Law in Andhra Pradesh

Water Law Reform

Over the last decade, the government of Andhra Pradesh (AP) and its former Chief Minister Chandrababu Naidu undertook a major

---

[48] Tushaar Shah *et al.*, 'Sustaining Asia's Groundwater Boom: An Overview of Issues and Evidence', 27 *Natural Resources Forum* 130 (2003).

[49] Vikas Rawal, 'Non-Markets Interventions in Water-Sharing: Case Studies from West Bengal, India', 2 *Journal of Agrarian Change* 545 (2002).

[50] Staff Reporter, '400 Tankers to Supply Drinking Water in City', *The Hindu*, 20 April 2003. *Source:* http://www.hinduonnet.com/2003/04/20/stories/2003042008680300.htm.

[51] Joël Ruet, Marie Gambiez and Emilie Lacour, 'Private Appropriation of Resource: Impact of Peri-Urban Farmers Selling Water to Chennai Metropolitan Water Board', 24 *Cities* 110 (2007).

water law reform. In 1997, the state passed the AP Farmers' Management of Irrigation Systems Act that has handed over water management responsibilities in irrigation canal systems to associations of water users. In the same year, the AP Water Resources Development Corporation Act was enacted sanctioning the setting-up of a corporation in charge of managing water resources in an integrated manner. Finally, in 2002 was promulgated the AP Water, Land and Trees Act that addresses in particular issues of ground water as well as surface water protection, and makes provision for the constitution of an authority in charge of performing a number of functions related to the provisions spelled out in the Act. The main features of these three Acts are discussed in the three following sections.

Andhra Pradesh Farmers' Management of Irrigation Systems Act (APFMIS), 1997

By passing the AP Farmers' Management of Irrigation Systems Act (APFMIS) in 1997, Andhra Pradesh has been the first state in India to introduce major reforms in the legal system providing for handing over the management of water resources in irrigation schemes sub-sections to Water User Associations. In 1996 the government issued a white paper on irrigation that had identified the following main shortcomings in the irrigation sector – decline in net irrigated area; low irrigation system efficiencies; low yields and farmer incomes; and low agricultural growth.[52] The blame for these weaknesses was put on different factors as government dominance and limited users' involvement, poor cost recovery, insufficient operations and maintenance allocations, deteriorating conditions of the irrigation and drainage network, low quality of agricultural extension, and weak incentives for government agencies to perform.[53]

---

[52] Government of Andhra Pradesh, 1996 White Paper on Irrigation, quoted in Roopa Madhav, Irrigation Reforms in Andhra Pradesh: Whither the Trajectory of Legal Changes? (Paper presented at the Conference Water, Law and the Commons organised by the International Environmental Law Research Centre, New Delhi, 8–10 December 2006). *Source:* http://www.ielrc.org/content/w0704.pdf.

[53] Keith Oblitas, J. Raymond Peter, Gautam Pingle *et al.*, Transferring Irrigation Management to Farmers in Andhra Pradesh, India (Washington: World Bank Technical Paper 449, 1999). *Source:* http://go.worldbank.org/686KSQBYN0.

The APFMIS was seen as a solution to redress the inefficiencies in the irrigation sector. The reform was inspired by international examples in which allowing for the active participation of farmers had had a positive impact on the performance of irrigated agriculture. The involvement of farmers in countries like the Philippines, Turkey, and particularly Mexico had permitted to improve operation and maintenance of infrastructures; rate of collection of water fees; and introduced an institutional mechanism for resolving water conflict.[54] Under the APFMIS, a total of 10,292 Water User Associations (WUA) were constituted on a war footing in Andhra Pradesh's minor, medium and major irrigation projects,[55] with the support of a $US 141 million loan provided by the World Bank under the Andhra Pradesh Economic Restructuring Project.[56]

Andhra Pradesh Water Resources Development Corporation Act (APWRDC), 1997

The enactment of the AP Water Resources Development Corporation Act (APWRDC) is particularly interesting in a multi-sector perspective since the Act deals with the broad spectrum of water uses, such as the 'promotion and operation of irrigation projects, command area development and schemes for drinking water and industrial water supply to harness the water of rivers of the state of Andhra Pradesh and for matters connected therewith or incidental thereto including flood control'. The setting-up of such an organisation for managing water resources in an integrated manner fits well in the river basin development model – as the

---

[54] Peter Mollinga, 'Power in Motion: A Critical Assessment of Canal Irrigation Reform in India', *in* Rakesh Hooja, Ganesh Pangare and K.V. Raju eds., *Users in Water Management* p. 265 (Jaipur and New Delhi: Rawat Publications, 2002). *See* also Douglas Vermillion, Impacts of Irrigation Management Transfer: A Review of the Evidence (Colombo: International Irrigation Management Institute, Research Report 11, 1997). *Source:* http://www.lk.iwmi.org/pubs/pub011/REPORT11.PDF.

[55] In Andhra Pradesh, irrigation projects are classified according to the area they serve. Minor irrigation projects have a command area which is less than 2,000 ha; medium projects between 2,000 and 10,000 ha; whereas major irrigation projects are designed so as to being able to irrigate more than 10,000 ha.

[56] J. Raymond Peter, 'Irrigation Reforms in Andhra Pradesh', *in* Rakesh Hooja, Ganesh Pangare and K.V. Raju eds., *Users in Water Management* 59 (Jaipur and New Delhi: Rawat Publications, 2002).

resource becomes scarce and intersectoral tensions develop, institutions overseeing the water use in large water units such as river basins are generally created.

Andhra Pradesh Water, Land and Trees Act (APWALT), 2002

The last reform in water law promoted by the former Chief Minister Chandrababu Naidu has been the enactment of the AP Water, Land and Trees Act (APWLT) in 2002. The APWLT Act puts a strong focus on water conservation, as well as on protection of surface and ground water from pollution. Moreover, it makes provision for the constitution of a Water, Land and Trees Authority for promoting water conservation, regulating surface and ground water exploitation in the state, and advising the Government on matters related to the Act.

Water Law Reform Implications for Water Allocation between Sectors

The devolution of a set of powers over water resources to WUAs through the enactment of the APFMIS Act bears a dramatic potential for smoothing intersectoral water transfers out of the agricultural sector. Farmers' organisations can constitute a counterpart to governments for negotiating the modalities of water reallocations, as points and timing of diversion, quantities to be withdrawn, or compensations to be provided to farmers for the water loss they experience. Examples of WUA having facilitated water reallocations from agriculture are the Chilean case and particularly the transfers in the Elqui Valley,[57] or Ditch Companies[58] in the United States.[59] A major difference between the WUAs constituted in Andhra Pradesh and those active in Chile or in the United States is the water rights system under which they evolve. Notably, WUA in Chile are entitled to transact water

---

[57] Robert R. Hearne and William Easter, Water Allocation and Water Markets. An Analysis of Gains-from-Trade in Chile (Washington: World Bank Technical Paper 315, 1995). *Source:* http://go.worldbank.org/VWIE3TBW40.

[58] A Ditch Company is a private water organisation, owned by farmers' associations, and incorporated under state charger, whose main purpose is to supply members with water.

[59] Loyal M. Hartman and Don Seastone, *Water Transfers: Economic Efficiency and Alternative Institutions* (Baltimore: Johns Hopkins Press, 1970).

according to market principles and forces, a right enshrined within the 1981 Chilean water code. In the United States, in some states Ditch Companies can trade water because it is allowed and encouraged by existing legislation.[60] Conversely, farmers in Andhra Pradesh have been vested with very limited powers over water resources through the enactment of the APFMIS Act. Though allocation and distribution of water within sub-sections of irrigation systems is now under the responsibility of an elected body of farmers' representatives, water rights allocation is exercised only within the respective sub-sections. No claims can be put forth in regard to receiving any specific quantity of water from the headwork, or to exchange the resource between sub-sections. Even if farmers had water transfer rights, inefficient irrigation infrastructure would impinge upon putting them into practice. Studies on surface water markets in irrigation systems in Australia for instance show that for markets to happen, relatively efficient physical irrigation infrastructure and irrigation management must be in place.[61] This can be easily understood – for marketing a good one must know how much of it is available – or will be available – for trading. Another crucial element for efficient trading is availing of information. In irrigation systems, exchanging good information on water availability and release patterns from the main reservoirs requires relatively well working gauging systems and other measurement devices, trained staff, and good governance practices. This degree of efficiency is generally not reached in irrigation schemes in most of the developing countries.[62]

---

[60] Although water transfers are generally subject to a number of conditions, notably protecting other water users from third-party effects. *See* e.g. David H. Getches, *Water Law in a Nutshell* (Minnesota: West, 1997).

[61] Hugh Turral, T. Etchells, Hector Malano *et al.*, 'Water Trading at the Margin: The Evolution of Water Markets in the Murray-Darling Basin', 41 *Water Resources Research* (2005) and Chris J. Perry, Michael Rock and David Seckler, Water as an Economic Good: A Solution, or a Problem? (Colombo, Sri Lanka: International Water Management Institute, Research Report 14, 1997). *Source:* http://www.iwmi.cgiar.org/Publications/IWMI_Research_Reports/PDF/PUB014/REPORT14.PDF.

[62] Robert Wade, 'The Information Problem of South Indian Irrigation Canals', 5 *Water Supply and Management* 31 (1981) and Robert Wade, 'The System of Administrative and Political Corruption: Canal Irrigation in South India', 18 *Journal of Development Studies* 287 (1982).

In addition to the APFMIS Act, the constitution of a Water Resources Development Corporation in 1997 through the Andhra Pradesh Water Resources Development Corporation Act might have provided the opportunity to constitute an organisation capable of regulating or providing advice on intersectoral water allocation. Unfortunately, the Water Development Corporation is not explicitly entrusted with the function and powers to plan intersectoral water allocation in an integrated manner, but only to look at each of the sectors separately. Though there is little written evidence allowing for the evaluation of the works of the corporation and of the effectiveness of the APWRDC Act, the existing literature suggests that the legislation has not been as strong in including transparency and accountability in its performance; nor in providing incentives and disincentives to staff and water users (in all sectors) to enhance water use efficiency.[63] The Act is not clear about water rights and emphasises controlling extraction only of surface water, and ground water is untouched. Furthermore, instead of setting up an independent body and actually delegating some of the government responsibilities on water allocation, the APWRDC Act further strengthens the role of the state, since the members of the corporation either belong to – or are elected by – the state government. In this regard, it is interesting to report the case of water law reforms in the neighbouring state of Maharashtra, where in 2005 was passed the Water Resources Regulatory Authority Act that provides for the constitution of a Water Regulatory Authority (WRA). Unlike in Andhra Pradesh, the membership structure of the WRA excludes political leaders, so as to try to reduce political interferences and government influence.[64] Among the powers, functions, and duties of the WRA there is the clear specification that it has to 'determine the distribution of entitlements for various categories of use'[65] including domestic, agricultural, irrigation, industrial and commercial.

---

[63] Ashok Gulati, Ruth Meinzen-Dick and K.V. Raju eds., *Institutional Reforms in Indian Irrigation* (New Delhi: Sage Publications, 2005).
[64] For a discussion on water law reforms in Maharashtra and Andhra Pradesh, see Philippe Cullet, 'Water Law Reforms. Analysis of Recent Developments', 48 *Journal of the Indian Law Institute* 206 (2006).
[65] Government of Maharashtra, Maharashtra Water Resources Regulatory Authority Act, 2005, Chapter 3, Section 11(a). *Source:* http://www.leadjournal.org/content/05080.pdf.

The Andhra Pradesh Water, Land and Trees Act of 2002 only tangentially accounts for the regulation of intersectoral water use, notably when it gives priority to drinking water through banning the sinking of tube wells within a pre-established distance from underground drinking water sources.[66] It has thereby only a very marginal relevance for intersectoral water allocation.

Water Administration

The Hyderabad Water Supply and Sewerage Board

The Hyderabad Metropolitan Water Supply and Sewerage Board (HMWSSB – hereafter also referred to as 'Water Board') established in 1989 is in charge of Hyderabad water supply and sanitation.[67] The HMWSSB was constituted under the Hyderabad Metropolitan Water Supply and Sewerage Act by consolidating two existing public departments, namely the Public Health Engineering Department formerly in charge of water supply, and the Municipal Corporation of Hyderabad which was responsible for the sewerage services. The constitution of the Water Board, encouraged by the World Bank, was seen as a means of establishing a water and sewerage authority with great financial autonomy, as well as heightened accountability to the consumers.[68] Financial autonomy was notably perceived as a means to entrust the HMWSSB with great operational and decision-making control, thus insulating it from political interferences. In practice, it has never been able to achieve financial autonomy, largely because of extremely low rates of water fees collection. The Water Board eventually became highly dependent on – and controlled by – the government. This dependence on the political establishment is further accentuated

---

[66] Government of Andhra Pradesh, Andhra Pradesh Water, Land and Trees Act, 2002, Chapter 3, Section 10(1). *Source:* http://www.ielrc.org/content/e0202.pdf.

[67] Website of the Hyderabad Metropolitan Water Supply and Sewerage Board. *Source:* http://www.hyderabadwater.gov.in.

[68] Jennifer Davis and Sunil Tanka, The Hyderabad Metropolitan Water Supply and Sewerage Board (Massachusets: Harvard University, Paper Prepared for the Kennedy School of Government, undated). *Source:* http://ocw.mit.edu/NR/rdonlyres/Urban-Studies-and-Planning/11-479Spring-2005/3AFFDD6C-7096-4CC8-85ED-89839F8BAC15/0/pm2background.pdf.

by the fact that the Chief Minister of Andhra Pradesh acts as the chairman of the Board of Directors of the HMWSSB.

Administrative Arrangements for Allocating Water from Agriculture to Hyderabad

Water supply is responsibility of the HMWSSB by virtue of the Hyderabad Metropolitan Water Supply and Sewerage Act of 1989 which posits in particular the following: 'On and from the date of coming into force of this chapter, all public reservoirs, tanks, cisterns, fountains, wells, and bore wells, pumps, pipes, taps, conduits and other works connected with the supply of water to the Hyderabad Metropolitan area, including the head work, reservoirs ..., shall vest in the Board and be subject to its control'.[69] Contrasting this statement is the actual situation, in particular in regard to the control over the reservoirs that feed the city, notably Singur on the Manjira River and Nagarjunasagar on the Krishna. Both reservoirs are multipurpose and serve Hyderabad, agriculture, and hydropower. The operation of Singur and Nagarjunasagar is under the responsibility of the Irrigation and Command Area Development (CAD) Department of Andhra Pradesh, the rationale being that the Water Board does not have the know-how to operate water reservoirs, a task traditionally entrusted to the Irrigation and CAD Department. Not controlling the operation of the reservoir has eventually curtailed the control of the Water Board upon Hyderabad water sources.

The case of Singur reservoir on the Manjira River is illustrative. The dam is operated by the Irrigation and CAD Department, though it serves Hyderabad and then agriculture on a priority basis. Water sharing between the two sectors is governed by a set of allocation rules laid down by the government through two administrative orders issued in 1989,[70] respectively in 1990.[71] Each and every

---

[69] Government of Andhra Pradesh, Hyderabad Metropolitan Water Supply and Sewerage Act, 1989, Chapter IV, Section 17.
[70] Andhra Pradesh Housing, Municipal Administration and Development (A2) Department, Government Order Ms No. 190, 12 April 1989.
[71] Andhra Pradesh Irrigation and CAD (Irrgn. V) Department, Government Order Ms No. 93, 24 February 1990.

release of water from Singur for agricultural purposes has to be sanctioned against abidance by the allocation rules, and has to be jointly agreed upon by the Principal Secretary for Irrigation and the Managing Director of the HMWSSB.[72] In case of discordance of opinion between Principal Secretary and Managing Director, the matter is brought to the attention of the Chief Minister who eventually takes the decision. In an attempt to establish its authority over Singur, the HMWSSB has put forth some proposals to the Government for being entrusted with the control of the dam, but these have been turned down because of the stiff opposition of the Irrigation and CAD Department.[73]

A comparable situation in which control over Hyderabad water sources is shared between the Water Board and the Irrigation and CAD Department is the transfer of water from the Krishna River, presently carried out by utilising pre-existing structures that are under the control of the Irrigation and CAD Department, and notably the so-called Akkampally balancing reservoir. The HMWSSB is now trying to obtain government approval and financial assistance for constructing an independent structure over which it would have total control.[74]

The duality in the administrative set up that characterise Hyderabad water supply had consequences for the allocation of water from agriculture to the city. By mandate, the Water Board and the Irrigation and CAD Department have divergent interests, and 'clients': the former has to ensure sufficient water supply to Hyderabad water users, whereas the latter to provide farmers with the maximum water possible so as to satisfy crops requirements. Though not proved with empirical evidence, one can expect that the existing organisational configuration has led to tensions and internal struggles between Water Board and Irrigation and CAD Department.

---

[72] Mattia Celio and Mark Giordano, 'Agriculture-Urban Water Transfers: A Case Study of Hyderabad, South-India', 5 *Paddy and Water Environment* 229 (2007).
[73] HMWSSB Staff, Personal Communication, 14 April 2006.
[74] Ibid.

Discussion and Conclusions: Water Institutions and Water Reallocation from Agriculture to Hyderabad in Andhra Pradesh

As stated by Williamson, water policy prescriptions have moved from 'getting the prices right', to 'getting the property rights right', and eventually to 'getting the institutions right'.[75] Water law, policy and administration are pivotal since they provide statutory or customary rules, formal and informal guidelines, as well as the administrative scaffolding for managing water resources. The institutional approach to water management is also reflected in recent trends towards the integrated management of water resources, an approach best exemplified by the Integrated Water Resources Management (IWRM) concept. Particular emphasis is given to river basin-wide approaches, active participation of the water users in water-related decision-making processes, and organisations for administering water use at different scales and preventing and resolving water conflicts.[76]

The study of Hyderabad water supply has shed light on multiple aspects of water institutions and reallocation. Water institutions vest the State Government of Andhra Pradesh with overwhelming powers over surface water resources. In an opposing trend to water law reforms undertaken in other countries grounded upon the IWRM principle,[77] an outstanding characteristic of the reform promoted by the government in Andhra Pradesh is the refusal to effectively relinquish water rights to water users. The constitution of Water User Associations through the enactment of the Andhra Pradesh Farmers' Management of Irrigation Systems Act in 1997 cannot be taken as an example of effective devolution

---

[75] Oliver E. Williamson, Institutions and Economic Organization: The Governance Perspective (Paper presented at the Annual Bank Conference on Development Economics organised by the World Bank, Washington, 1994), quoted in Maria Saleth and Ariel Dinar, *The Institutional Economics of Water: A Cross-Country Analysis of Institutions and Performance* (Cheltenham: Edward Elgar, 2004).

[76] Asit K Biswas, Olli Varis and Cecilia Tortajada eds, *Integrated Water Resources Management in South and South-East Asia* (New Delhi: Oxford University Press, 2005).

[77] For South Africa, see e.g. Desheng Hu, *Water Rights: An International and Comparative Study* (London: IWA Publishing, 2006).

of powers. Water management by farmers only applies to sub-distributaries of irrigation schemes and the scope for transacting water is very limited.[78] Also the set-up of a Water Resources Corporation in 1997 via the enactment of the Water Resources Development Corporation Act has not veritably represented a devolution of powers, since the members of the corporation are all appointed by – or representative of – the Government. Such a centralised system that excludes users' participation in decision-making and implementation of water projects has created the basis for the political opposition against the reallocation of water to Hyderabad from the agricultural sector, and can be considered as one of the factors explaining the low efficiency in the utilisation of water resources in the state.[79] For the Government, it has also been difficult to make decisions in regard to water reallocations to Hyderabad from agriculture because of the poor guidance provided by the 2002 National Water Policy (NWP). Though the NWP specifies water use priorities and makes reference to intersectoral water planning and utilisation, this article has shown that these provisions do not make up a solid ground to policy makers for justifying water transfers to cities.

The analysis of public expenditures in Andhra Pradesh from 1956–57 to 1996–97 done in this paper provides for interesting insights on water sharing between agriculture and urban areas. Expenditure data shows that over the period considered water supply and sanitation has not received so much of attention when compared against the irrigation and flood control sector. One reason for this is that 60 per cent of the population in Andhra Pradesh is employed in the agricultural sector.[80] Pro-irrigation water policies economically sustain a large share of the state population, secure food production, and indirectly provide for electoral support to the standing government. Lacking of a clear policy for transferring water from agriculture, the Government of Andhra Pradesh has

---

[78] Rakesh Hooja, Ganesh Pangare and K.V. Raju eds., *Users in Water Management* (Jaipur & New Delhi: Rawat Publications, 2002).

[79] On the low efficiency of irrigation, *see* e.g. V. Ratna Reddy, 'Irrigation: Development and Reforms', *in* C.H. Hanumantha Rao and S. Mahendra Dev eds, *Andhra Pradesh Development – Economic Reforms and Challenges Ahead* p. 170 (Hyderabad: Centre for Economic and Social Studies, 2003).

[80] According to the 2001 Census of India.

introduced a number of demand-oriented water initiatives in Hyderabad. Notably, water fees have been successively increased, rainwater harvesting was done mandatory for new buildings, and water tampering and leakages in the distribution system have been addressed. Though there is no evidence at hand on the actual impact of these measures, the grim situation of Hyderabad water supply suggests that their scope has been limited. Even if the last proposal of the Government in date to reuse treated wastewater will be actually implemented, Hyderabad will still need water from the agricultural sector. It was reckoned that even with the maximum reuse of wastewater, supply Hyderabad with water would engender an average net loss of irrigated agriculture of 40,000 ha by 2030.[81]

This chapter has also shown that administrative responsibilities over Hyderabad water supply are partially ambiguous. Management responsibilities and control over the city water sources are notably shared between the Hyderabad Metropolitan Water Supply and Sewerage Board (HMWSSB) on the one side, and the Irrigation and CAD Department on the other, though the former should hold overall responsibilities according to the provisions of the Hyderabad Metropolitan Water Supply and Sewerage Act (1989). Because of this situation of partially overlapping responsibilities, in times of drought water sharing between agriculture and Hyderabad at the Manjira and the Krishna rivers is subject to harsh negotiations, in which the intervention of the Chief Minister is often required.

The findings presented in this article have clearly highlighted the limitations and gaps of the existing institutional framework in Andhra Pradesh for transferring water from agriculture to urban areas. If one considers the growing need of cities for water and the prevailing conditions of water scarcity in south India and other parts of the world, it becomes clear that reforming institutions so as to allow for smoother and more efficient intersectoral water transfers should be among the top priorities of policy makers in the next decades.

---

[81] Daan J. van Rooijen, Hugh Turral and Trent Wade Biggs, 'Sponge City: Water Balance of Mega-City Water Use and Wastewater Use in Hyderabad, India', 54 *Irrigation and Drainage* 1 (2005).

UNIT IV

Environment and Human Rights

12

Balancing Development and Environmental Conservation and Protection of the Water Resource Base: The 'Greening' of Water Laws

STEFANO BURCHI[1]

1. Introduction

Many references to the term 'sustainable development' pervade discourses on the management of natural resources, particularly the diminishing supply of freshwater reserves worldwide. This entails reconciling the seemingly different goals of socio-economic development and environmental protection and conservation as essentially two sides of the same coin. Ultimately, efficient development strategies are those that sufficiently consider the finite nature of the water resource base and its dependent ecosystems. Addressing this concern, reforms in the water sector have increasingly mainstreamed environmental considerations into the elaboration of new laws or the review of existing legislation. This paper elucidates how environmental concerns are contained, reflected or given prominence within national water laws through

---

[1] The views expressed in this chapter are personal to the author and do not in any way engage, or reflect the official position of the Food and Agriculture Organisation of the United Nations. The author would like to gratefully acknowledge the contribution of Ms Ambra Gobena.

a number of regulatory and other mechanisms – essentially, it examines the 'greening' of modern water laws.

The mechanisms examined in detail in this paper include minimum environmental flows of rivers, environmental impact assessment (EIA) requirements, the national 'Reserve' and protected areas and zones, environmental water trades and water trusts, ecosystem service payment schemes and the specific protection of aquifers in recognition of their ecosystem support function. This list is not exhaustive of the ways in which environmental protection is accommodated in the water law framework; indeed, many provisions contained in legislation which regulate water use incorporate the protection and conservation of surface or underground water bodies to some degree.

The first section of the chapter provides an overview of the manner in which international water instruments address environmental issues, before taking note of the varied environmental and conservationist approaches integrated into national legislation, which are of a more general nature. The selected regulatory mechanisms highlighted above are then discussed in detail in the subsequent section. This is followed by a short discussion that traces the challenges in the legislative water reform process in Spain, and demonstrates the importance of incorporating an environmental dimension into water policy and the legislative drafting process.

2. Environment-friendly Provisions in Water Treaties and Laws

2.1 International Trends

As a starting point, a quick overview of international trends and the way in which environmental conservation provisions are framed at the international level is useful, even though these instruments often concern transboundary and shared water sources. Nevertheless, the principles can influence and guide the direction taken by national law-makers and are reflected in national legislation. Only a few water regulation treaties have been selected for discussion for the sake of brevity, but it should be borne in mind that many other international and regional treaties embrace (but are not restricted to) water management. International texts, it will be seen, often provide framework or skeletal provisions leaving a wide berth for states to apply such provisions through an appropriate mechanism. Often, this flexibility is provided by wordings such as 'state parties shall take all necessary measures', but this can be disadvantageous as the provisions are too general in nature. While water protection laws are by their very nature designed to protect water sources, in the past, water management was dominated by economic, technological and infrastructural priorities. Increasingly, both at international and national level, the focus has deviated somewhat to accommodate ecological and not solely human needs, with a greater emphasis on the protection of water sources, given their role in supporting aquatic and related ecosystems. These ecological needs have meant that the environment itself has become a consideration as a water 'user'. This shift in discourse and focus of water law and policy is in accordance with integrated water management principles, and it has been seen that, increasingly, water laws encompass a range of factors and issues that were not traditionally found within the realm of water management.

'Equitable and reasonable utilisation' was first enshrined in the Helsinki Rules on the Use of Waters of International Rivers 1966, laid down by the International Law Association (ILA) (the Helsinki Rules), and reiterated in subsequent texts on water management. This term is also included in Article 5(1) of the draft United Nations Convention of the Law of the Non-navigable Uses of International Watercourses 1997 (the UN Convention).[2] Although the UN Convention is still subject to ratification and therefore, strictly speaking, it is not binding, it has become one of the primary sources for water management on the international plane. Its tenets are mirrored in numerous other regional and bi-lateral texts.

Although environmental protection is not the main thrust of the UN Convention, this consideration has been accommodated in the text. The factors relevant to determining equitable and reasonable utilisation *inter alia* include 'the ecological nature of the water source' and the 'conservation protection, development and

---

[2] Convention on the Law of the Non-navigational Uses of International Watercourses, New York, 21 May 1997, *reprinted in* P. Cullet and A. Gowlland-Gualtieri eds., *Key Materials in International Environmental Law* (Aldershot: Ashgate, 2004), 481.

economy of use of the water resources of the watercourse and the costs of measures taken to that effect'.[3] A short statement of obligation to protect and preserve the ecosystems of international watercourses is found in Article 20 and a prohibition on the introduction of alien or new species which may detrimentally affect the ecosystem of the watercourse is included in Article 22. More than ten years later, in the recent Articles on the Law of Transboundary Aquifers (the Articles),[4] the environmental protection element is somewhat stronger than in the UN Convention, even though a similar structure has been adopted. Article 3 strikes a compromise between State sovereignity and the exercise of it in transboundary aquifer circumstances.

In the Articles, the equitable and reasonable utilisation standard means that among other considerations, the role of the aquifer in the related ecosystem must be taken into account.[5] States are also required to regularly exchange readily available data and information of an ecological nature, among other types of data, on the condition of the transboundary aquifer or aquifer system.[6] States should also identify recharge and discharge zones of their transboundary aquifer or aquifer system and minimise any potential detrimental effects on these processes.[7] One of the types of cooperation between developed and developing states, through competent international organisations where necessary, is the preparation of environmental impact assessments.[8]

Regional instruments also echo these provisions and the two instruments, which have been selected for analysis in this chapter, enjoy legally binding force. The United Nations Economic Commission for Europe (UNECE) Convention on the Protection and Use of Transboundary Watercourses and International Lakes 1992 (UNECE Convention) mandates states to ensure that transboundary waters are used with the 'aim of ecologically sound and rational water management, conservation of water resources and environmental protection', and to ensure the 'conservation

---

[3] Ibid., Articles 6 (1) (a) and 6 (1) (f).
[4] *See* UN General Assembly Resolution on the Law of Transboundary Aquifers, UN doc. A/RES/63/124 (2009).
[5] Ibid., Article 5(1)(i).
[6] Ibid., Article 8(1).
[7] Ibid., Article 11.
[8] Ibid., Article 16.

and, where necessary, restoration of ecosystems'.[9] The importance of ensuring sustainability is incorporated in Article 2(5)(b), which directs that water resources be managed in such a way as to meet the needs of the present generation 'without compromising the ability of future generations to meet their own needs'. The UNECE Convention proposes various measures for the protection of water resources, such as environmental impact assessments and sustainable water resources management through the ecosystems approach, and it promotes the development of environmentally sound technologies, and production and consumption patterns.[10] In Annex II, the instrument fleshes out guidelines containing graduated measures for the development of best environmental practices that are mostly related to maintaining good water quality and preventing water pollution.

The objectives of the Southern African Development Community's Protocol on Shared Watercourses 1995 (the SADC Protocol) clearly elucidate the balancing act between development, sustainability and environmental goals. It adopts the equitable and reasonable utilisation approach but injects the concept of sustainability. It also advocates integrated, environmentally sound development and management of shared watercourses. Equitable and reasonable utilisation *inter alia* entails social, economic and environmental needs of the states concerned, and 'the conservation, protection, development and economy of use of the water uses of the shared watercourse and the costs of measures taken to that effect'.[11] It more explicitly frames the equilibrium to be achieved thus,

> State Parties shall maintain a proper balance between resource development for a higher standard of living for their people and

---

[9] Article 2(2)(b) and (d), Convention on the Protection and Use of Transboundary Watercourses and International Lakes, Helsinki, 17 March 1992. *Source:* http://www.ielrc.org/content/e9210.pdf.

[10] Ibid., Articles 3 and 5.

[11] Article 3(8), Protocol on Shared Watercourses in the Southern African Development Community Region, Johannesburg, 28 August 1995. The Protocol was subsequently replaced by the Revised Protocol on Shared Water Resources in the Southern African Development Community, Windhoek, 7 August 2000. *Source:* http://faolex.fao.org/docs/texts/mul15902.doc/

conservation and enhancement of the environment to promote sustainable development.[12]

The SADC Protocol contains standard provisions on the duty to protect and preserve the ecosystems of a shared watercourse, prohibits the introduction of harmful alien or new species and includes a general stipulation to take measures to protect and preserve the aquatic environment.[13]

While only international texts that are legally binding require compliance and adherence by states, the shape and direction of trends in water management is often fleshed out to a greater extent in 'soft' law (non-binding) instruments such as declarations and statements of principles. These instruments often adopt a bolder approach and serve to advocate important new concepts; these types of texts may eventually achieve recognition and voluntary adherence which renders their contents to be of a customary law nature.

The Berlin Rules on Water Resources 2004 put forward by the ILA (the Berlin Rules) are an example of the collective embodiment of the current concepts of sustainable development, environmental protection and integrated water management techniques.[14] The Berlin Rules promote 'conjunctive management', which denotes a holistic approach and integration of underground, surface and atmospheric management of water sources (Article 5), and advocate the management of other natural resources along with water (Article 6). This is integrated water management from a narrow environmental perspective. The Berlin Rules also endorse the fundamental tenet of sustainable management (Article 7). States are required to take all appropriate measures to prevent 'environmental harm', which is to include '(a) injury to the environment and any other loss or damage caused by such harm; and (b) the costs of reasonable measures to restore the environment actually undertaken or to be undertaken'.[15]

---

[12] Ibid., Article 3(4).
[13] Ibid., Article 4(2).
[14] *See* Berlin Rules on Water Resources, *in* International Law Association, Report of the Seventy-first Conference (2004). *Source:* http://www.asil.org/ilib/WaterReport2004.pdf.
[15] Ibid., Article 3(8).

Chapter V is dedicated to the protection of the aquatic environment and contains provisions aimed *inter alia* at the protection of 'ecological integrity' (and the flows necessary to protect the ecological integrity (Article 24), which refers to the 'natural condition of waters and other resources sufficient to assure the biological, chemical, and physical integrity of the aquatic environment',[16] which are necessary to sustain ecosystems dependent on particular waters. It is noteworthy that in Article 23(2), the Berlin Rules promote the precautionary principle, which has become a staple of environmental laws. Thus, States are required to prevent, eliminate, reduce, or control harm to the aquatic environment when there is a serious risk of significant adverse effect or to the sustainable use of waters even without conclusive proof of a causal relation between an act or omission and its expected effects.

Chapter VI is devoted to impact assessments – the obligation to examine expected effects on the environment and sustainability of water use of a particular activity, rights of parties to participate in procedures related to impact assessments in other states, the delineation of some of the elements for such a process, and the parameters of responses to emergency situations involving the environment.

The wide scope of the Berlin Rules indicates that integrated water management encompasses human rights principles, development priorities and protection of the environment. Various definitions of integrated water management exist – a narrower interpretation in the environmental context refers to management at the river basin level taking into account non-aquatic resources. A broader view of integrated water management incorporates an underpinning principle of sustainable development, which itself implies the equal prioritisation of economic, social and environmental exigencies. The term 'Integrated Water Resources Management' (IWRM) has been defined as 'a process, which promotes the coordinated development and management of water, land and related resources in order to maximise the resultant economic and social welfare in an equitable manner without compromising the sustainability of vital ecosystems'.[17]

---

[16] Ibid., Article 3(6).
[17] Technical Advisory Committee of the Global Water Partnership, Johannesburg World Summit on Sustainable Development (WSSD) 2002.

The degree of successful incorporation of sustainable development and integrated water management techniques can be gathered from the increasingly broad scope of primary water laws and the range of social economic and environmental issues that have been included therein. Another balancing act required by legislative drafters is ensuring that the legal text produced is precise and comprehensive – detailing mechanisms to be adopted, the rights and obligations of various state or non-state actors and the functions and responsibilities of institutions – while still allowing for flexibility in implementation and enforcement. For example, the Chilean water law reform has been heralded as an example of a step by step approach tailored to a country's current stage of development, with adaptable strategies that facilitate the identification and correction of problems according to changing circumstances.[18] Progress has been made in the Chilean context towards 'economic efficiency, social equity and environmental sustainability' referred to as the three 'Es of integrated water management'.[19]

The Dublin Statement on Water and Sustainable Development 1992 (the Dublin Statement) buttresses the argument for integrated water management and its guiding principles recognise the importance and vulnerability of water as a resource, placing sustenance of life, social and economic development and the environment on equal footing.[20] Principle No. 4 of the Dublin Statement, that is 'water has an economic value in all its competing uses and should be recognised as an economic good', highlights the theme of this chapter, and stipulates that:

> [p]ast failure to recognise the economic value of water has led to wasteful and environmentally damaging uses of the resource. Managing water as an economic good is an important way of

---

[18] Sandy Williams and Sarah Carriger, Water and Sustainable Development: Lessons from Chile, Global Water Partnership (GWP) Technical Committee (TEC), Policy Brief No. 2 (2006), p. 5, 6. *Source:* http://www.gwpforum.org/gwp/library/Policybrief2Chile.pdf.

[19] Ibid., p. 1.

[20] *See* Dublin Statement on Water and Sustainable Development, International Conference on Water and the Environment, Dublin, 31 January 1992. *Source:* http://www.ielrc.org/content/e9209.pdf.

achieving efficient and equitable use, and of encouraging conservation and protection of water resources.

It notes the importance of the role of water for the environment and for humans, and in this regard, it advocates the integrated management of river basins, which provides 'the opportunity to safeguard aquatic ecosystems, and make their benefits available to society on a sustainable basis'.

2.2 General Environmental Provisions in National Legislation

Environment-friendly provisions can be found in a range of sectors other than water and broad environmental laws, for example, land use planning and control, forestry and agriculture laws to name a few. However, the primary purpose of the laws analysed in this chapter, for the most part, is the regulation of water resources. There are several types of laws that can be employed by a national legal system, and a hierarchy of norms can be created depending on the type of law used. Non-derogable rights and obligations are found in national constitutions which are recognised as the supreme law of the land, and many countries have included the right to a clean and safe environment and water for its citizens in their constitutions.

The instruments examined in this chapter are predominantly primary water laws enacted by the legislature and are referred to as statutes, acts or codes. These primary texts are often fleshed out and implemented by subordinate legislation such as decrees, regulations and similar by-laws. While these latter types of laws are 'lower' in the hierarchy, they are nevertheless crucial to the implementation of primary laws as the details of managing institutions and in-depth directions on rights, obligations and processes are found here.

The trend towards integrated water resources management has meant that water protection provisions can be found in legislation governing other natural resources such as forest or land laws in addition to, or independent of, water resources legislation. The Andhra Pradesh Water, Land and Trees Act 2002[21] stipulates that

---

[21] India, Andhra Pradesh Water Land and Trees Act, Act No. 10 of 2002. *Source:* http://www.ielrc.org/content/e0202.pdf.

local authorities may formulate guidelines for landscaping and tree planting along canal banks and water bodies, and must ensure tree plantation in the 'fore-shore area of open water bodies'.[22] Tree felling or branch cutting in these areas is subject to permit conditions.[23] Similarly, land development and construction plans may be subject to vetting by water resource authorities who regulate and monitor activities, which may potentially damage nearby water bodies. In Florida, USA, land development activities such as the construction of buildings and roads, which exceed a prescribed size, fall within the remit of the 'environmental resource permitting' program.[24] This program which entails state-wide monitoring and control of activities which alter the surface flow of waters, receives statutory backing in the Florida Administrative Code.[25] An applicant for an approval permit must provide reasonable assurance that the activity will not cause adverse water quantity or quality impacts, nor 'adversely impact the value of functions provided to fish and wildlife and listed species by wetlands and other surface waters'.[26]

Anti-pollution provisions are also included in almost all water legislations as a crucial instrument for the protection of water quality, but the objective of this chapter is to elucidate how specific environmental purposes are factored into the law in addition to the protection of the water body itself. Statutes give effect to a range of strategies with a conservationist design based on local and national priorities. National water programs and planning, with which all government decisions must generally comply, increasingly take into account environmental protection exigencies, having gone through processes of public consultation and liaison between water management committees and other government bodies (including environmental agencies). The New South Wales (Australia) Catchment Management Authorities Act 2003 provides that the water use provisions of the master plan must ascertain the uses or activities which adversely affect water sources or their dependent ecosystems, and also land-based activities with negative

---

[22] Ibid., section 30.
[23] Ibid., section 28.
[24] *See* Joelle Hervic, 'Water, Water Everywhere?', 77/1 *Florida Bar Journal* 49 (2003).
[25] Florida Administrative Code, Rule Number 40C-4.301.
[26] Ibid., Rule Number 40C-4.301 §1(e).

ramifications for adjacent water bodies such as soil erosion, increase in salinity, or clearing of vegetation.[27] Environmental protection priorities are often underscored in the fundamental or guiding tenets of many water management laws, mandating that these principles are to be borne in mind in the reading, construction and implementation of the law. A case in point is the water management principles of the New South Wales (Australia) Water Management Act 2000 (the NSW Water Management Act),[28] which mandate that:

> 'water sources floodplains and dependent ecosystems (including groundwater and wetlands) should be protected and restored ...; habitats, animals and plants that benefit from water, or are potentially affected by managed activities should be protected and (in the case of habitats) restored; ... the cumulative impacts of water management licences and approvals and other activities on water sources and their dependent ecosystems, should be considered and minimised'.

Also, section 3 of the South Africa National Water Act 1998 (the South African NWA) elucidates the necessary balance that must be struck between development and environment protection goals:

> 3(1) As the public trustee of the nation's water resources the National Government, acting through the Minister, must ensure that water is protected, used, developed, conserved, managed and controlled in a sustainable and equitable manner, for the benefit of all persons and in accordance with its constitutional mandate.
>
> (2) Without limiting subsection (1), the Minister is ultimately responsible to ensure that water is allocated equitably and used beneficially in the public interest, while promoting environmental values.[29]

---

[27] Article 23(b) and (c), Australia New South Wales Catchment Management Authorities, Act 104 of 2003.

[28] Australia New South Wales Water Management Act, Act No. 92 of 2000 [hereafter the NSW Water Management Act].

[29] South Africa National Water Act, Act No. 36 of 1998 [hereafter South Africa, NWA].

Moreover, the NSW Water Management Act directs that the health of water systems – surface and underground – must come first in the allocation of available water resources for competing uses. This priority translates into a statutory obligation to meet the basic environmental requirements of designated water bodies, and into other complementary mechanisms earmarking water resources to meet the environmental requirements of other water bodies, based on a ranking of the state's water resources according to the level of environmental stress or risk and conservation value.[30]

3. Selected Mechanisms

3.1 Minimum Environmental Flow Requirements of Rivers

Minimum flow refers to the least amount of water required within a watercourse, which is necessary to maintain water quality and the survival of dependent ecosystem varieties. The statutory requirement of a protected minimum 'environmental' flow is most frequently used, where an explicit reference is made to the need to maintain the minimum flow requirement of watercourses, for example, to maintain fish populations and the health of riverine ecosystems.[31] Progressively, water laws have emphasised the importance of maintaining this minimum quantity largely through monitoring and regulating the number and volume of abstractions. In this regard, environmental minimum flow requirements of rivers may be given priority over available river flows.[32]

In certain countries, laws may stipulate the actual percentage of minimum flow requirements. For example, the Chilean legislation notes that this figure should not be greater than twenty percent of the average annual flow, or in exceptional cases as set by the President, not more than forty percent of the average annual flow.[33] Under this law, minimum requirements will only affect permits granted *after* the establishment of standard percentages.

---

[30] Australia New South Wales Water Management Act, Act No. 92 of 2000 [hereafter the NSW Water Management Act], section 8.

[31] Article 64, Kyrgyzstan, Water Code Law No. 8 of 2005.

[32] Article 26(1), Spain Law Concerning the National Water Master Plan, Law No. 10 of 2001 [hereafter Spain Law].

[33] Article 129bis 1, Chile Law Amending the Water Code Law No. 20.017 of 2005.

Considering stream flows vary naturally along the watercourse and at different times of the year, certain laws may legislate on the minimum flow for each individual stream type. The Swiss Federal Law on the Protection of Waters 1998 prescribes water protection targets and minimum flow figures for different average flow rates which take into account the ecological function of the water bodies.[34] While regulations at the federal level establish minimum flows, the cantons may flesh out these provisions depending on geographic, economic and ecological factors.[35]

3.2 Environmental Impact Assessments (EIAs)

Permits and licences are the primary instruments of water use regulation, with environmental considerations frequently injected into licensing criteria. Examples would include technical ecological evaluation or environmental impact analyses before the issuance of a permit; authorisations based on considerations such as the annual average availability of water, and existing water rights and uses;[36] and, in relation to waste disposal, setting levels for chemical and physical components, its volume and specifying requisite treatment measures.[37] In Namibia, the consideration of environmental effects is one of the specified criteria for successful applications for water abstraction and effluent discharge (taken into account when processing applications), and also forms part of the terms and conditions of such concessions.[38]

---

[34] Megan Dyson, Ger Bergkamp and John Scanlon eds., *Flow: The Essentials of Environmental Flows* (Gland: IUCN, 2003). *Source:* http://www.iucn.org/dbtw-wpd/edocs/2003-021.pdf.

[35] Stefan M.M. Kuks, The Evolution of National Water Regimes in Europe: Transitions in Water Rights and Water Policies (Paper presented at the Conference entitled Sustainable Water Management: Comparing Perspectives from Australia, Europe and the United States organised by the National Europe Centre at The Australian National University, Canberra, 15–16 September 2005).

[36] Article 22, Mexico, Law on National Waters of 1992 as amended by Decree of the President of the Republic on 29 April 2004.

[37] Section 29, South Africa, NWA, note 29 above.

[38] Sections 33, 34, 35 and 37, Namibia, Water Resources Management Act, Act No. 24 of 2004 [hereafter Namibia, WRM Act].

EIAs are most commonly found as part of the statutory set-up with respect to granting concessions for water use (most frequently, surface and ground water abstractions and waste disposal). Several national water laws follow the trend of including EIAs as prerequisites for licenses. In Cameroon, applications for water abstraction authorisations must be accompanied by an impact study of such proposed use, together with conclusions from the agency responsible for the environment.[39] Permits for the discharge of wastewater, as well as development and abstraction concessions, must be applied for together with an environmental impact statement, also prepared under the relevant environment protection legislation under the Mexican legislation.[40]

Similarly, the Kenyan statutory framework outlining the procedure for obtaining permits stipulates that an environmental impact assessment shall be carried out in line with relevant provisions detailed in the Environmental Management and Coordination Act 1999.[41] Such studies are also used in a number of other water-related areas. For example, under the Chinese Water Law of 2002, review and approval of an impact assessment report must be completed before the construction of sewerage outfall projects.[42] In South Africa, consultation and environmental impact assessment must be prepared before the construction of waterworks, the report summary of which must be published in the Gazette. Also, two years following the completion of such waterworks, the Minister must again consider the results of another environmental impact assessment.[43]

3.3 National Reserves and Protected Areas for Environmental Purposes

The South Africa NWA, which created the 'ecological' and 'human needs' reserves, effectively served as a prototype for numerous

---

[39] Article 5(4)(a), Cameroon, Décret No. 2001/164/PM.
[40] Article 21bis (iii), Article 22, Mexico, Law on National Waters of 1992 as amended by Decree of the President of the Republic on 29 April 2004.
[41] Article 29(3), Kenya Water Act (Cap 732), Act No. 8 of 2002.
[42] Article 34, China Water Law of 2002.
[43] Section 110, South Africa, NWA, note 29 above.

subsequent legislations which established such a category. The Armenia Water Code defines a 'national reserve' as 'the quality and quantity of water that is required to satisfy present and future basic human needs, as well as to protect aquatic ecosystems and to secure sustainable development and use of that water resource'.[44] Other statutes recognise the use of reserves for domestic and urban needs, but almost always incorporate the environmental protection dimension. The Spanish Law on the National Water Master Plan empowers the government to set aside entire rivers (or sections thereof), aquifers or other water bodies as part of the environmental reserve; it also indicates that one of the possible legal consequences of such a reservation is that new water abstraction rights and licenses in that area may be prohibited.[45] In Kenya, a determination of the reserve entails a component for protection of aquatic ecosystems 'in order to secure ecologically sustainable development and use of the water resource',[46] and a sufficient stipend is made available for each constituent of the reserve.[47] Laws frequently direct that the requirements of the 'Reserve' are taken into consideration in all water resource-related decisions by the government and also in the formulation of national and catchment-level strategies.

In Namibia, protected areas and zones are conceptually analogous to the notion of the Reserve, and those relevant to this study aim for the protection of:

> any water resource, riverine habitat, watershed, wetland, environment or ecosystem at risk of depletion, contamination, extinction or disturbance from any source, including aquatic and terrestrial weeds.[48]

The purposes of such a designation – and banned activities therein – are often included in the main statute, and the specific geographic boundaries within which they apply are indicated in

---

[44] Article 1, Armenia Water Code 2002.
[45] Article 25, Spain Law, note 32 above.
[46] Article 1, Kenya, Water Act, note 41 above.
[47] Ibid., Article 13(2).
[48] Section 72, Namibia WRM Act, note 38 above.

subsidiary legislation. The types of proscribed activities in the water body or its vicinity include the application or storage of pesticides or fertiliser chemicals, road construction, tree felling, mining, abstractions and effluent discharge. The government may be directed to establish procedures concerning the allocation of land use and forest use in water ecosystem protection zones; the construction of pipelines or other communication devices, and the extraction of biological resources and materials in water ecosystem protection zones.[49]

The procedure by which Reserves and protected areas are established is often included in the law which elaborates the steps for public consultation, for example, in the national law gazette, inviting written comments and stipulating time frames and deadlines for feedback. The Victoria Water (Irrigation Farm Dams) Act 2002[50] involves significant stakeholder input, thus acquiring extensive information bases, which include authorities charged with environmental protection.[51] Article 32(a) of the Act expressly identifies the objective of binding management plans for the water area produced by stakeholder 'consultative committees' as being to ensure the equitable and sustainable management of resources, and prescribes 'conditions relating to the protection of the environment, including the riverine and riparian environment'. Enforcement provisions for protected areas can take the form of registering and publicising the specific zones, as well as setting up monitoring systems and programs to ensure compliance.[52]

3.4 Water Trading and Trusts

3.4.1 Water Trading and the Environment

Traditionally used as a device to alleviate pressure on scarce freshwater sources through the efficient allocation of water resources for abstraction and use permits and concessions, the trade

---

[49] Article 121(5), Armenia Water Code 2002.
[50] Australia Victoria Water (Irrigation Farm Dams) Act, Act No. 5 of 2002.
[51] For example, the Consultative Committees in charge of drafting a management plan for the area. Ibid., Article 29.
[52] Article 6, EU Directive 2000/60/EC of 23 October 2000 Establishing a Framework for Community Action in the Field of Water Policy.

of water rights usually involves transfers of water for monetary compensation. The pre-requisites to such a framework necessitate legislation that recognises the limits on the availability of the resource, clearly defines water property rights and establishes the parameters of the trading structure,[53] notably conditions on transfers and use aimed at preventing adverse third party effects, particularly on the environment. The Mexican law, for instance, authorises the transfer of permits wholly or partially, permanently or temporarily during certain seasons.[54] Temporary transfers are subject to prior notification to the government,[55] whereas permanent transfers require a review before the exchange if the transfer does not entail modifications to the terms of the grant or if it may have third-party, environmental or hydrological effects.[56]

The California Water Code, which distinguishes between long-term and short-term water entitlement transfers, mandates that 'the change will not operate to the injury of any legal user' of the relevant water body.[57] Whereas provisions governing short term transfers call for the avoidance of 'unreasonable' effects on fish and wildlife, this proviso is missing from the provisions governing long-term transfers. To fill this lacuna, the State Water Resources Control Board relies on its responsibility under the public trust doctrine to judge whether the approval of such a long-term change is in public interest.[58] Environmental protection and conservation is gaining priority in what is in the public interest.

Elsewhere in the western United States, where the 'prior appropriation' doctrine of water allocation dominates, the upshot of the doctrine's requirement to use a water right or risk losing it,

---

[53] Megan Dyson and John Scanlon, 'Trading in Water Entitlements in the Murray Darling Basin in Australia – Realising the Potential for Environmental Benefits', 1 *IUCN ELP Newsletter* (2002). *Source:* http://www.iucn.org/themes/law/pdfdocuments/Neuchatel_Trade%20MDBC%20ELP%20Article%20Final.pdf.
[54] Article 22, Mexico, Law on National Waters of 1992 as amended by Decree of the President of the Republic of 29 April 2004.
[55] Ibid., Article 23bis.
[56] Ibid., Article 33.
[57] California Water Code, § 1702.
[58] Division of Water Rights, State Water Resources Control Board, California Environmental Protection Agency, A Guide to Water Transfers (Draft, July 1999).

and of water salvage laws intending to prevent waste and encourage full water use, is the tendency to over-use abstraction rights, which contributes to environmental degradation.[59] This particular problem has been addressed in Oregon state legislation,[60] which directs the salvager to return, to the state, 25 per cent of the conserved water to maintain stream flow levels, in exchange for granting the water user the right to reallocate (sell or lease) the remaining portion of saved water.

The Spanish transition towards the use of water transfers as an efficient allocation mechanism together with the creation of water banks is examined in section 3.7 below.

3.4.2 Water Trusts

The pioneering Oregon Trust in the US is illustrative of an alternative method to restore the flows of water sources, by acquiring 'out-of-stream rights' and converting them to 'instream flows'. The former comprise resource-intensive water rights such as for irrigation purposes, while the latter connotes non-consumptive use; such a conversion seeks to rehabilitate stream flows during non-consumptive periods, which as a result of contributing factors such as over-abstraction under the prior appropriation doctrine, were often diminished below the minimum flow requirement.[61] Oregon's Instream Water Rights Act[62] recognised instream flows protecting aquatic habitats as a beneficial use of water, providing three ways to create instream rights. One such method is the creation of trust rights, which are those the state has acquired through purchase, lease – wholly or partially – or donation (which may be subject to conditions such as, for example, that the trust be used for environmental purposes, and thus must be administered in accordance with that condition).

The strategy adopted by the Oregon Trust was to leave the larger rivers and water bodies to the federal budget, and instead

---

[59] A. Dan Tarlock ed., *Water Transfers in the West: Efficiency, Equity, and the Environment* (Washington: National Academy Press, 1992).
[60] The Conserved Water Program, Oregon Revised Statute 537.455, 1987.
[61] Andrew Purkey and Clay Landry, 'A New Tool for New Partnerships: Water Acquisitions and the Oregon Trust Fund,' 12 *Journal of Water Law* 5 (2001).
[62] Instream Water Rights Act (Oregon Revised Statute 537.348).

concentrate its lesser state budget on acquiring rights attached to smaller water sources, which affords a greater ecological advantage as this accounts for a higher proportion of the water in smaller rivers.[63] The benefits of such a mechanism are especially perceptible in legal systems which accord priority of water use according to the 'first in time, first in right' rule; trust rights retain the date of the original right, thus maintaining seniority in terms of use.[64]

As shown by the Oregon Water Trust, 'environmental water transactions have gained a prominent role as an important tool in protecting and restoring water-dependent ecosystems in a way that minimises disruption and controversy'.[65] The recourse to the market technique in the pursuit of water-related environmental protection goals aptly illustrates the juxtaposition of development and conservation. In essence, the environment has become a market player, and transfers can be based on environmental considerations.

3.5 Environment or Ecosystem Service Payments

Financial mechanisms can also be employed towards ecological and conservation purposes through payment for services that confer water-related environmental benefits. Ecosystem services refer to the natural 'interactions of living organisms with their environment', which provide important benefits to society such as purifying water or detoxifying waste;[66] more commonly, they fall within the purview of environmental statutes but are now finding their way into some modern water laws as well. Payment systems thus offer financial incentives for land owners or managers to carry out or refrain from certain activities, which ultimately reflect on the quality and dependability of freshwater systems.

A movement towards accommodating this mechanism in the

---

[63] Andrew Purkey and Clay Landry, 'A New Tool for New Partnerships: Water Acquisitions and the Oregon Trust Fund,' 12 *Journal of Water Law* 5 (2001).
[64] Steven Malloch, Liquid Assets: Protecting and Restoring the West's Rivers and Wetlands through Environmental Water Transactions (Arlington, Canada: Trout Unlimited, 2005). *Source:* http://www.nature.org/initiatives/freshwater/files/liquid_assets.pdf.
[65] Ibid.
[66] James Salzman, 'Creating Markets for Ecosystem Services: Notes from the Field', 80 *New York University Law Review* 870 (2005).
[67] Ibid.

legislative framework has been observed,[67] and is gaining momentum in water resources statutes as evidenced in Costa Rica, with Ecuador and Guatemala having local regulations to this effect. In line with the theme of this chapter, Costa Rica instituted a water tariff structure which highlights the economic, social and environmental importance of water.[68] Water charges comprise a 'use' element and an 'environmental' element;[69] half of the proceeds of the collection of water charges are allocated for national water management and for specific projects, and the remainder to conserve, maintain and restore the basin unit ecosystem which includes surrounding forests. As part of the National Forestry Fund that finances the Environmental Services Payment Programme, this is used to remunerate private property holders within forests for the services rendered therein, which result in water resource conservation and protection. A part of these funds can also go to the municipalities to fund the purchase of private land for the protection of ground water recharge areas, and for the protection of water sources of local significance.[70]

3.6 Controlling the Exploitation of Groundwater Resources in Recognition of their Ecosystem Support Function

Water laws often contain discrete and self-standing provisions on ground water protection, notably licensing requirements of borehole drilling and well construction, in response to the importance of these resources as a source in their own right, their connection to surface water bodies, and their support function to neighbouring wetlands and forests. The NSW legislation provides for an aquifer interference activity approval by the government, and in any event, the activity must avoid land degradation such as the decline of native vegetation, increased acidity, and soil erosion. The management plan for the relevant area where such controlled activity occurs must identify the nature of the aquifer interference having any effect, including 'cumulative impacts on water sources

---

[68] Costa Rica Decree of the President of the Republic No. 32868 of 24 August 2005.
[69] Ibid., Article 3.
[70] Ibid., Articles 14 and 15.
[71] Section 5(8), NSW Water Management Act, note 28 above.

or their dependent ecosystems, and the extent of those impacts'.[71] Plans for such controlled activity also deal with undertaking work with a view to rehabilitating the water source or its dependent ecosystems and habitats.[72] The Namibia WRM Act empowers the Minister to establish the 'safe yield' of aquifers when making determinations regarding its use, where 'safe yield' refers to the amount and rate of abstraction which would not cause damage to the aquifer, the quality of the water or the environment.[73]

3.7 Spanish Water Management Experience: The Evolution of Water Law and Policy

While Spain has pioneered the incorporation of integrated management of water sources at river basin level and boasts an intricate water distribution system, some of its more environmentally-conscious policies have lagged behind. The shift in Spain's water supply paradigm to water demand management and protection of resources has been slow. But through trial and error, it has come to light that water management techniques that fail to account for ecological and environmental aspects negate any efforts through other methods for the efficient management of water. This glimpse into the Spanish water sector reform is useful to illustrate the challenges faced by drafters and policy makers in managing the water sector in line with sound environmental principles.

The impetus for the evolution of water law in Spain has derived in part from its European Union obligations, but water sector reform was also inevitable following the drought of 1990–1995, the prominence of a 'supply pattern of management' to cater for irrigation and industry, an overexploitation of aquifers and the deterioration of water quality.[74] This situation was exacerbated by the decline of ecological integrity in rivers and inadequate fiscal policies, which prevented the improvement of Spain's hydraulic

---

[72] Ibid., Section 33.
[73] Section 51, Namibia WRM Act, note 38 above.
[74] Antonio Fanlo Loras, 'Water Resources Management in Spain', *in* S. Marchisio, G. Tamburelli and L. Pecoraro eds., *Sustainable Management and Rational Use of Water Resources: A Legal Framework for the Mediterranean* 149–167 (Italy: Instituto di Studi Giuridici Internazionali, 1999).

works construction program and limited the ability to manage resources adequately.[75] Spain's traditional emphasis on the construction of hydraulic infrastructure, technology and irrigation systems is a product of its low natural water availability (one of the lowest in Europe) and the need for an extensive irrigation system (the largest in Europe). Its law and policy reflected these physical and climatic exigencies.[76]

The Water Law No. 29 of 1985 (the Law) was overdue – the last national water law was in 1879. It brought about radical changes in the water management system and was the basis for subsequent amendments. Increasingly, the environmental dimension gained priority alongside economic and technological considerations and the Law demonstrated a shift in emphasis from a supply management pattern to demand and quality based management, but it was not without flaws. Although environmental conservation considerations of the resource and its surroundings were required to be taken into account, the Law set out water uses in order of priority as population supply, irrigation, energy production, other industrial uses, aquaculture, recreation, navigation and transport and a catch all category of 'other uses'.[77] Notably, no mention of environmental use or minimum flow preservation is made in the Law. This provision has been kept intact in the Legislative Decree No. 1 of 2001 (the Decree) and its subsequent amendments.

Spain was one of the pioneers in integrated water management along hydrographic basin lines with the hydrologic basin as an administrative unit (*Confederaciones hidrográficas*) since the 1920s, later confirmed in Article 1.2 of the Law. The general objectives of hydrological plans under the Law were to 'achieve a better fulfilment of water demands, and balance and harmonise regional and sectoral development, increasing the availability of the resource,

---

[75] Alberto Garrido, 'Analysis of Spanish Water Law Reform', *in* B.R. Bruns, C. Ringler and R. Meinzen-Dick eds., *Water Rights Reform: Lessons for Institutional Design* 219 (Washington, DC: International Food Policy Research Institute, 2005).

[76] Antonio Fanlo Loras, 'Water Resources Management in Spain', *in* S. Marchisio, G. Tamburelli and L. Pecoraro eds., *Sustainable Management and Rational Use of Water Resources: A Legal Framework for the Mediterranean* 149–167 (Italy: Instituto di Studi Giuridici Internazionali, 1999).

[77] Article 58.3, Spain, Water Law No. 29 of 1985.

protecting its quality, both economising on and rationalising its use in harmony with the environment and the other natural resources'.[78] This integrated water management approach is reflected in Article 40, which contains the mandatory contents for hydrological plans which include *inter alia* resource allocation and reserve for current and future uses, for the conservation and repair of the environment (including protected areas), hydrological-forest for soil conservation and guidelines for aquifer recharge and protection.

Article 13.3 of the Law mandated the management of water in accordance with environmental principles such as increasing water quantity and quality as a guide to the performance of public institutions, and in granting water permits. The legislation incorporates some of the techniques discussed in this chapter such as the use of environmental impact assessments before permit authorisation and before the construction of hydraulic works projects;[79] and the establishment of minimum flow for common uses, sanitary or ecological purposes. The increasing profile and priority of the environmental perspective has resulted in the adoption of strategies such as water savings through maximum quotas for different usages, improvement of irrigation techniques and water reuse.[80] The Law mandated the fulfilment of other requirements it considered 'environmental' obligations such as the maintenance of hydraulic works and grids, emission and pollution limits, etc.

While the Water Law seemed to contain some important principles, it ignored the concept of including environmental planning as a part of hydrological planning. Considering Spain's numerous appearances before the European Commission relating to its management of water sources (and before the European Court of Justice as a result of a deterioration of its wetlands), the country was experiencing a general decline in its water quantity and quality, which has often been attributed to problems relating to implementation of the law.

---

[78] Ibid., Article 38.1.
[79] Ibid., Articles 90, 128, and 130.4 respectively.
[80] Loras, note 74 above.

Numerous subsequent amendments demonstrated flaws in the policy behind this law and were necessary to bring the Spanish regime in line with the EU Water Framework Directive.[81] The 2001 Decree, which consolidates the amendments to the Law, seeks to strike a new balance between the hitherto prevailing utilitarian thrust of water resources management and the innovative environmental propulsion brought about by the EU Directive, with the pendulum tending since to swing towards the latter. Compelling evidence of this trend is the statutory treatment of new water rights (abstraction concessions and authorisations), and the regulation of water rights trading.

Under the above-mentioned consolidated Water Law, the government water resources administration is under a statutory direction to ensure that new abstraction rights are compatible with 'respect for the environment' and, in particular, with ecological flow requirements of watercourses and with the environmental demands crystallised in the national and basin water resources plans.[82] This provision sets a clear limit on the government's water allocation authority, in the interest of the environmental dimension of water resources management. In addition, and in line with the above-mentioned EU Water Directive, the new dispensation inaugurated by the consolidated Water Law opens up opportunities for the trading of water rights. Trades, however, are subject to prior government consent, with a detrimental impact on ecological flow requirements of watercourses, and on the state of aquatic ecosystems, being grounds for denial of government consent to a proposed trade.[83] Moreover, recent case law seems to suggest that the trading of water rights opens up fresh opportunities for a pro-environment re-assessment of existing water rights by government, as it reviews proposed trades under the above-mentioned Water Law provision.[84] Also, this provision, and its application by some

---

[81] EU Directive 2000/60/EC of 23 October 2000 Establishing a Framework for Community Action in the Field of Water Policy.
[82] Spain Real Decreto Legislativo 1/2001, que aprueba el Texto Refundido de la Ley de Aguas, Article 98.
[83] Ibid., Article 68.3.
[84] See Pedro Brufao Curiel, *La revisión ambiental de las concesiones y autorizaciones de aguas* 50–56. (Zaragoza: Colección Nueva Cultura del Agua No 18, 2008.)

courts, points in the direction of an emerging new pro-environment paradigm on the Spanish water management scene.

4. Discussion and Conclusion

The foregoing analysis draws from contemporary laws from a range of jurisdictions with differing water policy priorities, but nevertheless incorporating similar environmentally conscious regulations. The mechanisms identified in this chapter are not exhaustive, but are indicative of recent trends towards a 'greening' approach to water law. However, these 'greening' strategies are not without practical difficulties. For example, the emergent ecosystem payment schemes are advantageous in their low-cost and straightforward implementation, but their environmental efficiency has nevertheless been questioned. Also, the utility of considering minimum flows for permit issuance is qualified by the fact that laws cannot have retrospective effect – while it is often a primary consideration for granting new permits and licenses or renewing existing ones, those that were issued prior to the passing of the law (which introduces such provisions) are precluded from its ambit. Furthermore, some provisions which seek to protect instream flows have been criticised for having overly-limiting language or a narrow scope, for example, protecting 'fish' instead of the more comprehensive formulation of 'ecosystems'.[85]

It should also be highlighted that the purpose of some environment-friendly water law provisions may be frustrated by other laws. For example, the Philippines Biofuels Act 2006[86] stipulates that, as an incentive to the production of biofuels, water effluent from the production of biofuels is exempt from wastewater charges – the raison-d'être of the latter being to discourage effluent discharges as much as possible. Clearly, this is a case of two environment-friendly statutes working at cross-purposes, possibly reflecting inconsistent policy direction – or a conscious decision to sacrifice one environmental goal for another.

---

[85] A. Dan Tarlock ed., *Water Transfers in the West: Efficiency, Equity, and the Environment* (Washington: National Academy Press, 1992).
[86] Philippines Biofuels Act, Act No 9360 of 2006. *Source:* http://www.lead-journal.org/content/07368.pdf.

On a concluding note, the 'greening' of water laws which emerges from a comparative review of the more recent generation of water resources statutes can be regarded as a tangible manifestation of mounting concern for the *sustainability* of resource development and utilisation, with sustainability turning from an elusive distant goal to precise rights and obligations accruing to the government in its capacity as a custodian and manager of the resource, and to members of the public as resource developers and users.

13

The Right to Water as a Human Right or a Bird's Right: Does Cooperative Governance Offer a Way Out of a Conflict of Interests and Legal Complexity?

JONATHAN VERSCHUUREN

1. Introduction

The right to water has been recognised as a human right under various international human rights instruments. On the other hand, various international legal instruments to protect nature have led government institutions to reserve enough water for protected areas, for instance wetlands designated under the Convention on Wetlands of International Importance especially as Waterfowl Habitat 1971 (the Ramsar Convention).[1] These international legal obligations may conflict with each other, giving rise to legal problems not only within a country, but also between countries. The latter is the case when two or more countries use the same river as a source of drinking water and for ecological purposes (such as protection of a wetland). In theory, the principle of reasonable and equitable use and the concept of common river basin management, as laid down in the Convention on the Law of

---

[1] Convention on Wetlands, Ramsar, Iran, 2 February 1971, *reprinted in* P. Cullet and A. Gowlland-Gualtieri eds., *Key Materials in International Environmental Law* (Aldershot: Ashgate, 2004), 248 [hereafter Ramsar Convention].

the Non-navigational Uses of International Watercourses 1997 (UN Water Convention),[2] are considered to offer a way out of this potential conflict. However, the question is whether, in practice, these principles really are the solution to the conflict between the right to water and the duty to protect wetlands of international importance, and if so, under what conditions do they function adequately.

This chapter consists of a theoretical and an empirical part. The theoretical part (Section 2) will start with an analysis of the international legal instruments on the right to water and the obligation to protect wetlands. Then, I will turn to the principle of reasonable and equitable use and the concept of common river basin management. How have these concepts been regulated, what is their purpose, and do they (in theory) offer a way out of the conflict between the right to water and the obligation to protect wetlands, especially in multilateral situations? All of these questions will be answered on the basis of a desk study into relevant legal texts and literature, thus concluding the theoretical part.

In the empirical part of the chapter (Section 3), a case study will be presented. This case study of the Orange River, which runs through four countries in southern Africa, and into a protected Ramsar wetland on that river's estuary located on the border between Namibia and South Africa, will show whether the principle of reasonable and equitable use and the concept of common river basin management actually solve the conflict between man and nature in a transboundary context. The case will be studied within the theoretical framework of co-operative governance. The second part of the chapter is structured as follows: first the case will be laid out; then, the findings will be presented. The final conclusions are drawn in Section 4.

---

[2] Convention on the Law of the Non-navigational Uses of International Watercourses, New York, 21 May 1997, *reprinted in* P. Cullet and A. Gowlland-Gualtieri eds., *Key Materials in International Environmental Law* (Aldershot: Ashgate, 2004), 481 [hereafter UN Water Convention].

2. The Right to Water Versus the Obligation to Protect Wetlands

2.1 The Right to Water in International and National legal Instruments

Although the Ministerial Declaration adopted at the 4th World Water Forum in Mexico in 2006 (the Declaration) does not mention the right to water, the issue whether such a right exists in international law was heavily debated during the conference.[3] The fact that, despite these debates, there is no mention at all of the right to water in the Declaration shows that the right is disputed. States sometimes fear that the rights-based approach not only forces them to change their national legislation, but also conflicts with the current trend of privatisation and the increasing role of the market mechanism, which reduces government intervention.

Still, the right to water has already been acknowledged as a human right in various international human rights documents, including the International Convention on Economic, Social and Cultural Rights 1966 (ICESCR). Article 12(1), ICESCR encompasses everyone's right to the enjoyment of the highest attainable standard of physical and mental health. To achieve the benefits of this right, states should improve all aspects of environmental and industrial hygiene, and provide for the healthy development of children [Article 12(2) (a) and (b)]. In 2000, the Committee on Economic, Social and Cultural Rights (the Committee) adopted a General Comment on this human right, stating that Article 12 not only deals with health care, but also with all other factors that determine the enjoyment of good health, such as access to safe drinking water, personal hygiene, sufficient safe food and shelter.[4] Before that, the availability of water had been acknowledged as being part of the human right to an adequate standard of living (Article 11).[5]

---

[3] P. Martinez Austria and P. van Hofwegen eds., *Synthesis of the 4th World Water Forum* 90 (Copilco El Bajo: Comisión Nacional de Agua, 2006).

[4] UN Committee on Economic, Social and Cultural Rights, General Comment No. 14, UN Doc. E/C.12/2000/4 (2000).

[5] UN Committee on Economic, Social and Cultural Rights, General Comment No. 4, UN Doc. E/1992/23 (1991) and Human Rights Committee (CCPR) General Comment No. 6 (1982) *in* Compilation of General Comments and General Recommendations Adopted by Human Rights Treaty Bodies, UN Doc. HRI/GEN/1/Rev.1 (1994) at 6.

In 2002, the Committee adopted General Comment No. 15 that deals entirely with the right to water, and that is considered to be the most influential document on this right.[6] In this document, the right to water has been defined as follows:[7]

> The right to drinking water entitles everyone to safe, sufficient, affordable and accessible drinking water that is adequate for daily individual requirements (drinking, household sanitation, food preparation and hygiene).

The right to water has been specifically mentioned in several other binding human rights documents as well, such as the Convention on the Rights of the Child 1979 (CRC) and the Convention on the Elimination of All Forms of Discrimination against Women 1989 (CEDAW).[8] Article 24, CRC has a formulation that is comparable with Article 12, ICESCR, although it adds, in Article 24(2), that measures are to be taken to combat disease and malnutrition, including within the framework of primary health care, through, inter alia, the application of readily available technology and the provision of adequate nutritious foods and clean drinking-water, taking into consideration the dangers and risks of environmental pollution. Article 14(2)(h), CEDAW states that states shall ensure the right of rural women to enjoy adequate living conditions, particularly in relation to housing, sanitation, electricity and water supply, transport and communications.

These human rights documents only mention the right to water as an individual human right without referring to the tension between this right and water management aimed at more than just

---

[6] Committee on Economic, Social and Cultural Rights, General Comment 15: The Right to Water (Articles 11 and 12 of the International Covenant on Economic, Social and Cultural Rights), UN Doc. E/C.12/2002/11 (2002). *Source:* http://www.ielrc.org/content/e0221.pdf [hereafter General Comment 15].

[7] The various elements of this definition have been further worked out in General Comment 15.

[8] An overview of all texts that implicitly and explicitly refer to the right of water in human rights conventions is available at http://www.righttowater.org.uk. For more texts that refer to the right to water, for instance those of the International Labour Organisation (ILO), *see* Birgit Toebes, *The Right to Health as a Human Right in International Law* (Antwerp: Intersentia 1999).

the provision of drinking water, for instance, protection of the environment and nature. Only the CRC explicitly states that the dangers of environmental pollution should be taken into account, but the word 'pollution' hints at the fact that dehydration of wetlands as a possible side effect of water supply for human purposes had not been considered while drafting the Convention. General Comment No. 15 has the same flaw, although it does mention that the realisation of the right to water has to take place in a sustainable manner, so that the right can be exercised by present and future generations.[9]

The World Health Organisation goes a few steps further by observing, in a recent report, that the right to water should be balanced in an integrated catchment policy with all the other water needs, such as irrigation, power generation, and nature conservation.[10] Integrated river basin management indeed may accomplish much, although in extremely dry areas it will probably prove to be impossible to serve all water needs at the same time. In international river basins, the situation is even more complicated because the various conflicting needs are present on both sides of the border. An upstream state may, for instance, because the right to water is invoked by its citizens, either on the basis of the international human rights documents mentioned above or on the basis of national law, have a legal duty to supply most of the water to its citizens, leaving too little for the citizens of the downstream state, or to a downstream wetland.

The 1999 UNECE Protocol on Water and Health (the UNECE Protocol)[11] to the 1992 Convention on the Protection and Use of Transboundary Watercourses and International Lakes (the UNECE Convention or the Helsinki Convention),[12] takes the same approach. Although this Protocol does not explicitly recognise a right to water, it does state, in Article 6(1), that the Parties shall pursue the aims

---

[9] General Comment 15, note 6 above para. 20.
[10] World Health Organisation, *Right to Water* 18–21 (Geneva: WHO, 2003).
[11] Protocol on Water and Health to the 1992 Convention on the Protection and Use of Transboundary Watercourses and International Lakes, London, 17 June 1999, Doc. MP.WAT/2000/1. *Source:* http://www.ielrc.org/content/e9910.pdf.
[12] Convention on the Protection and Use of Transboundary Watercourses and International Lakes, Helsinki, 17 March 1992. *Source:* http://www.ielrc.org/content/e9210.pdf. *See* section 2.3 below [hereafter UNECE Convention].

of access to drinking water for everyone and provision of sanitation for everyone:

> within a framework of integrated water management systems aimed at sustainable use of water resources, ambient water quality which does not endanger human health, and protection of water ecosystems.

National constitutions sometimes explicitly acknowledge the right to water as well, especially in Africa. The constitutions of Gambia, Uganda, Zambia, South Africa and Ethiopia recognise the right to water. Let us have a closer look at the right to water as laid down in the South African Constitution:

(i) Everyone has the right to have access to (a) health care services, including reproductive health care; (b) sufficient food and water; and (c) social security, including, if they are unable to support themselves and their dependants, appropriate social assistance.

(ii) The state must take reasonable legislative and other measures, within its available resources, to achieve the progressive realisation of each of these rights.[13]

Section 27(2) makes it clear that the right to water is a social right, which needs government intervention in order to be realised. This does not mean that this right cannot be enforced by individual citizens. The South African Constitutional Court has, on various occasions, judged that a certain government policy was contrary to certain socio-economic rights. The first case in which such a judgment was given was the *Grootboom* case on the right to housing (Section 26) and the right to shelter for children (Section 28).[14] The

---

[13] Constitution of the Republic of South Africa N0. 108 of 1996, 18 December 1996, Bill of Rights, Section 27.

[14] Constitutional Court of South Africa, CCT 11/00, 2001 (1) SA 46 (CC), published in 9 *Tilburg Foreign L.R.* 417–445 (2002) (annotated by Raymond Bos). *Source:* http://www.constitutionalcourt.org.za. There is also a Constitutional Court case on Section 27, although this case did not concern the right to water as such, but the right to health care. *See* Constitutional Court of South Africa, CCT 8/02, 2002, published in 11 *Tilburg Foreign L.R.* 671–702 (2003) (annotated by Danie Brand). *Source:* http://www.constitutionalcourt.org.za.

Constitutional Court judged that these rights can be invoked before a court. However, this does not mean that the Government has to supply a house to anyone who asks for one, but that the authorities have to be able to show that they have, within available means, a coherent programme with which the socio-economic rights actually can and will be effectuated.

The constitutional right to water has been further elaborated in specific South African water legislation, such as the National Water Act 1998 (the NWA) and the Water Services Act 1997.[15] For instance, under the latter Act, it is determined that every person must be able to get at least twenty-five litres of safe drinking water within two hundred metres of his or her home.[16] The policy is aimed at providing this amount of water through the regular waterworks. Those who cannot afford to purchase these services get water from the Government for free (six thousand litres per household annually). Fifty-seven per cent of the South African population thus receives free drinking water, supplied by the state.[17]

Contrary to the human rights documents mentioned above, the South African NWA does, to some extent, regulate the balancing of water needs. Section 3 will deal in detail with the same.

Finally, it is worthwhile to note that sometimes a right to water is recognised in statutory law rather than in the constitution. This, for instance, is the case in Namibia, where the new Water Resources Management Act 2004 (the WRM Act)[18] states as a general principle that safe drinking water is a basic human right.[19]

---

[15] N. Gabru, 'Some Comments on Water Rights in South Africa', 1 *Potchefstroom Electronic L.J.* 1–33 (2005). *Source:* http://www.puk.ac.za/fakulteite/regte/per/issue05v1.html. *See* further on these statutes Section 3.3 below.

[16] According to the definition of 'basic water supply', laid down in Section 2 of the 2001 Compulsory National Standards and Measures to Conserve Water Regulations, based on the 1997 Water Services Act.

[17] N. Gabru, Some Comments on Water Rights in South Africa', 1 *Potchefstroom Electronic L.J.* 1–33 (2005). *Source:* http://www.puk.ac.za/fakulteite/regte/per/issue05v1.html. *See* further on these statutes Section 3.3 below, 26–27.

[18] Act No. 24 of 2004, Namibian Government Gazette No. 3357 of 23 December 2004 [hereafter Namibia WRM Act].

[19] Ibid., Section 3.

2.2 The Obligation to Protect Wetlands in International Legal Instruments

The most renowned international convention with regard to wetlands, including transboundary wetlands, is the Ramsar Convention.[20] According to the Ramsar Convention, the contracting parties are obliged to formulate and implement their planning law so as to promote the conservation of wetlands designated under the Convention ('Ramsar sites'), and as far as possible the wise use of wetlands in their territory,[21] without prejudice to the exclusive sovereign rights of the contracting party in whose territory the wetland is situated.[22] 'Wise use' of wetlands is defined as 'the maintenance of their ecological character, achieved through the implementation of ecosystem approaches, within the context of sustainable development'.[23] In addition, 'ecological character' is defined as the combination of the ecosystem components, processes and benefits/ services that characterise the wetland at a given point in time', for listed wetlands being the time of designation of the wetland for the Ramsar list.[24]

Furthermore, the establishment of nature reserves on wetlands should be promoted.[25] Deletion or restricting the boundaries of a designated site is only allowed in the urgent national interest of the state involved.[26] Finally, the contracting parties have to endeavour, through management, to increase the waterfowl

---

[20] The other relevant conventions are the 1979 Berne Convention on the Conservation of European Wildlife and Natural Habitats, *UKTS* 56, (1992), the 1979 Bonn Convention on Migratory Species of Wild Animals, 19 *Int'l Leg. Mat.* 15 (1980), the 1992 Rio Convention on Biological Diversity, 31 *Int'l Leg. Mat.* 851 (1992), and the 2003 (revised) African Convention on the Conservation of Nature and Natural Resources (signed in Maputo, Mozambique, 11 June 2003). *Source:* http://www.african-union.org.

[21] Ramsar Convention, note 1 above, Article 3(1).

[22] Ibid., Article 2(2).

[23] Resolution IX.1 Annex A, adopted during the 9th Meeting of the Conference of the Contracting Parties to the Convention on Wetlands, Kampala, Uganda, 8–15 November 2005, Para. 22. *Source:* http://www.ramsar.org/res/key_res_ix_01_ annexa_e.pdf.

[24] Ibid., Para. 15.

[25] Ramsar Convention, note 1 above, Article 4(1).

[26] Ibid., Article 4(2).

populations on appropriate wetlands.²⁷ For transboundary wetlands ('shared wetlands' or 'international wetlands'), there is a specific provision in the Convention stating that parties shall consult with each other about implementing obligations arising from it, as well as endeavour to coordinate and support present and future policies and regulations concerning the preservation of wetlands and their flora and fauna.²⁸

Since 1971, a number of resolutions, handbooks and guidelines have been adopted that further define the general provisions cited above. The so-called 'Ramsar Toolkit' is a set of no less than seventeen Handbooks for the wise use of wetlands, including those on the drafting of national wetlands policies, wise use in general, the designation process, river basin management, participation of local communities, etc. The concept of integrated management has been promulgated in the Handbook on management of wetlands.²⁹ Site management plans must be integrated into the public development planning system at local, regional or national level. According to this Handbook, 'the integration of site management plans into spatial and economic planning at the appropriate level will ensure implementation, public participation and local ownership'.³⁰ In addition, a multi-scalar approach to wise use planning and management should be adopted and 'linked with broad-scale landscape and ecosystem planning, including at the integrated river basin ..., because policy and planning decisions at these scales will affect the conservation and wise use of wetland sites'.³¹ The Handbook recites the part of Agenda 21 in which the multi-interest utilisation of water resources was recognised,³² and states that integrated river basin management aims at bringing

---

[27] Ibid., Article 4(4).
[28] Ibid., Article 5(1).
[29] The third and latest edition of the Ramsar Convention Handbooks (2007) is available at http://www.ramsar.org/lib/lib_handbooks2006_e.htm.
[30] Ramsar Convention Secretariat, *Ramsar Handbook for the Wise Use of Wetlands, Handbook 16: Managing Wetlands* 11–12 (3rd ed. 2007). *Source:* http://www.ramsar.org/lib/lib_handbooks2006_e16.pdf.
[31] Ibid.
[32] Ibid., 14. *See* sections 18.8 and 18.9 of Agenda 21, *in* Report of the United Nations Conference on Environment and Development, Rio de Janeiro, UN Doc. A/CONF.151/26/Rev.1 (Vol. 1), Annex II (1992). *Source:* http://www.ielrc.org/content/e9211.pdf.

together stakeholders at all levels, from politicians to local communities, and at considering water demands for different sectors within the basin. To be able to do so, the benefits of wetlands have to be determined in order to justify the required allocation.

For transboundary wetlands, there is another Handbook with more detailed advice on how to pursue international cooperation on the management of such areas.[33] Referring to the 1992 Helsinki Convention, the Handbook indicates that multi-state management commissions should be established to promote international cooperation,[34] and urges states to harmonise wetland management with the obligations arising from watercourse agreements.[35]

Generally, it can be observed that over the last few years, wetland management has been integrated into river basin management, recognising the fact that wetlands usually are only a part of a bigger catchment area and largely depend, for their conservation, on the quality of the entire catchment.[36] To this end, the Ramsar Convention Bureau and the Secretariat of the Convention on Biodiversity have joined hands in a River Basin Initiative.[37] In 2005, the 9th Conference of Parties (COP) of the Ramsar Convention adopted a resolution laying down practical guidelines for the integration of wetland management into river basin management.[38] Attention is focused on: (i) improving communication between the wetland management sector and the water management sector, (ii) improving cooperation between the water sector and the wetlands sector through cooperative governance, for instance, by formally harmonising policy and

---

[33] Ramsar Convention Secretariat, *Ramsar Handbook for the Wise Use of Wetlands, Handbook 17: International Cooperation* (3rd ed. 2007). *Source:* http://www.ramsar.org/lib/lib_handbooks2006_e17.pdf. *See also* section 2.3 below.

[34] Ibid., 12–13.

[35] Ibid., 13.

[36] Resolution VII.18, reprinted in Ramsar Convention Secretariat, *Ramsar Handbook for the Wise Use of Wetlands, Handbook 7: River basin management* (3rd ed. 2007). *Source:* http://www.ramsar.org/lib/lib_handbooks2006_e07.pdf.

[37] The website for River Basin Initiative is available at http://www.riverbasin.org.

[38] Resolution IX.1, Kampala 2005, Annex C i ('River Basin Management: Additional Guidance and a Framework for the Analysis of Case Studies'). *Source:* http://www.ramsar.org/res/key_res_ix_01_annexci_e.htm.

legislation or by other, less far-reaching forms of cross-sectoral cooperation, and (iii) up scaling wetlands management to the river basin level.

2.3 The Principle of Reasonable and Equitable Use and the Concept of Common River Basin Management

In international law, the use of water in a transboundary context is governed by the principle of reasonable and equitable use, and by conventions such as the UNECE Convention. The UN Water Convention addresses the same topic on a global level,[39] although the wording of this Convention is less strong than the UNECE Convention.[40] From a substantive point of view, there are similarities between the two conventions, most notably the establishment of a joint body in order to achieve a common management of the international watercourse. Both conventions also support an ecosystem approach, in which all consequences of human activities on the entire ecosystem are considered, respecting the integrity of the ecosystem as a whole.[41]

The UNECE Convention most elaborately defines the measures that have to be taken to protect transboundary water systems, such as (a) prevention and control of pollution, (b) ecologically and rationally sound water management, conservation of water resources and environmental protection, (c) reasonable and

---

[39] Until February 2009, only sixteen countries had ratified the UN Water Convention, including Namibia and South-Africa. The Convention will enter into force after ratification by 35 countries.

[40] Patricia W. Birnie and Alan E. Boyle, *International Law and the Environment* 305 (Oxford: Oxford University Press, 2002). For more detail, *see* Attila Tanzi, 'The Relationship between the 1992 UNECE Convention on the Protection and Use of Transboundary Watercourses and International Lakes and the 1997 UN Water Convention on the Law of the Non Navigational Uses of International Watercourses (2000). *Source:* http://www.unece.org/env/water.

[41] Jutta Brunnée and Stephen J. Toope, 'Environmental Security and Freshwater Resources: A Case for International Ecosystem Law', 5 *Yb. Int'l Envtl. L.* 41, 55 (1994). The obligation to do so is more strictly formulated in the UNECE Convention than in the UN Water Convention. *See* Owen McIntyre, 'The Emergence of an 'Ecosystem Approach' to the Protection of International Watercourses under International Law', 13 *Rev. Europe. Community Int'l Envtl L.* 1, 13 (2004).

equitable use taking into account the transboundary character,[42] (d) conservation, and, where necessary, restoration of ecosystems.[43] Several legal principles, such as the precautionary principle and the polluter-pays principle, apply.[44]

Under the UNECE Convention, the joint bodies have a wide range of tasks, including elaborating monitoring programs concerning water quality and water quantity, exchanging information, elaborating emission limits for waste water and joint water-quality objectives, developing concerted action programs for the reduction of pollution loads and implementing environmental impact assessments.[45] The UN Water Convention only has a short provision on management, stating that consultations between watercourse states may include the establishment of a joint management mechanism, whose task it is to plan the sustainable development of an international watercourse and provide for the implementation of any plans adopted, and otherwise promote the rational and optimal utilisation, protection and control of the watercourse.[46]

A large number of guidelines are available for the application of the UNECE Convention, such as the Guidelines on Monitoring and Assessment of Transboundary Rivers (the Guidelines).[47] Again, these Guidelines stress the need for an integrated approach. The state of the river and related ecosystem should be assessed in an

---

[42] Using both the principle of equitable use and the principle of prevention of harm has been criticised for their inherent upstream/downstream conflict; some authors advocate a 'needs based' approach, rather than a 'rights based' approach. *See* Heather L. Beach *et al.*, *Transboundary Freshwater Dispute Resolution. Theory, Practice, and Annotated References* 74 (Tokyo: United Nations University Press, 2000).

[43] UNECE Convention, note 12 above, Article 2(2). Again, from an environmental protection point of view, the UN Water Convention is much weaker. There, equitable and reasonable utilisation is the only principle, in which the ecological factor just seems to be a minor one. *See* Article 6(1).

[44] Ibid., Article 2(3). These principles are absent in the UN Water Convention.

[45] Ibid., Article 9(2). *See also* Malgosia Fitzmaurice and Olufemi Elias, *Watercourse Co-operation in Northern Europe – A Model for the Future* (The Hague: T.M.C. Asser, 2004).

[46] UN Water Convention, note 2 above, Article 24.

[47] UNECE Task Force on Monitoring & Assessment, Guidelines on Monitoring and Assessment of Transboundary Rivers (Lelystad: UNECE, 2000).

integrated manner, based on criteria that include water quality and quantity for different human uses as well as flora and fauna.[48] The Guidelines also identify three sources of conflicts: (a) the competition for water (consumptive use versus non-consumptive use), (b) conflicts between human intervention and nature, and (c) different interests of riparian countries.[49] These (potential) conflicts have to be acknowledged when formulating an integrated management plan.

Outside the UNECE region, bilateral or multilateral conventions establishing joint river basin management commissions have also been concluded, often based on the UN Water Convention, even though this Convention has not entered into force yet. A good example of such a commission that has to deal with various claims on a transboundary river basin is the Orange-Senqu River Commission (the ORASECOM), consisting of representatives from Botswana, Lesotho, Namibia and South Africa.[50] The Council serves as a technical advisor to the authorities of the states involved on matters relating to the development, utilisation and conservation of the water resources of the river system.[51] The Parties to this Agreement, which was not only based on the UN Water Convention, but also on a Protocol of the Southern African Development Community (SADC),[52] agree, *inter alia*, to:

> Utilize the resources of the river system in an equitable and reasonable manner with a view to attaining optimal and sustainable utilization thereof, and benefits there from, consistent with adequate protection of the river system [Article 7(2)],
>
> Take all appropriate measures to prevent the causing of significant harm to any other Party [Article 7(3)],
>
> Individually and jointly take all measures necessary to protect and preserve the river system from its sources and headwaters to its common terminus [Article 7(12)], including the estuary of

---

[48] Ibid., 10.
[49] Ibid., 14.
[50] Agreement on the Establishment of the Orange-Senqu River Commission, signed in Windhoek on 3 November 2000.
[51] Ibid., Article 4.
[52] Shared Watercourse Systems Protocol, Windhoek, 7 August 2000. This Protocol replaces the 1995 version. It entered into force on 22 September 2003.

the river system and the marine environment taking into account generally accepted international rules and standards [Article 7(14)],

Individually and jointly prevent, reduce and control pollution of the river system that may cause significant harm to one or more of the Parties, including harm to the environment, or to human health or safety, or to the ecosystem of the river system [Article 7(13)].

Again, reference is made to important principles of international law, such as reasonable and equitable use (or equitable utilisation). This principle is considered to be the most important principle in international freshwater law.[53] According to the principle, states may not use water in such a manner so as to prevent or otherwise limit other riparian states from making full use of their equitable and reasonable entitlements in relation to the shared river.[54]

The question that arises is whether this principle can limit the realisation of the human right to water. Suppose that the realisation of the right to water upstream leads to a serious decline of the available water, to such an extent that there cannot be a reasonable and equitable use downstream. According to the principle of reasonable and equitable use, this is not allowed. The consequence then would be that the upstream state can only partially guarantee its citizens' right to water. Thinking further along the lines of the principle, the right to water of all people in the river basin should be equally restricted, so that everyone has an equal share of the (scarce) available water. Suffice to say that ecological water uses, for instance, for the conservation of wetlands in the river basin, will be under extreme pressure in such a situation. This was the case in the case study presented in the later half of the paper.

---

[53] Patricia W. Birnie and Alan E. Boyle, *International Law and the Environment* 305 (Oxford: Oxford University Press, 2002). For more detail, *see* Attila Tanzi, 'The Relationship between the 1992 UNECE Convention on the Protection and Use of Transboundary Watercourses and International Lakes and the 1997 UN Water Convention on the Law of the Non Navigational Uses of International Watercourses (2000). *Source:* http://www.unece.org/env/water at 302 and Philippe Sands, *Principles of International Environmental Law* 461–462 (Cambridge: Cambridge University Press, 2003).

[54] Philippe Sands, *Principles of International Environmental Law* 461–462 (Cambridge: Cambridge University Press, 2003) at 462.

2.4 Conclusion

In international water law, the ecosystem approach, through integrated river basin management, encompasses the obligation to balance all water uses within the river basin. The main goal should be to protect the river-ecosystem as a whole, including wetlands located within the river basin. In human rights law, the right to water is gaining more and more acceptance as a basic human right. However, the human rights documents pay little attention to the necessity to integrate water supply for basic human needs into a wider water policy, taking into account 'ecosystem needs' as well. The only exception is the UNECE Protocol, which is not a human rights document *per se*, although it does impose a duty on states to pursue the objective of access to drinking water for everyone.

3. Case Study: Orange River and Orange River Mouth Wetland

3.1 Approach

In this section, a case study into the Orange River and the transboundary protected wetland 'Orange River Mouth' will be presented, especially focussing on the practical application of the principle of reasonable and equitable use and the concept of common river basin management, as worked out in the UN Water Convention and the UNECE Convention on international watercourses.[55] The methodology used for this case study is as follows. First, an extensive desk study was carried out into the relevant legislation, policy documents, evaluation studies, minutes of relevant meetings, and case law. In addition, meetings were attended, interviews with relevant key persons were held, and site visits were conducted.[56] The data collected has been analysed using the concept of multi-level governance as a theoretical framework.[57]

---

[55] The findings of the case study are only briefly presented here. A much more detailed report of this and another case study have been published in a comprehensive article on the Ramsar Convention. *See* Jonathan Verschuuren, 'The Case of Transboundary Wetlands Under the Ramsar Convention: Keep the Lawyers Out!', 19 *Colo. J. Int'l Envtl. L. & Pol'y* 49 (2008).

[56] A total of seventeen interviews were held. The data are on file with the author.

[57] The theoretical framework will not be worked out here because of a lack of space (*see* note 55 above).

3.2 Introduction to the Case

The Orange-Senqu river basin is a large river basin covering an area of approximately one million square kilometres in Lesotho, Botswana, South Africa and Namibia, with a total population of 14.27 million. In downstream Namibia, there is an average annual rainfall of only 185 mm, and so this area depends heavily on the surface and ground water available in the river basin. Upstream uses (for power generation, irrigation, households, industries and mineral mining) determine the fate of the dry areas downstream, and thus the fate of the people living there, as well as the fate of a transboundary wetland of international importance, located at the estuary of the river, in the dry Namibian/ South African desert. Upstream use is very high, especially for irrigation purposes.[58] Large water transfer schemes have been developed throughout the South African and Lesotho section of the river basin.[59] Release of water is determined by the operator of several dams constructed for the generation of hydropower.[60] A further growth in the water requirement for urban, industrial and mining use is expected.[61]

At the very end of this river lies a wetland of international importance, designated as such under the Ramsar Convention by both Namibia and South Africa – the Orange River Mouth wetland (ORM). The wetland consists of a dynamic estuary ecosystem. During high tide, water from the Atlantic Ocean enters the river mouth; during floods, the Orange River transports fresh water well into the ocean. The water level of the river varies with seasonal changes. Sometimes, the water level is so low that the mouth closes. Shifting sandbanks and mudflats, small islands, channel bars, as well as littoral salt marshes are the result of the 'rhythmic tidal inundation'.[62] The Orange River Mouth is the only wet and green

---

[58] In 2000, 88 per cent of the total gross water use was for irrigation purposes. *See* Dept. of Water Affairs and Forestry, Internal Strategic Perspective for the Orange River System: Overarching 2–8 (Pretoria: DWAF, 2004) [hereafter DWAF – Internal Strategic Perspective].

[59] Alan H. Conley and Peter H. van Niekerk, 'Sustainable Management of International Waters: The Orange River Case', 2 *Water Policy* 131, 137 (2000).

[60] Dept. of Water Affairs and Forestry, Internal Strategic Perspective for the Orange River System: Overarching 2–8 (Pretoria: DWAF, 2004) 2–9.

[61] Ibid., 2–12.

[62] Northern Cape Province (NDEC), 3rd Draft Orange River Mouth Management Plan 1 (2004).

area in an arid environment, both on the Namibian and the South African side of the river. The nearest coastal wetlands are 400 and 500 kilometres to the South and North respectively. Therefore, the area is important for waterfowl, both migratory and breeding birds, including several rare and endangered species.[63]

Since the early 1990s, the area has been degraded.[64] The salt marshes have dried up as a consequence of the construction of a road cutting off the salt marshes from the river and as a result of a general scarcity of water. The river lost its seasonal water level changes as a consequence of several upstream dams that control the amount of water in the river.[65] The number of birds dropped from around 25,000 in the mid-1980s to 6,200 in 2001. The population of the Cape Cormorant (*phalacrocorax capensis*) totally disappeared.[66] In 2004, part of the salt marsh was restored. The number of water birds present in the area is stable now at around 7,000 birds.

3.3 Findings

A. Usefulness of the principle of reasonable and equitable use and the concept of common river basin management for transnational decision-making

The principles of international law, such as the principle of reasonable and equitable use, as well as more specific basic concepts of international water law and environmental law, such as the concepts of common river basin management and wise use of wetlands, provide adequate support and guidance to talks and negotiations in general, and decision-making in particular. Since these overarching principles are the same for all stakeholders

---

[63] D. Lincoln *et al.*, *Important Bird Areas in Africa and Associated Islands. Priority Sites for Conservation* 820 (Newbury: Pisces Publications/BirdLife International 2001) and Abe Abrahams, 'Orange River Mouth Transboundary Ramsar Site: Green Scene', 55 *African Wildlife* 46 (2001).

[64] G.I. Cowan and G.C. Marneweck, South African National Report to the Ramsar Convention 1996 at 7–8 (Pretoria: Dept. of Environmental Affairs and Tourism, 1996).

[65] Hans Beekman, I. Saayman and S. Hughes, Vulnerability of Water Resources to Environmental Change in Southern Africa 32 (Pretoria: CSIR, 2003).

[66] Mark D. Anderson *et al.*, 'Waterbird Populations at the Orange River Mouth from 1980–2001: A Re-assessment of its Ramsar Status', 74 *Ostrich* 159 (2003).

involved and since they are broadly accepted, they form a common ground for initiating talks. At the same time, they leave much discretion and thus provide ample room for the development of new ideas, policies and projects.

Although the scope of the Ramsar Convention on the one hand and the UN Water Convention and the UNECE Convention on international watercourses on the other hand differs, the adoption of the concept of integrated river basin management in the Ramsar Convention has strongly influenced laws and policies dealing with wetland conservation. This has enabled an integration of policies with regard to water management and wetland management. The concepts of integrated river basin management and wise use have also been laid down in regional water law in southern Africa, more specifically, the SADC Shared Watercourses Protocol and the SADC Protocol on Wildlife Conservation and Law Enforcement. We have seen a similar development in other regions of the world, for instance in Europe, where the EU Water Framework Directive was clearly inspired by the UNECE Convention, and the EU Wild Birds Directive is used as an instrument to protect Ramsar sites. In addition, bilateral and multilateral agreements on specific river basins have been concluded, again based on the same principles and concepts (see further below, under Section C).

All relevant conventions and protocols thus stimulate states to adopt an integrated perspective to protected areas, integrating wetland management and (general) water management. They also offer a framework for close cooperation between states in the management of transboundary sites.

However, the case study also shows that the effect of international water law remains limited to the general guidance that these principles and basic concepts thus offer. More specific obligations arising, for instance, from the Ramsar Convention, or from soft law documents, such as the various Handbooks that accompany the Convention, hardly play a role in legal practice.[67]

The main reason is that most of the people involved are only concerned with the overarching concepts. They fear that a very detailed framework will hamper their discussions on the relevant

---

[67] Only the listing in the Montreux record of the ORM area is an important factor. It serves as an impetus for the South African national authorities to be involved in the management of the area.

water issues. Negotiations and talks on delicate issues such as the distribution of the available water only flourish when there is enough space for manoeuvering. In fact, the distribution of water is not the only delicate issue that is at stake. The case shows that there are other complicated and sensitive legal issues that have to be dealt with as well, such as a border dispute between South Africa and Namibia,[68] and a land claims issue on the South African side of the wetland.[69] Difficult enough as these issues are, there is no need for legal norms that only further complicate things, hence also the problems with national law, dealt with under Section B. The talks on all of these issues mainly take place within bilateral and multilateral commissions. These commissions that play a crucial role in decision-making processes on transboundary water issues have been dealt with under Section C.

B. Complications brought about by national law

Problems particularly arise at the national level where different legal systems exist on each side of the border and, more importantly, where a variety of competent authorities have their own specific legal domains.

Let us first take a brief look at South African and Namibian water laws. In South Africa, water legislation is mainly set out in the National Water Act 1998 (NWA).[70] The NWA is aimed at water management in a broad sense and introduces Catchment

---

[68] Gerhard Erasmus and Debbie Hamman, 'Where Is the Orange River Mouth? The Demarcation of the South African/Namibian Maritime Boundary', 13 *South Afr. Yb. Int'l L.* 49 (1987–1988) and Heidi Currie, The Namibian-South African boundary question in terms of the modern law of equitable maritime boundary delimitation (LLM Dissertation 2005). *Source:* http://www.feike.co.za.

[69] Hanri Mostert and Peter Fitzpatrick, 'Law Against Law: Indigenous Rights and the Richtersveld Cases', 2 *Law, Social Justice & Global Development J.* 1 (2005).

[70] Act No. 36 of 1998 [hereafter NWA]. The other main water statute is the Water Services Act No. 108 of 1997 [hereafter WSA]. Both Acts replace over one hundred previous statutes dealing with water. *See* Jan Glazewski, *Environmental Law in South Africa* 427 (Durban: LexisNexis Butterworths, 2005). The WSA provides the regulatory framework for local authorities to supply water and sanitation services in their area and is not so relevant for wetlands management.

Management Agencies as the competent authorities for an entire river basin. The quantity and quality of the water that is needed to satisfy basic human need,[71] *and* to protect the aquatic ecosystem of river and wetland is specified in the 'Water Reserve'.[72] Once the Reserve has been determined, it must be observed when exercising any power or performing any duty in terms of the NWA,[73] such as granting licenses for the use of water (where 'use' includes taking of water, discharging substances, altering the bed, banks, course or characteristics of a watercourse, using the water for recreational purposes, etc.[74]). This means that other allocations, such as the use of water for irrigation purposes or for domestic use beyond basic human need, can only be granted to the extent that water remains after the Reserve has been set aside. In theory, this is a very interesting principle balancing the individual human right and the duty to protect valuable ecosystems, such as wetlands.[75] However, in practice, it appears to be difficult, if not impossible, at least in some parts of the country, to set a sufficient 'Reserve', let alone to have the remaining water to be distributed for other purposes. It appears that in such a situation of scarcity, ecosystems, despite their good legal position, are the worst affected.[76] The statute also regulates the prevention and control of pollution through a system of individual and general authorisations.

---

[71] Twenty-five litres of safe drinking water per person. *See* section 2.1 of the text.

[72] NWA, note 70 above, Section 16. The other main water statute is the Water Services Act No. 108 of 1997 [hereafter WSA]. Both Acts replace over one hundred previous statutes dealing with water. *See* Jan Glazewski, *Environmental Law in South Africa* 427 (Durban: LexisNexis Butterworths, 2005). The WSA provides the regulatory framework for local authorities to supply water and sanitation services in their area and is not so relevant for wetlands management.

[73] NWA, note 70 above, Section 18.

[74] Ibid., Section 21.

[75] The creation of a 'Reserve' is considered to be a formidable innovation by Robyn Stein, 'Water Law in a Democratic South Africa: A Country Case Study Examining the Introduction of a Public Rights System', 83 *Texas L. Rev.* 2167, 2181 (2005).

[76] Interview at Dept. of Water Affairs and Forestry, Pretoria, 18 May 2006 [hereafter Interview at DWAF]. The two legislations open up the possibility to expropriate any property in the public interest (Section 65 of NWA). This may be a final resort to reduce existing (historic) water uses in areas where the reserve cannot be met.

In Namibia, the WRM Act[77] mainly deals with allocation of water for human use. As already stated above, this Act acknowledges the right to water as a basic human right. However, the WRM Act also introduces river basin management and the establishment of Basin Management Committees.[78] Once these have been established, it is believed that Namibian and South African water management in the Orange River basin can be better aligned.[79] The differences between Namibian and South African water legislation are considered to be obstacles for water management cooperation in the Orange River basin.[80] The WRM Act forms the basis for joint water management in line with the SADC Protocol on Shared Watercourses.[81] It has comparable provisions to the South African NWA. Interestingly, the WRM Act states as one of its fundamental principles:

The harmonisation of human needs with environmental ecosystems and the species that depend upon them, while recognising that those ecosystems must be protected to the maximum extend.[82]

According to the WRM Act, water has to be reserved to meet domestic household needs *and* to protect aquatic and wetland ecosystems.[83] The abstraction of water can be subject to environmental impact analysis,[84] and the impact on aquatic ecosystems has to be taken into account when granting licenses to abstract and use water.[85]

---

[77] Act No. 24 of 2004, Namibian Government Gazette No. 3357 of 23 December 2004 [hereafter Namibia WRM Act].

[78] Section 12. *See* also Maria Amakali and Loise Shixwameni, River Basin Management in Namibia (Paper presented at the 3rd WaterNet/Warfsa Symposium 'Water Demand Management for Sustainable Development', Dar es Salaam, 30–31 October 2002). *Source:* http://www.waternetonline.ihe.nl/aboutWN/pdf/Amakali&Shixwameni.pdf.

[79] Interview at DWAF.

[80] Ibid.

[81] Ibid., Section 54(b).

[82] Ibid., Section 3(d).

[83] Ibid., Section 27(1).

[84] Ibid., Section 33(3) (c).

[85] Ibid., Section 35(b).

Although the overarching principles and concepts of international water law have been taken as a starting point in national legislation in both countries, on a more detailed level, national legislation in both countries still is vastly different, which makes it difficult to manage the area in an integrated manner.[86] The Namibian WRM Act, for instance, does not contain a distinction between a river and an estuary, whereas the South African NW Act does. Another example is the important role that the provinces play in nature conservation matters in South Africa in addition to that of the national authorities whereas, in Namibia, nature conservation is completely centralised. In addition, within each country, there are systematic differences between water legislation on the one hand and nature conservation legislation on the other. The NW Act explicitly recognises wetlands as a type of water for which specific requirements are set, whereas the South African Protected Areas Act 2003 does not recognise this habitat type. Under that new law, Ramsar sites are only protected after they have been explicitly designated as protected areas.[87] Therefore, from a strictly legal point of view, the ORM remains largely unprotected in South Africa. Fortunately, the SADC Protocols, especially the one on shared watercourses, enable cross border cooperation on these issues, although this is a slow process.[88]

Legal complexities like these have to be overcome within the joint management commissions that deal with the water issues concerning a transboundary river basin. These commissions have been dealt with under Section C. The way the commissions try to overcome the legal complexities originating not from the international, but from the national, levels have been dealt with under Section D.

C. Existence of several joint commissions

Over the past fifteen years, various bilateral and multilateral water commissions have been established, such as the abovementioned ORASECOM based on the UN and UNECE conventions on

---

[86] Interview at DWAF.
[87] National Environmental Management: Protected Areas Act No. 57 of 2003, South African Government Gazette No. 26025 of 18 February 2004.
[88] Interview at DWAF.

international watercourses and the SADC Protocol on Shared Watercourses. In addition, already in 1992, Namibia and South Africa established a Permanent Water Commission (PWC),[89] acting, like ORASECOM, as a technical advisor for the competent authorities in both countries on trans-frontier water-related issues. At the same time, the Vioolsdrift and Noordoewer Joint Irrigation Authority was established,[90] to administer a joint irrigation scheme, allowing both countries to divert water from the Orange river for irrigation purposes.[91]

A bilateral committee that only deals with the Orange River Mouth wetland is the Orange River Mouth Interim Management Committee (ORMIMC). This informal committee meets twice a year, and consists of all the stakeholders involved in the area, such as various divisions of the Namibian Ministry of Environment and Tourism, the Namibian Department of Water and Agriculture, the Namibian Ministry of Fisheries and Marine Resources, various divisions of the South African Department of Environmental Affairs and Tourism, the South African Department of Water Affairs and Forestry, the Northern Cape Provincial Department of Tourism, Environment and Conservation, the Alexkor and Namdeb mining companies. The committee also includes the Namibian zinc mining company Skorpion Zinc, the Richtersveld community and the Richtersveld municipality, the South African Coastal Working Group NGO, the South African National Biodiversity Institute's Working for Wetlands Programme, and estuarine researchers of South Africa's University of Port Elizabeth. The Committee serves as an advisory body to the respective competent authorities. The Committee has no formal legal basis, although it is frequently mentioned in policy documents, such as the South African National Environmental Management and Implementation Plan.[92] The ORMIMC is considered to be the driving force behind current initiatives at the Central Government level in South Africa to rehabilitate the area, get it removed from the Montreux record, to get the area designated as a protected area under South African

---

[89] Signed in Noordoewer on 14 September 1992.
[90] Agreement on the Vioolsdrift and Noordoewer Joint Irrigation Scheme, signed in Noordoewer on 14 September 1992.
[91] Ibid., Article 3(3).
[92] Para. 3.2.1.6 of the Plan, General Notice No. 354 of 2002.

law,[93] and to draft a management plan for the Ramsar site to be used by Alexkor and the Richtersveld community. When the area has been formally declared a provincial nature reserve, the ORMIMC will probably be replaced by a formal management organisation.[94]

D. The introduction of a co-operative governance approach does not guarantee success

Within such committees and commissions, a cooperative governance approach, with all stakeholders involved, is applied. Since the national government alone does not have decisive power over the river basin or the Ramsar site, the two (or more, if you take the Orange River basin into account) national governments depend on each other. In addition, there are provincial and local authorities that have a say in the management of the river basin too, as well as functionally organised authorities, such as the water authorities. Obviously, these exist on both sides of the national border. Co-operation between all of these authorities is achieved through various commissions and committees, some of which have been described above. Some of these, especially the ORMIMC, have a broad network-like structure without a clear legal basis, involving not only public authorities, but private actors as well, such as private companies, local communities and NGOs from both Namibia and South Africa.

The case study shows that the involvement of such stakeholders is essential. Once the differences between the various parties involved are overcome, the road is paved for the national governments to reach a common position on water use. Informal and non legalistic structures, such as the ORMIMC, offer a platform where such agreements can be reached.

The involvement of stakeholders also allows the establishment of 'co-management' of the wetland. Co-management has been defined as the active participation in the management of a wetland by the community of all individuals and groups having some

---

[93] Interview at Department of Environment and Tourism, Pretoria, 17 May 2006 'The ORMIMC picks us up and drives us. Without them, probably nothing would have happened'. Also, 'We rely on IMCs because they are our eyes and ears at a local level'.
[94] Ibid.

connection with, or interest in, that wetland.⁹⁵ The ultimate goal of co-operative management is to achieve sustainable utilisation of the wetland's resources through sharing authority and responsibility with the people who work and live in and near the wetland.⁹⁶ In addition, voluntary compliance will be stimulated. It was concluded from the case that voluntary compliance should be the prime option, rather than government monitoring and enforcement.

According to some, stakeholder involvement should not be restricted to smaller areas, such as the Ramsar site, but extend to the entire river basin.⁹⁷ In my view, this is a rather theoretical option in river basins as large as the Orange River with more than fourteen million inhabitants.⁹⁸ Therefore, I think that stakeholder involvement will still have to take place at the level of a protected wetland within such a body as the ORMIMC. In addition, some stakeholder involvement at the river basin level will have to be organised by the catchment authority, such as ORASECOM in the case of the Orange River, but the sheer size of the area should lower the expectations of the outcome of such a stakeholder process.⁹⁹

Obviously, it is then essential that when several cooperative governance processes take place within a river basin, there is a good, communicative and open relationship between the various

---

⁹⁵ Gordon Claridge and Bernard O'Callaghan, *Community Involvement in Wetland Management: Lessons from the Field* 19 (Kuala Lumpur: Wetlands International, 1997).

⁹⁶ Ibid., 25 and 30.

⁹⁷ UNEP/Wetlands International, Integrated River Basin Management – Experiences in Asia and the Pacific 135 (Kuala Lumpur: Wetlands International, 1997).

⁹⁸ Savenije and Van der Zaag argue that some decisions are to be taken at the river basin level, while others should be taken at a much lower level such as the sub-catchment. *See* Hubert H. G. Savenije and Pieter van der Zaag, 'Conceptual Framework for the Management of Shared River Basins, With Special Reference to the SADC and EU', 2 *Water Policy* 9, 26 (2000).

⁹⁹ Research shows that integrated river basin management by more than two states is extremely difficult. *See* Richard E. Just and Sinaia Netanyahu, 'International Water Resource Conflicts: Experience and Potential', *in* Richard E. Just and Sinaia Netanyahu eds, *Conflict and Cooperation on Trans-boundary Water Resources* 1, 24 (Boston: Kluwer Academic Publishers, 1998).

bodies, so that the cooperative governance processes are well coordinated. In addition, it is important that the general public in the area is well informed by the stakeholders that are involved in the process. There is a danger that the general public that is not involved in the stakeholder process is not able to keep up with developments within the inner circle of the stakeholders. Resistance from the general public against the results of the stakeholder process may cause serious setbacks, once politicians feel that they cannot ignore this resistance when reaching final decisions to implement the outcome of the stakeholder process.

When looking at the content of the work of these commissions and committees from a lawyer's perspective, it must be concluded that the law is intentionally excluded from this process as much as possible. The actors involved try to overcome the legal complexity by abstracting or withdrawing somewhat from the law. They enter into talks and negotiations in order to discover the best way to deal with water issues, or, in the ORM case, to manage the protected wetland, taking into account the interests of all parties involved. In my observation, all the participants had the best intentions to conserve the area, but during these talks, they did not want to be impeded by the legal details. They simply temporarily withdrew from the complex legal situation in order to discover what they actually want to achieve from the management of the protected area.

Has the problem concerning the battle for water between Namibians, South Africans and the cormorants been solved? The answer is no, at least not yet. Agreements have been reached to reserve water for the wetland by the dam operators, especially by the new Vioolsdrift dam. However, the water flow remains constant (thus, not allowing for seasonal changes) and scarce, due to other necessary uses, such as irrigation and the provision of drinking water. There are promises to increase the amount of water released upstream and to further open the mouth, allowing more seawater into the wetland.[100] However, there is a danger of further disruption of seasonal changes and of the area being totally flooded; there

---

[100] Artificially opening or closing the estuary mouth, if carried out injudiciously, may have a detrimental effect on the estuary. Hence, there now are guidelines for such a process. *See* Lara van Niekerk and Piet Huizinga, *Guidelines for the Mouth Management of the Orange River Estuary* (Stellenbosch: CSIR, 2005).

have to be tidal flows in the salt marsh in order to retain it. The focus of future talks is on this issue.

In general, it was concluded that the stakeholder process is not the end, but merely an (important) first step towards an equitable and wise use of the scarce water. Once the goals have been laid down in such a process, legal procedures will have to be followed to mould the various agreements into policy plans, permits and other decisions taken by governmental authorities, and in company management plans and other decisions at the level of business corporations. The conversion of the agreements into legal decisions by a variety of institutions (government agencies in all states involved on various levels, individual business corporations, and NGOs) appears to be a difficult and dangerous task.

This task is difficult because of the complicated legal situation described above. Several authorities will have to apply different pieces of legislation to implement the outcome of the cooperative governance process. Since, as is shown, the stakeholders have some idea of the existing legal requirements but are not concerned with the details, a certain outcome may very well prove to be difficult to convert into legal decision-making. Sometimes, the norms in the various acts and regulations simply coexist, but applying them to the same area could result in a contradictory outcome.

The task of implementing the outcome into legal decision-making is also dangerous, because the stakeholders may not recognise the outcome of the talks in the decisions that the competent authorities take. This may result in disappointment about the entire process and in a withdrawal from co-management of the wetland and resort to other means to achieve their goals, for instance, going to court.[101] There is a risk that the competent authorities at various levels in both countries think that after an agreement is reached, the various joint bodies are no longer needed. They sometimes even seize the opportunity that the law offers to take a decision that is contrary to the outcome of the stakeholder process. After the talks are over and legal decisions are to be taken, competent authorities tend to revert to their old position, using their own specific legal domain to 'do it their way'.

---

[101] Piet Gilhuis *et al.*, 'Negotiated Decision-Making in the Shadow of the Law', *in* Boudewijn de Waard ed., *Negotiated Decision-Making* 219, 225 (The Hague: BJu, 2000).

4. Conclusion

The right to water has been recognised as a human right under various international human rights instruments, such as the International Convention on Economic, Social and Cultural Rights and the Convention on the Rights of the Child. These legal instruments primarily focus on access to safe drinking water. On the other hand, various international legal instruments to protect nature have led government institutions to reserve enough water for protected areas, for instance, wetlands of international importance designated under the Ramsar Convention. These international legal obligations may conflict. A wetland situated in the estuary of a river can be severely damaged or even destroyed when too much water is used for human purposes upstream (not only for drinking water, but also for irrigation purposes, generation of energy, industrial uses, etc). The situation gets even more complicated when the river is located in more than one country.

Therefore, the first conclusion is that it is important to include the notion of integrated river basin management into the debate on the human right to water. The 1999 Protocol on Health and Water to the UNECE Convention on the Protection and Use of Transboundary Watercourses and International Lakes, and the South African National Water Act provide good examples of how such integration should be provided for in legal texts.

Integration in legal texts is relatively easy compared to integration in legal practice. In theory, the principle of reasonable and equitable use and the concept of common river basin management, as laid down in the Convention on the Law of the Non-navigational Uses of International Watercourses, offer a way out of the potential conflict between human uses and ecosystem uses. They should allow for a fair and reasonable distribution of the available amount of water in the entire river basin for all relevant purposes, discussed in a transnational commission in a co-operative setting, and involving all relevant stakeholders.

However, these theoretical concepts are not easy to implement in practice for several reasons. Obviously, one reason is that in some areas, there simply is too little water to reconcile the realisation of the human right to water and the protection of wetlands. Another reason is the legal complexity of cases like these.

A large number of legal rules apply to any given area – international law, regional law (EU law, or in southern Africa, SADC law), national law and local or provincial law in all countries involved, not only on water, but also on other issues such as environmental protection. National legislation should, as is the case both in Namibia and South Africa, regulate the balancing of the various interests involved, especially the right to water and the duty to protect aquatic ecosystems.

A cooperative governance approach, where all relevant stakeholders together try to figure out how the available water is to be reasonably and equitably shared, is an important mechanism to achieve an outcome that is acceptable for all. To achieve such an outcome, the stakeholders temporarily withdraw from the legal specifics and focus on the main principles of the relevant international law. The case study presented here shows that a co-operative governance approach involving all relevant stakeholders is successful. Conflicts of interests are overcome, paving the way for long-term integrated and sustainable management of the site, avoiding legal conflicts within or between the states involved. However, such a stakeholder process is time consuming and slow, and should be undertaken carefully, keeping a close eye on all sensitive positions.

Although, often successful at first, the process may very well run into legal complexities once the carefully concluded agreements are to be consolidated into legal decision-making at all levels of government, in all the countries involved. This complexity can be seized by those within the government that want to do it their way. Therefore, it is important that the co-operative governance process continues during the translation of the agreements into legal decisions, and that all relevant government institutions are actually involved in the process and committed to its outcome.

14

South Africa's Water Law and Policy Framework: Implications for the Right to Water

ALIX GOWLLAND-GUALTIERI

Introduction

The post-apartheid reforms in South Africa which put into place the existing water framework were intended to redress the disparities inherited from the prior racial segregation policies. Apartheid had entrenched stark inequalities between black and white communities also in the face of access to water, while the natural scarcity of national freshwater resources in South Africa also contributes to diminishing availability of water and increasing competition between the various users.[1] Consequently, water reform policy and water justice became a central aspect of the new Government's policy of reconstruction and development[2] and

---

[1] On the history of water in South Africa, *see* R. Francis, 'Water Justice in South Africa: Natural Resources Policy at the Intersection of Human Rights, Economics, and Political Power', 18 *Georgetown Intl Envtl L.Rev.* 149 (2005); D.D. Tewari, 'A Brief Historical Analysis of Water Rights in South Africa', 30 *Water International* 184 (2005).

[2] In particular, *see* South Africa, White Paper on Water Supply and Sanitation Policy (1994) [hereafter 1994 White Paper]; South Africa, White Paper on Reconstruction and Development (1994) [hereafter 1994 White Paper on Reconstruction and Development]; South Africa, White Paper on a National Water Policy for South Africa (1997) [hereafter 1997 White Paper].

indeed remain very topical issues today. The right to water was entrenched in the constitution adopted in 1996 and in subsequent legislation, and its implementation was furthered a few years later by means of a policy of 'free basic water' adopted by the government.

The South Africa experience is interesting in that it sheds light on developments taking place in the context of renewed interest for the formalisation of a right to water in international law as well as in the national legal orders of a growing number of countries. While on the one hand the implementation of the right has resulted in the development of a policy of free entitlement to water for consumption and domestic use, there remain in South Africa huge disparities in access to basic water services and allocation of water, mostly as a legacy from the apartheid regime but also as the result of the application of an essentially economic approach to water policy. The integration of such concepts as cost-recovery and privatisation in water policy have been viewed as maintaining the poorest segments of the population with little or no access to water for household needs and sanitation, and limited water infrastructure. This creates tensions that underpin the definition and implementation of the right to water.

Part 1 of this chapter describes the law and policy framework for the right to water in South Africa, where the constitutional right has been concretised in a number of legislative and policy documents including the Water Services Act (WSA), the National Water Act (NWA) and the Free Basic Water Policy. Parts 2, 3 and 4 turn to some of the challenges observed in the realisation of the right to water which relate more specifically to the application of economic policies to water. The chapter ends with concluding remarks.

1. Law and Policy Framework for the Right to Water

A Constitutional Protection

South Africa is remarkable in that it formally recognises the right of access to water at the constitutional level, where it underpins the whole law and policy water framework. The constitution adopted on 8 May 1996 represents the cornerstone of the sweeping water policy reform that was undertaken in the period of transition following the end of the apartheid regime.[3] It contains a comprehensive bill of rights including *inter alia* the right to a healthy environment, housing, health care, food and social security, education, and culture.[4] The right of access to water is set forth in Section 27, which reads:

'(1) Everyone has the right to have access to
a. [...]
b. sufficient [...] water; and
[...]

(2) The state must take reasonable legislative and other measures, within its available resources, to achieve the progressive realisation of each of these rights [...]'.

The content of the right found in Section 27 relates both to physical and economic access to water. This obligation is tempered by the fact that the state has to take only 'reasonable' legislative and other measures 'within its available resources' to achieve the 'progressive realisation' of the right of access to water.[5] The Constitution does not directly provide for the right of individuals to access water, but rather places an obligation on all three spheres of government to take reasonable action to give effect to the general rights of the population. While the national government is required to establish a framework to ensure the realisation of this right, local governments have the responsibility to ensure the delivery of water to their communities. The 1996 Constitution provides as a corollary that constitutional rights may only be limited 'to the extent that the limitation is reasonable and justifiable in an open

---

[3] South Africa, Constitution of 1996 (Constitution Act 108, 1996) [hereafter 1996 Constitution].

[4] Ibid., at Sec. 24, 26, 27, 29 and 31. On the inclusion of environmental rights in the South African Constitution, *see, e.g.,* J. Glazewski, 'Environmental Rights and the New South African Constitution', *in* A. Boyle and M. Anderson, *Human Rights Approaches to Environmental Protection* 177 (Oxford: Clarendon Press, 1996).

[5] Sec. 27 (2), 1996 Constitution, note 3 above. *Note* that Section 28 (1)(c), which concerns the right of 'every child ... to basic nutrition, shelter, basic health care services and social services', does not include such a qualification.

and democratic society based on human dignity, equality and freedom, taking into account all relevant factors …'.[6] Relevant factors include *inter alia* the nature of the right and the importance and purpose of the limitation.

The question of whether the constitutional social and economic rights are justiciable has been a central one when addressing the implications of the right to water.[7] While the Constitutional Court has not yet ruled on a case concerning the right to water, a lower court has found that an alleged violation of the right is indeed a justiciable matter.[8] In 2000, the Constitutional Court adopted the so-called 'Grootboom' decision, which concerned the justiciability of the right of access to housing.[9] The case addressed more specifically what is entailed by the obligation of the state to take reasonable legislative and other measures within the available resources of the state so as to progressively fulfil socio-economic rights.[10] The case is important in describing how state policies can be reviewed by a court on the basis of reasonabless. The reasonabless inquiry examines first whether responsibilities and tasks have been allocated to the different spheres of government and whether appropriate financial and human resources are available. Second, it dictates that programmes for socio-economic rights obligations must be balanced and flexible, and include the appropriate provision for responding to crisis situations. While the Constitutional Court has found that socio-economic rights are

---

[6] Ibid., Sec. 36 (1).

[7] Traditionally, only civil and political rights have been considered justiciable. *See*, S. Liebenberg, 'The Value of Human Dignity in Interpreting Socio-Economic Rights', 21 *South African J. Human Rts* 1; M. Swart, 'Left Out In The Cold? Crafting Constitutional Remedies For The Poorest Of The Poor', 21 *South African J. Human Rts* 215 (2005).

[8] *See Highveldridge Residents Concerned Party v Highveldridge Transitional Local Council* [2002] 6 SA 66 (T) [hereafter *Highveldridge case*].

[9] *See South Africa v Grootboom* [2000] 11 BCLR 1169 (CC) [hereafter *Grootboom case*]. Discussing different interpretations that have been assigned to the judgment, *see* M. Wesson, '*Grootboom* and Beyond: Reassessing the Socioeconomic Jurisprudence of the South Africa Constitutional Court', 20 *South African J. Human Rts* 284 (2004).

[10] *See* C. Steinberg, 'Can Reasonabless Protect the Poor? A Review of South Africa's Socio-economic Rights Jurisprudence', 23 *South African L.J.* 264 (2006).

justiciable, its case-law shows that it is difficult to prove a violation of the Constitution, in particular because the plaintiff bears the burden of proving that the government's actions are unreasonable. This might constitute a significant obstacle to bringing a case based on alleged violations of the constitutional right to water.

The constitutional recognition of the right of access to water in South Africa is symptomatic of a growing movement in international law to view access to safe drinking water as a fundamental human right.[11] The adopted or planned constitutions and legislation of many states have also recognised water as a human right.[12] The existence of a stand-alone right to water has long been debated in international human rights law since it is not explicitly stated in the main instruments, in particular the International Covenant on Civil and Political Rights (ICCPR)[13] and the International Covenant on Economic, Social and Cultural Rights (ICESCR).[14] In 2002, the adoption by the Committee on Economic, Social and Cultural Rights of General Comment No. 15 on the right to water formally confirmed its status as a stand-alone human right.[15] General Comment No. 15 regards the right to water

---

[11] Using the South African constitutional recognition of the right to water as supporting the existence of an international right, *see* P.H. Gleick, 'The Human Right to Water', 1 *Water Policy* 487, 494 (1998). *See* further A. Gowlland-Gualtieri, 'International Human Rights Aspects of Water Law Reforms', *in* P. Cullet, A. Gowlland-Gualtieri, R. Madhav and U. Ramanathan eds., *Water Law for the Twenty-first Century: National and International Aspects of Water Law Reforms in India* (Abingdon: Routledge, forthcoming 2009).

[12] These include Belgium, Ecuador, Ethiopia, Kenya, the Democratic Republic of Congo, Uganda, Uruguay and Zambia.

[13] International Covenant on Civil and Political Rights, UN Doc. A/6316 (1966), 999 *UNTS* 171 [hereafter ICCPR].

[14] International Covenant on Economic, Social and Cultural Rights, UN Doc. A/6316 (1966), 993 *UNTS* 3 [hereafter ICESCR].

[15] Committee on Economic, Social and Cultural Rights, General Comment 15: The Right to Water (Articles 11 and 12 of the International Covenant on Economic, Social and Cultural Rights), UN Doc. E/C.12/2002/11 (2002). *Source:* http://www.ielrc.org/content/e0221.pdf [hereafter General Comment No. 15]. *See* UN Sub-Commission on the Promotion and Protection of Human Rights, Relationship between the enjoyment of economic, social and cultural rights and the promotion of the realisation of the right to drinking water supply and sanitation, Preliminary report submitted by Mr El Hadji Guissé,

as a fundamental one because a necessary component of the right to an adequate standing of living and to the right to health found in articles 11 and 12 of the ICESCR. In September 2007, the UN Human Rights Council (HRC) held its first session dedicated to the human right to water. The report prepared by the UN High Commissioner for Human Rights (UNHCHR) declared that 'it is now time to consider access to safe drinking water and sanitation[16] as a human right, defined as the right to equal and non-discriminatory access to a sufficient amount of safe drinking water for personal and professional uses – drinking, personal sanitation, washing of clothes, food preparation and personal and household hygiene – to sustain life and health.'[17] It provides a detailed analysis of the human right to water, and concludes that it is now clearly an explicit, stand-alone right, as supported by numerous international instruments.[18] On 28 March 2008, the HRC adopted by consensus a resolution establishing a new Independent Expert on the issue of human rights obligations related to access to safe drinking water and sanitation.[19] The mandate of the Independent Expert is first to identify, promote and exchange on best practices related to access to safe drinking water and sanitation, and, in that regard, to prepare a compendium of best practices; and second to carry out further clarification of the content of human rights obligations, including non-discrimination obligations, in relation to access to safe drinking water and sanitation.

UN Doc. E/CN.4/Sub.2/2002/10 (2002) [hereafter Sub-Commission Preliminary Report] and Report of the Special Rapporteur, Draft Guidelines for the Realization of the Right to Drinking Water and Sanitation, UN Doc. E/CN.4/Sub.2/2005/25 (2005). *Source:* http://www.ielrc.org/content/e0501.pdf [hereafter Sub-Commission Draft Guidelines].

[16] While this paper focuses on the human right to water, this right is often linked to the right to adequate sanitation.

[17] Para. 66, Report of the United Nations High Commissioner for Human Rights on the scope and content of the relevant human rights obligations related to equitable access to safe drinking water and sanitation under international human rights instruments, UN Doc. A/HRC/6/3 (2007) [hereafter UNHCHR Report].

[18] Ibid., Annexes I and II.

[19] HRC, Promotion and Protection of all Human Rights, Civil, Political, Economic, Social And Cultural Rights, Including the Right to Development, Human rights and access to safe drinking water and sanitation, UN Doc. A/HRC/7/L.16, A/HRC/7/L.16 (20 March 2008).

B The National Water Act (1998)

The preamble to the NWA,[20] adopted in 1998, embraces the human rights principles found in the 1996 Constitution. It creates a comprehensive legal framework for the management of water resources, that is, rivers, streams, dams and groundwater while recognising that 'the ultimate aim of water resource management is to achieve the sustainable use of water for the benefit of all users.' This implies amongst other things taking account of 'the basic human needs of present and future generations', 'promoting equitable access to water' and 'promoting the efficient, sustainable and beneficial use of water in the public interest'.[21] The management of water resources remains the responsibility of the national government, while local governments (municipalities) are competent for the management of water and sanitation services under the WSA.[22] Since the present chapter focuses on the latter, only some brief remarks on the NWA will be made here.

It is firstly noteworthy that national water resources are managed through a public trust rather than private ownership,[23] with the national government acting through the Minister of Water Affairs and Forestry as the public trustee.[24] As the trustee, the government must ensure that water is protected, used, developed, conserved, managed and controlled in a sustainable and equitable manner, for the benefit of all persons, and in accordance with its constitutional mandate. In addition, the NWA de-links water rights and land ownership by replacing the previous riparian system of allocation, which linked water rights to land ownership, with a compulsory licensing regime to achieve more equitable water

---

[20] South Africa, National Water Act 36 (1998) [hereafter NWA].

[21] Ibid., Art. 2.

[22] South Africa, Water Services Act 108 (1997) [hereafter WSA]. *See further* Part I (C) in the text.

[23] Art. 3, NWA, note 20 above.

[24] According to the DWAF, '[p]ublic trustee means that the Minister has authority over water throughout the country. Water is a natural resource that belongs *to all people*. As the public trustee of the nation's water resources, the Minister is responsible for public interest and must ensure that all water everywhere in the country is managed for the benefit of all people, including future generations.' [emphasis in text]. *See* South Africa, DWAF, Guide to the National Water Act, at 12.

redistribution in the population.[25] The de-linking of water use claims and land ownership is viewed as necessary to ensure that those not owning or controlling land have equal access and use of water.[26] Second, the NWA prioritises basic human needs over other water uses, which represents an important component of the realisation of the right to water.[27] For this purpose, it establishes the 'Reserve' which consists of a basic human needs reserve and of an ecological reserve.[28] This is the only right to water found in the NWA and it has priority over all other water uses; in other words, the amount of water required for the Reserve must be ensured before water resources are allocated to other water users.

C The Water Services Act (1997)

The supply of safe drinking water commonly falls under the competence of local authorities.[29] Accordingly, municipalities in South Africa are responsible for the accessibility of drinking water services for households and other municipal water users under the 1997 WSA. The WSA codifies Section 27, Paragraph 1(b), of the Constitution, which calls for the right of access to basic water supply and basic sanitation necessary to ensure sufficient water

---

[25] See Chap. 4, Part 1, NWA, note 20 above.

[26] Ibid., Chap. 4, Part 3.

[27] See, e.g., Para. 26(c), Plan of Implementation of the World Summit on Sustainable Development, UN Doc. A/CONF.199/20 (2002) [hereafter Johannesburg Plan of Implementation]. Paragraph 6 of General Comment No. 15, note 15 above, notes that priority in the allocation of water must be given to the right to water for personal and domestic uses. Priority should also be given to the water resources required to prevent starvation and disease, as well as water required to meet the core obligations of each of the ICECSR's rights. See also United Nations Development Programme (UNDP), Human Development Report 2006 (UNDP, 2006), which distinguishes between water for life in the household and water to sustain ecological systems and for livelihoods.

[28] Chap. 3, part 3, NWA, note 20 above. See also Art. 1 (1)(xviii). The concept of the reserve can be found in other nation's legislations. See, e.g., Part I, Art. 2(1), Kenya Water Act (Cap 732) No. 8 of 2002.

[29] On the responsibilities of local authorities in providing access to water, see World Health Organisation (WHO), Guidelines for Drinking-Water Quality (WHO, 3rd ed., 2006), at 11–12.

and an environment not harmful to health or well-being.³⁰ The preamble to the WSA also underscores its relationship with the NWA in that 'the provision of water supply services and sanitation services, although an activity distinct from the overall management of water resources, must be undertaken in a manner consistent with the broader goals of water resource management'. Other governmental regulations have been adopted to give effect to the right of access to water, in particular the 2003 Strategic Framework for Water Services which sets out the national framework for the water services sector.³¹

The WSA addresses questions of water quantity, quality and access. 'Basic water supply' is defined as 'the prescribed minimum standard of water supply services necessary for the reliable supply of a sufficient quantity and quality of water to households, including informal households, to support life and personal hygiene'.³² The contours of the notion of basic supply have been determined in later regulations issued by the Department of Water Affairs and Forestry (DWAF). These provide that the minimum standard for basic water supply services subsumes *inter alia* a minimum quantity of potable water of twenty-five litres per person per day or six kilolitres per household per month, available within 200 metres of a household and with an effectiveness such that no consumer is without a supply for more than seven full days in any year.³³ The government has determined that this basic amount of water should be available for free for each individual.³⁴ While it is up to each country to determine the minimum reasonable amount of water needed to cover personal and domestic uses, the World Health Organization (WHO) has stated that between 50 and 100 litres of water per person per day are needed to ensure that all health concerns are met.³⁵ The threshold of 25 litres per person

---

³⁰ *See* Preamble and Sec. 3(1), WSA, note 22 above. *See further* Part 1 (C) in the text.

³¹ South Africa, Strategic Framework for Water Services (2003) [hereafter 2003 Strategic Framework].

³² Ibid., Sec. 1(iii).

³³ *See* Para. 6.5.3. 1997 White Paper, note 2 above; Para. 3, DWAF, Regulations Relating to Compulsory National Standards and Measures to Conserve Water (2001) [hereafter 2001 Regulations].

³⁴ *See further* Part I (D) in the text.

³⁵ WHO, The Right to Water (2003), at 13 [hereafter WHO Right to Water].

per day represents the lowest level to maintain life, but this amount raises health concerns because it is insufficient to meet basic hygiene and consumption requirements. The UNHCHR Report does for itself not settle on a precise amount.[36] Physical accessibility to water means that water be provided within or in close proximity to the home in a way that provides regular water and prevents excessive collection time.[37] The WHO specifies that a water source with capacity to provide sufficient, safe and regular water should normally be within less than one thousand metres of the household and collection time should not exceed thirty minutes in order for around 20 litres a day of water to be collected.[38]

That water must be of sufficient quality implies according to the WHO that the water must represent no health risks, and must be free of toxic chemical substances and microbiological pathogens.[39] Regulations under the WSA require that water service authorities include a suitable programme to sample the quality of potable water provided by it to consumers in its development plan.[40] The WSA also requires that no person may dispose off industrial effluent without approval from the requisite authority[41] and empowers the national government to set compulsory national standards relating to the quality of the water discharged into any water services or water resource system.[42] Regulations also address responsibilities of water services institutions to carry out measures to prevent entry of objectionable substances into drains and watercourses.

The WSA provides that water service authorities have the duty to all consumers or potential consumers in their area of jurisdiction to progressively ensure efficient, affordable, economical and sustainable access to water services.[43] This duty is subject to, *inter alia*, the availability of resources, equitable allocation of resources to all current and potential consumers and the duty of consumers

---

[36] Para. 15, UNHCHR Report, note 17 above.
[37] Ibid., Para. 25.
[38] G. Howard and J. Bartram, Domestic Water Quantity, Service Level and Health (WHO, 2003), at 22.
[39] WHO Right to Water, note 35 above at 15.
[40] *See* Sec. 5(1), 2001 Regulations, note 33 above.
[41] *See* Sec. 7(2), WSA, note 22 above.
[42] Ibid., Sec. 9(1).
[43] Ibid., Secs. 3(2)–(3) and 11(1).

to pay reasonable charges.⁴⁴ A water service authority may not unreasonably refuse or fail to give access to water services to a consumer or potential consumer in its area of jurisdiction, but may impose reasonable limitations on the use of water services.⁴⁵ It however provides that in emergency situations, an authority must take reasonable steps to provide basic water supply and sanitation to all persons and may do so at the cost of the authority.⁴⁶ On equity and non-discrimination, while General Comment No. 15 highlights the principle of equity that demands that poorer households should not be disproportionately burdened with water expenses as compared to richer households,⁴⁷ the UNHCHR Report underlines that considerations of equity are often broader than the question of how to distribute water expenses and that equitable access considered within a human rights framework refers to equal and non-discriminatory access.⁴⁸ The principles of equality and non-discrimination require that no population group is excluded and that priority in allocating limited public resources is given to those who do not have access or who face discrimination in accessing safe drinking water and sanitation.⁴⁹

The question of the economic accessibility of water is a crucial one. It is addressed in more detail in the following sections.

D The Free Basic Water Policy (2001)

General Comment No. 15 stresses that the most critical target for the implementation of the right to water is to ensure the affordability of the minimum amount of water sufficient for essential personal and domestic uses to prevent disease.⁵⁰ Governments have a wide margin of discretion in choosing the most appropriate means to make water affordable. One of these means is to set a basic amount of water available for free for each individual. The concept of free basic water is not prevalent on the domestic level, and not readily recognised in international

---

⁴⁴ Ibid., Sec. 11(2).
⁴⁵ Ibid., Sec. 11(6).
⁴⁶ Ibid., Sec. 11(5).
⁴⁷ *See* Para. 27, General Comment No. 15, note 15 above.
⁴⁸ *See* Para. 22, UNHCHR Report, note 17 above.
⁴⁹ Ibid., Para. 24.
⁵⁰ *See* Paras. 12(c)(ii) and 27, General Comment No. 15, note 15 above.

documents. In addressing this issue, the UNHCHR Report does acknowledge that affordability requires that direct and indirect costs related to water and sanitation should not prevent a person from accessing safe drinking water and should not compromise his or her ability to enjoy other rights, such as the right to food, housing, health and education.[51] However, it provides also that while no one should be deprived of access to drinking water because of an inability to pay, the international human rights framework does not imply a right to free water. The possibility that safe drinking water and sanitation should be provided for free in certain circumstances is thus contemplated but not set as a rule.

South Africa is one of the few countries that has determined that a basic amount of water should be available for free for each individual in line with the constitutional requirement to progressively realise access to water for all citizens. Its policy of Free Basic Water was formally adopted in February 2001.[52] Hailed as part of the government's strategy to alleviate poverty and improve public health, the policy was a response to the significant problems that remained with respect to access to basic water and sanitation services of large parts of the population.[53] It has however been strongly opposed by private operators and multilateral financial institutions in South Africa, and has been subject to broader criticism as to its effectiveness in furthering the realisation of the right.

---

[51] *See* Para. 28, UNHCHR Report, note 17 above. These costs include both connection and delivery costs.

[52] *See* South Africa, DWAF, Free Basic Water Implementation Strategy Document (2001) [hereafter FBW Implementation Strategy].

[53] Although strides had been made since the end of apartheid in providing citizens with basic water supplies, government figure showed that in 2001, 11 per cent of the population still had no access to safe water supply, a further 15 per cent did not have defined basic service levels and 41 per cent did not have adequate sanitation services. *See* 2003 Strategic Framework, note 31 above, at iii. The 2000–2001 massive outbreak of cholera in Kwazulu-Natal and other parts of the country which killed several hundred people also brought the critical water situation faced by millions of citizens to the forefront of national and international attention. *See* D. Hemson *et al.*, *Still Paying the Price: Revisiting the Cholera Epidemic of 2000–01 in South Africa* (Human Sciences Research Council, Occasional Papers Series Number 8, February 2006. *Source:* http://www.hsrc.ac.za/research/output/outputDocuments/4077_Hemson_Stillpayingtheprice.pdf.

The Free Basic Water policy targets the water needs of the most impoverished citizens by guaranteeing each household a free minimum quantity of potable water set at six kilolitres per household per month. These regulations are based on the assumption that each individual person needs 25 litres of water per day. The amount of free water is the same for every household, irrespective of wealth and number of persons comprising it. Although it is a policy of the national government, the responsibility for implementing the policy rests with local governments.[54] The national government however provides support to local governments towards implementation. Free basic water services are to be financed from the local government equitable share, which is a constitutionally required portion of the annual national budget allocated to local governments, as well as through cross-subsidisation between users within a system of supply or water services authority area where appropriate.[55] In order to ensure the financial sustainability of the provision of free water, municipalities are required to adopt a block tariff system according to which the cost of water increases with usage, subject to the requirement that the first block of water for up to six kilolitres per household per month should be provided free. The price of water then increases for every additional block of water used by a household to ensure that those who use large amounts of water subsidise to some extent the free provision of six kilolitres of water for all households.

The idea behind the Free Basic Water policy is an ambitious and progressive one. Its implementation has nevertheless faced serious obstacles. The first shortcoming that has been pointed out concerns the lack of funding for local governments. Cross-subsidisation has not appeared to be a viable source of funding especially in rural communities where there are not enough high volume water users to cross-subsidise the provision of free water. Neither do private water companies consider providing a minimum amount of water for free as economically viable. Local governments are facing serious problems in providing for water and sanitation

---

[54] *See* Para. 2, 2003 Strategic Framework, note 31 above.
[55] *See* Para. 4.4.1, 2003 Strategic Framework, note 31 above. The Strategic Framework notes that the equitable share should have been temporarily increased for the 2003–2004 period specifically to assist local governments implement free basic services. This has not been the case so far.

services in general, which have led them to take drastic cost-recovery measures such as disconnections that deprive their residents of any access to water.[56] This in turn means that people are deprived of their free basic amount of water altogether. It is noteworthy that the UNHCHR Report underscores that when water services and facilities are operated by local authorities, states retain the obligation to ensure that these have at their disposal sufficient resources, authority and capacity to maintain and extend the necessary water and sanitation services.[57]

Second, there remain very important infrastructural problems in many areas of South Africa which means that water delivery of any kind is simply not possible. The implementation of the policy to provide free basic water therefore requires a rapid improvement in water infrastructure, especially for the rural poor.

A third problem concerns the quantity of free water that has been determined by the government as the minimum quantity necessary for survival. In a household of eight people, the six kilolitre per household per month amount translates as 25 litres per person per day.[58] The amount of 25 litres of water per person per day is considered insufficient to meet basic human needs, particularly for the urban poor, and thus has been considered not to fulfil the requirements found in Section 27, Paragraph 1(b), of the Constitution.[59] The 2003 Strategic Framework accordingly encourages water service authorities to increase the basic quantity of water provided free of charge to at least 50 litres per person per day to poor households, although this has not happened to date.[60] The limitation applicable to the amount of free water constitutes a heavy impediment to particularly vulnerable households, including those headed by women or children, and those affected by HIV/AIDS.[61] The constitutionality of the level of free basic water has

---

[56] On disconnections, *see further* Part III C(1).

[57] *See* Para. 39, UNHCHR Report, note 17 above.

[58] As the allocation of free basic water is made on a household basis and not an individual one, and since the average poor household is typically comprised of more than eight individuals, large, poor households are penalised.

[59] M. Kidd, 'Not a Drop to Drink: Disconnection of Water Services for Non-Payment and the Right of Access to Water', 20 *South African J. Human Rts* 119 (2004). WHO, The Right to Water (2003), at 13.

[60] Para. 4.4.1, 2003 Strategic Framework, note 31 above.

[61] Ibid., Para. 4.4.1.

been contested in an application submitted in July 2006 by five residents of Phiri, Soweto, against the City of Johannesburg, Johannesburg Water (PTY) Ltd and the DWAF.[62] In particular, the applicants' motion includes an affidavit by Peter Gleick maintaining that a flat level of six kilolitres of water per household per month is insufficient to meet minimum basic requirements in the urban context of Phiri for all households.[63] The Court is consequently being asked to order Johannesburg Water to provide a free basic water supply of 50 litres per person per day, which is viewed as the minimum starting point to provide people in the applicants' position with access to sufficient water as guaranteed under the 1996 Constitution.

Finally, as developed in the following sections, the Free Basic Water policy is meant to be implemented in a framework that has encouraged economic approaches to water management. Coupled with a policy of cost-recovery, this means that once a household goes over the amount of free water allocated and cannot pay, it will face having its water supply disconnected. Indeed, once consumption exceeds the free amount, charges are levied for the *full* amount. Disconnection of course means that the household will have no water at all including the free basic amount. This constitutes a severe impediment to the realisation of access to water for all.

2. Cost Recovery

While on the one hand the South African water framework includes a human rights approach to water, including the provision of a basic amount of free water, it has also been seen as embracing the economic approaches to water management actively promoted by international donors including the World Bank and International Monetary Fund (IMF).[64] While several international documents

---

[62] High Court of South Africa (Witwatersrand Local Division), *In the matter between: Lindiwe Mazibuko, Grace Munyai, Jennifer Makoatsane, Sophia Malekutu, Vusimuzi Paki (Applicants) and The City of Johannesburg, Johannesburg Water (Pty) Ltd., the Minister of Water Affairs And Forestry (Respondents)* (July 2006) [hereafter *Johannesburg Water case*].

[63] P.H. Gleick, Supporting Affidavit, Para. 8 (on file with the author).

[64] *See*, A. Baietti *et al.*, Characteristics of Wellperforming Public Water Utilities, World Bank, Water Supply & Sanitation Working Notes (May 2006).

do provide that the affordability requirement is not incompatible with the principle of cost recovery for water and sanitation services, they do define limits to cost recovery and highlight the fact that it should not become a barrier to access to safe drinking water and sanitation, notably by the poor.[65] For instance, the Johannesburg Plan of Implementation underlines that cost-recovery objectives should not become a barrier to access to safe drinking water by poor people.[66]

In South Africa, the applicable law and policy framework which has put in place a policy of cost recovery can be viewed as creating challenges to the realisation of the right to water by impeding the access of the poorer segments of the population to a basic quantity of clean water. A policy of cost recovery implies that the full cost of the operation and maintenance of water utilities should be financed through fees paid by water consumers.[67] The idea is that water usage should be priced in order to reflect the true societal cost of consuming the resource and to finance the cost of managing and delivering it to end-users. The other side of the coin is that accessibility of water services is contingent upon ability to pay. This has led in South Africa to rising use of water service disconnections, the installation of pre-paid water meters, and a certain level of privatisation of the water services sector.

The WSA puts in place a pricing scheme for water intended for domestic use.[68] Full cost recovery is tempered by the right of access to water, which implies that the cost of accessing water must be set at a level that ensures that people can have access to water without having to forgo access to other basic needs. While the WSA does make provisions for affordability, it does not explicitly set tariffs according to ability to pay. Accordingly, norms and standards for water tariffs may differentiate on an equitable basis between different users of water services, the types of water services and geographic areas, taking into account amongst other factors the socio-economic and physical attributes of each area. In setting

---

[65] *See* Para. 28, UNHCHR Report, note 17 above.

[66] *See* Para. 26(b), Johannesburg Plan of Implementation, note 22 above.

[67] *See,* Para. 6.5.3, 1997 White Paper, note 2 above.

[68] Water pricing also occurs under the NWA with regard to the cost of developing and managing water resources so that they are protected and conserved for beneficial use. These costs are recovered from water users by means of water use charges. *See* Chap. 5, Part 1, NWA, note 20 above.

these standards, the government is required to consider among other imperatives social equity, the financial sustainability of the water services and the recovery of reasonable costs.[69] Water tariffs are based on block tariffs, which are aimed at allowing for redistribution of water resources from richer to poorer areas through cross-subsidisation. The WSA moreover prescribes that the government can establish compulsory provisions and requirements for any contracts with a water service provider so as to ensure that water services are provided on a fair, efficient, equitable, cost-effective and sustainable basis and comply with the Act.[70] However, while it gives competence to the Minister to raise funds, including from Parliament, to provide subsidies to a water service institution,[71] the WSA does not provide specific guarantees of funding to local governments without an adequate tax base to support affordable water supply services. The 2001 Regulations on Water Tariffs provide that a water service institution must consider the right of access to basic water supply and the right of access to basic sanitation when determining which water services tariffs are to be subsidised.[72] When setting tariffs, the institution must differentiate between both the category and the level of services provided. Tariffs on water services designed to provide an uncontrolled volume of water must include a volume based charge which supports the viability and sustainability of water supply services to the poor, discourages inefficient water use and takes into account the incremental cost of increasing the capacity of the water supply infrastructure.

The 2003 Strategic Framework confirms that over and above basic water services and sanitation, consumers will have to pay for water services.[73] Tariffs must take into account the affordability of water services for the poor and the 'subsidies necessary to ensure the affordability of water services to poor households.' The Framework also provides that the approach of water services

---

[69] *See* Sec. 10(3), WSA, note 22 above.

[70] Ibid., Sec. 19 (5).

[71] Ibid., Sec. 64.

[72] *See* Sec. 3(2), South Africa, Norms and Standards in Respect of Tariffs for Water Services in terms of Action 10(1) of the Water Services Act (Act No. 108 of 1997) (2001).

[73] *See* Para. 4.5.3, 2003 Strategic Framework, note 31 above.

authorities must be guided by a number of principles, the first of which is 'compassion' and that consequently local governments must develop and implement credit control policies that are 'compassionate, especially towards poor and vulnerable households'.[74]

Although the WSA and other documents require of water service authorities to provide consumers in their jurisdiction with affordable access to water and the corresponding duty of consumers to pay reasonable charges for water use, cost recovery is used as a guiding principle in water services management. National policy has been to price water at a level reflecting the full cost of providing water and sanitation services to households; there has been only minimal cross-subsidisation from rich to poor households. This evidences the tensions that exist between application of full cost recovery policies, and of more progressive and equitable social policies. According to the 2003 Strategic Framework, '[t]he prices of water and sanitation services reflect the fact that they are both social and economic goods ....'[75] The application of a policy of cost recovery has created serious obstacles in the realisation of the right of access to water.[76] It has firstly led to dramatic increases in the price of water, leading to substantial debt in low-income households.[77] Since during apartheid white South Africans and the industrial sector benefited from heavily subsidised municipal services, charging communities the full cost of service delivery has led to higher rates in poor, black neighbourhoods which require the installation of basic water supply infrastructure. At the same time, provisions for financial assistance have not been sufficient or not implemented in many regions. A second issue linked to cost-recovery has been that of arrears on water bills. Great emphasis has been placed by local governments on recovering the massive arrears debt that exist in the poorest communities, despite the

---

[74] Ibid., Para. 4.5.8.
[75] Ibid., Para. 2.
[76] UN, Department of Economic and Social Affairs, Interagency Task Force on Gender and Water, A Gender Perspective on Water Resources and Sanitation, Background Paper No. 2, DESA/DSD/2005/2, submitted at the Twelfth Session of the UN Commission on Sustainable Development, 14–30 April 2004, at 16 (Proposals on the application of a sustainable cost-recovery policy).
[77] See Francis, note 1 above at 172.

evident impossibility of consumers to afford current service bills. Many households have very high municipal services arrears, which include electricity, water and waste removal.[78] A policy of cost-recovery in the water sector has also led to increases in disconnections of water services as well as the establishment in some communities of a system of prepayment for water. These latter two aspects are further developed below.

3. The Provision of Water Through the Private Sector

A further factor that has been considered an obstacle to the realisation of the right to water is the growing tendency towards the involvement of the private sector in water management, whether through what is referred to as 'corporatisation' of institutions or through more direct privatisation mechanisms.[79] The potential adverse effects that the private provision of water services can have on access to safe drinking water can lead to the suggestion that human rights obligations should prevent the private provision of these basic services. The dominant approach of UN treaty bodies and special procedures in this regard is however that the human rights framework does not dictate a particular form of service delivery and leaves it to states to determine the best ways to implement their human rights obligations. According to the UNHCHR Report, when water or sanitation services are operated by the private sector, states should ensure that such private provision does not compromise equal, affordable and physically accessible water and sanitation of a good quality.[80] They must therefore regulate and control private water and sanitation providers through an effective regulatory system which includes independent monitoring, participation, and imposition of penalties

---

[78] P. McInnes, 'Entrenching Inequalities: The Impact of Corporatization on Water Injustices in Pretoria', in D.A. McDonald and G. Ruiters (eds.), *The Age of Commodity: Water Privatization in Southern Africa* 99, 99 (London: Earthscan, 2005).

[79] In the first case scenario, water services are owned and operated by the local government but are restructured following market principles in order to increase their efficiency. In the second, the management of state-owned water services is delegated to private corporations.

[80] *See* Para. 38, UNHCHR Report, note 17 above.

in case of non-compliance. The consultation process which led to the adoption of the Report also raised the question of the responsibilities of transnational corporations and other business enterprises in relation to access to safe drinking water and sanitation.[81] The nature and scope of the responsibilities of business enterprises under international human rights law is currently being explored by the HRC through the mandate of the Special Representative of the Secretary-General on the issue of human rights and transnational corporations and other business enterprises.[82] Further elaboration is certainly needed regarding the human rights requirements concerning the private provision of water and the type of regulatory system that states must put in place in that respect.

South Africa is increasingly involving the private sector in the delivery and management of services, and for this purpose municipalities have adopted business models for water services.[83] Corporatisation of services is commonly the first step towards direct involvement of the private sector. Whether water systems are fully state-run but commercialised, or whether they have been taken over by private corporations, the focus is on the promotion of cost recovery and other market principles often at the detriment of more human rights-oriented considerations. It is thus important to note that the institutional arrangement is not necessarily the most

---

[81] *See* Office of the High Commissioner for Human Rights, Consultation on Human Rights and Access to Safe-Drinking Water and Sanitation, Summary of Discussion (May 2007), at 4.

[82] *See,* HRC, Report of the Special Representative of the Secretary-General on the issue of human rights and transnational corporations and other business enterprises, John Ruggie, UN Doc. A/HRC/4/35 (2007).

[83] *See* S. Flynn and D. Mzikenge Chirwa, 'The Constitutional Implications of Commercializing Water in South Africa', *in* D.A. McDonald and G. Ruiters (eds.), *The Age of Commodity: Water Privatization in Southern Africa* 99, 99 (London: Earthscan, 2005) at 59. In South Africa, there is no full privatisation, or divestiture of public water service infrastructure to private companies. *See* Paras. 3.4.7 and 4.1, Strategic Framework, note 31 above. A general shift towards private sector participation and privatisation of network utilities can also be observed in other African states. *See,* A. Jerome, Infrastructure Privatization and Liberalization in Africa: The Quest for the Holy Grail or Coup de Grace? (Palma de Mallorca, Spain: 4th Mediterranean Seminar on International Development, University of Balearic Islands, September 2004).

important factor in terms of application of human rights and equity principles in water service delivery. The WSA and the NWA entrenched the opportunities for private sector involvement in post-apartheid South Africa[84] and private investment represents one of the key principles buttressing the 2003 Strategic Framework.[85] Since 1999, several local governments have entered into long-term contracts with international water corporations. These include Nelspruit,[86] Dolphin Coast and Johannesburg.

Opposition to privatisation in the water sector has been active in South Africa, particularly from NGOs and unions which point to the detrimental effects on health and safety resulting from a focus on economic profit and the incentive by private service providers to provide water to wealthier areas. The question has also arisen of whether the policy of privatisation of essential services, and in particular water, is consistent with constitutional obligations relating to social and economic rights.[87] The Bill of Rights found in the 1996 Constitution is indeed not limited to state action, since its Section 8, paragraph 2, binds natural and juristic persons also. This would imply that some constitutional duties apply directly to private entities, although in the absence of related judicial cases it is unclear how a court would treat the applicability of the constitutionally-based right to water to private actors.[88] In any case, the delegation by the state of the provision of basic services to private actors does not mean that the state can delegate its human right obligations; thus, a policy to privatise or corporatise

---

[84] *See*, Sec. 19, WSA, note 22 above.

[85] *See* Para. 3.1, 2003 Strategic Framework, note 31 above. Examining the financial role of Great Britain in this push towards privatisation, *see* G. Monbiot, 'Exploitation on tap: Why is Britain using aid money to persuade South Africa to privatise its public services?' in The Guardian Unlimited (19 October 2004). *Source:* http://www.guardian.co.uk/comment/story/0, 1330405,00.html#article_continue.

[86] *See* L. Smith *et al.*, Testing the limits of market-based solutions to the delivery of essential services: the Nelspruit Water Concession (Johannesburg: Centre for Policy Studies, September 2003).

[87] There have been few studies in South Africa on the effects of privatisation from a human rights perspective.

[88] *Note* that the Constitutional Court has suggested that the duty to respect socioeconomic rights binds private actions. *See Grootboom case*, note 9 above at Para. 34.

water services to any extent must still comply with the duty to progressively realise socio-economic rights. In particular, the duty to respect and fulfil the right to water requires that the state must ensure that pricing will not make water unaffordable and that efforts are made to realise access to services for all. Decisions to restructure basic service delivery should also be based on participatory processes.

4. Disconnections and Pre-payment

A Limitations and Disconnections of Water Services

In the context of a policy of cost-recovery, limitations and disconnections of water services appear as logical options for water providers (whether public or private) in case of non-payment by users. Human rights law addresses the issue from the point of view that procedural safeguards are needed in case of water and sanitation disconnections rather than prohibiting such procedures as such.[89] While human rights obligations in relation to access to safe drinking water do not prohibit as such disconnections of water services, disconnection procedures must comprise (a) timely and full disclosure of information on the proposed measures; (b) reasonable notice of the proposed action; (c) legal recourse and remedies for those affected; and (d) legal assistance for obtaining legal remedies.[90] Additionally, the UNHCHR Report specifies that the quantity of safe drinking water a person can access may be reduced, but full disconnection may only be permissible if there is access to an alternative source which can provide a minimum amount of safe drinking water needed to prevent disease.

In South Africa, there has been a rise in water disconnections as a response to a household or neighbourhood's inability to pay for water services. The question of whether the provision of such an important resource, and indeed one that is protected in the Constitution, can legally be interrupted has therefore become very

---

[89] Water cut-offs are, however, prohibited by law in many countries, including Argentina, Australia, Austria, Belgium, Brazil, Ireland, Luxembourg, Mexico, New Zealand, Norway, Spain, Sweden, Switzerland, the United Kingdom and Ukraine.

[90] *See* Paras. 57–59, UNHCHR Report, note 17 above. *See also* Paras. 44(a) and 56, General Comment No. 15, note 15 above.

pertinent, and has been the object of several judicial decisions.[91] The WSA sets forth legal procedural and substantive criteria applicable to limitations and disconnections of water services by water providers. Overall, such procedures must be fair and equitable.[92] They must provide for reasonable notice of intention to apply the measure and for an opportunity to make representations. Section 4, paragraph 3(c) of the Act provides that a person may not be denied access to basic water services for non-payment where that person proves to the satisfaction of the relevant water services authority that he or she is unable to pay for basic services. The WSA does however not provide for the situation in which the individual suffering from the disconnection of the water supply is not the same as the person responsible for paying the bill, for instance children in schools or renters whose rent includes the provision of water. The 2001 Regulations further provide that where services are interrupted for more than 24 hours for reasons other than the user's non-compliance with conditions of service, a water service institution must ensure that the consumer has access to alternative water service comprising at least 10 litres of potable water per person per day.[93] The 2003 Strategic Framework refers more explicitly to water disconnections for domestic users. It grants service providers the right to disconnect water services to domestic consumers, although service cut-offs should only be used as a last resort.[94]

While the criteria applicable to limitations or disconnections of water services found in these documents are in general similar to those outlined in international human rights instruments, they do not go as far as to include the essential condition that '[u]nder no circumstances shall an individual be deprived of the minimum essential level of water.'[95] Indeed, when water services are disconnected, individuals are deprived from even a basic amount

---

[91] *See Manquele v Durban Transitional Metropolitan Council* [2001] JOL 8956 (D); *Residents of Bon Vista Mansions v Southern Metropolitan Local Council* [2002] (6) BCLR 625 (W); *Highveldridge case*, note 8 above; and *Johannesburg Water case*, note 62 above.

[92] *See* Sec. 4(3), WSA, note 22 above.

[93] *See* Sec. 4, 2001 Regulations, note 33 above.

[94] *See* Para. 4.5.8, 2003 Strategic Framework, note 31 above.

[95] Para. 56, General Comment No. 15, note 15 above. (Emphasis added).

of water, thereby seriously comprising the implementation of the constitutional right to water and of the Free Basic Water policy. As a result, the DWAF has called upon municipalities to refrain from complete disconnection and that even when consumers do not respect payment orders, water supply should be reduced to a 'trickle supply' to provide the free basic amount rather than being disconnected.[96] This has not appeared to be widely implemented by local governments.

B Pre-payment of Water Services

As another consequence of the application of a policy of cost recovery, the installation of pre-paid water meters mainly in the poorest neighbourhoods is becoming a means employed to ensure payment for water use. Pre-paid meters represent a 'convenient' tool for public or private water providers because since they charge for water up-front, they allow for full cost recovery with little administrative paperwork.[97] The system however creates significant hurdles for the poor and contributes to impeding their access to basic water.[98]

First, the system implies that people have to pay for water before they use it. Since in case of non-payment water is immediately disconnected, there is no room for application of the criteria found in the WSA, which require inter alia that reasonable notice of disconnections be provided and ability to pay taken into consideration. Second, the availability of water is made dependent upon the correct functioning of the devices, which in reality have proven to be complex, unreliable and faulty.[99] Third, the system

---

[96] *See* R. Kasrils, Minister of Water Affairs and Forestry, Pre-paid water meters serves peoples rights, 13 April 2004 (on file with the author).

[97] The production of pre-paid water meters has also been identified as a lucrative export market in South Africa.

[98] *Note* that pre-paid meters have been illegal in the United Kingdom since 1998 due to the adverse effects on health for the poor. They are however still used in other countries, including Brazil, Egypt, Lesotho, Namibia, Sudan, the Philippines and the United States.

[99] *See*, Public Citizen, Orange Farm, South Africa: The Forced Implementation of Prepaid Water Meters (June 2004), at 7 [hereafter Orange Farm Case Study] at 28–29.

of pre-paid water meters prevents communication between communities and water providers and thus does not allow for adequate public participation in water management. The experience of the main applicant in a judicial application involving the system of pre-paid meters illustrates well the absence of a human dimension in the context of access to water:

> 'When the free 6 kilolitres of water is finished, the water supply is discontinued without any notice. There is no person to whom I can explain the reason why I cannot pay, or why I need the water to remain connected. The pre-payment meter automatically cuts off the water.'[100]

Fourth, pre-paid meters are often installed without the provision of correct information to and consultation with local communities, and even without their consent or knowledge.[101] As a result, the installation of pre-paid water meters has forced people in the most deprived neighbourhoods to look for other, often contaminated, sources of water when they cannot afford to pay for the resource. Despite several failed experiences,[102] the installation of water meters has continued unabated and without adequate public consultation. In particular, they have been introduced in Johannesburg's surrounding townships in parallel with the privatisation of delivery of water services.[103]

The effects of the application of a policy of cost-recovery, particularly the practice of water prepayment and of disconnections, have in effect prevented the realisation of the right of access to water found in the Constitution in impoverished communities. Service cut-offs have had disastrous health consequences and have caused social unrest and violence in many communities, including the Johannesburg townships of Soweto and Orange Farm.[104]

---

[100] L. Mazibuko, Founding Affidavit, Paragraph 102 (on file with the author), in the case referred to at note 62 above. The application asks the Johannesburg High Court to declare the decisions of Johannesburg Water to unilaterally install prepayment meters in Phiri unlawful and unconstitutional.

[101] *See*, Orange Farm Case Study, note 99 above at 12–15.

[102] Ibid., 11.

[103] Although in September 2001 the City of Cape Town announced its decision not to implement pre-paid water meters in the city.

[104] P. Bond, *The Battle over Water in South Africa* (AfricaFiles). *Source:* http://www.africafiles.org/article.asp?ID=4564.

Moreover, the high administrative costs of performing service cut-offs and meter installations, or hiring collections agencies and lawyers, has meant that the provision of water has operated at a net economic loss.[105]

Concluding Remarks

The law and policy framework for water established after the apartheid era in South Africa is noteworthy particularly because the main thrust of the reforms undertaken was to entrench the right to water at the constitutional level. The constitutional right has laid the basis for legislation on water related both to the management of water resources at the national level and the management of water and sanitation services at the local one, as well as the adoption of the Free Basic Water policy. In parallel, there has been a growing recognition on the international level that access to safe drinking water is a fundamental human right and must be addressed within a legal framework, as reflected most recently in the work of the HRC. The increasing references to the right to water in human rights instruments both as a component of other human rights and as a self-standing right highlights the growing importance of this issue to the international community.

Both on the national and the international levels, however, tensions have emerged between the enunciation of a human rights approach to water on the one hand, evidenced by explicit reference to the right, and on the other by the economic approaches to water which have become prevalent particularly in the policies of international trade and financial institutions. Governments are incrementally withdrawing from direct responsibility for providing water to its citizens, and are instead divesting responsibility to private providers or local governments that have been corporatised. The application of economic policies to water in South Africa have resulted in increased commodification of the resource and have contributed in effect to posing significant challenges to the realisation of the constitutional right of access to water especially for the poorer segments of the population. Despite the recognition that '[t]he cost associated with providing free basic water to poor households is not large for a country of our economic and size',[106]

---

[105] See Francis, note 1 above at 170.
[106] See Para. 4.4.1, 2003 Strategic Framework, note 31 above.

there remain persistent inequalities in the face of access to water services and infrastructure, and the implementation of the Government's Free Basic Water policy has met with serious obstacles in addressing problems of accessibility and affordability of water. Local governments are increasingly resorting to disconnection of water services for non-payment and to the installation of pre-paid water meters which allow people to access water only if they pay for it. These measures have dramatic health consequences as people are forced to resort to polluted rivers, streams and even open pits to draw water for daily survival.

There remains an urgent need in international law to determine more precisely the contours of states' obligations in relation to access to safe drinking water, and more particularly with regard to the regulation of the private sector in the context of the private provision of drinking water; the increasing commercialisation of the water sector; the criteria needed to protect the right to water in the face of disconnection of service; and more generally the obligations of local authorities.

15

Respect, Protect, Fulfill: the Implementation of the Human Right to Water in South Africa

INGA T. WINKLER*

Introduction

Water services in South Africa have started a hot debate. On one hand, the state is often regarded as being at the forefront in terms of water services provision particularly due to its explicit acknowledgement of the human right to water and its Free Basic Water (FBW) Policy.[1] On the other hand, South Africa is a country suffering from extreme inequalities. Its GINI index[2] is extremely high with 57.8.[3] Also, access to water is extremely uneven, a legacy of the apartheid era.[4] Historically 'white' suburbs account for more

---

* The author would like to thank Dr. Lena Partzsch, Anna Zimmer, Dr. Björn Lüssem and Anna Oehmichen as well as the editors for their helpful comments on an earlier draft of this contribution.
[1] United Nations Development Programme, *Human Development Report 2006, Beyond Scarcity: Power, Poverty and the Global Water Crisis* 64 (New York: Palgrave Macmillan, 2006).
[2] In the GINI index, a value of zero represents perfect equality and a value of 100 perfect inequality.
[3] United Nations Development Programme, note 1 above at 337.
[4] Rose Francis, 'Water Justice in South Africa: Natural Resources Policy at the Intersection of Human Rights, Economics, and Political Power', 18 *Georgetown International Environmental Law Review* 149, 154 (2005). On the hydrology of apartheid cf. Ken Conca, *Governing Water: Contentious Transnational Politics and Global Institution Building* 322 et seqq. (Cambridge: MIT Press, 2006).

than 50 per cent of residential water use with 'whites' just comprising roughly 10 per cent of the population.[5] A great number of people still lack access to water services, mostly marginalised and vulnerable groups of society. Many more suffer from high prices and the use of unsafe water.[6] These inequalities in access to safe and affordable water have caused much resistance against water service commercialisation, privatisation and cost recovery on which the government mainly relies to expand services.

This chapter tries to achieve a balanced view from a human rights perspective recognising the country's achievements, but also taking a close look at the challenges in the implementation of the right to water. It analyses the implementation of the human right to water in South Africa in regard to the obligations borne by the state that correspond to the right to water under the common tripartite distinction between obligations to respect, protect and fulfill. The chapter starts by outlining the normative content of the right to water in order to determine the standard against which the implementation is to be assessed and by presenting the set of different obligations as a framework for analysing how South Africa aims to meet these obligations. This analysis starts by presenting the legislative framework and policies aimed at the implementation of the right to water. In the remaining part of the chapter, challenges to the implementation of the right are examined. Several concerns are raised which refer to the different obligations ranging from affordability concerns over widespread disconnections to the complete lack of access.

The Human Right to Water

In international law, the human right to water is guaranteed under the International Covenant on Economic, Social and Cultural Rights (Social Covenant) in particular as being derived from the right to an adequate standard of living and the right to the highest attainable

---

[5] Francis, note 4 above at 150. *See also* Jaap de Visser, Edward Cottle and Johann Mettler, 'Realising the Right of Access to Water: Pipe Dream or Watershed', 7 *Law, Democracy and Development* 27 (2003).
[6] South African Human Rights Commission, The Right to Water 4 (5th Economic and Social Rights Report Series, Financial Year 2002/2003, Johannesburg, 2004).

standard of health.[7] However, South Africa is not party to that Covenant so its provisions are not binding onto the state. Yet, the South African Constitution itself explicitly recognises the right to water. The Constitution is regarded as being one of the most progressive in the world,[8] in particular due to its far-reaching commitment to socio-economic rights.[9] Section 27(1)(b) guarantees the right to have access to sufficient food and water. The second paragraph acknowledges that the full realisation of socio-economic rights such as the right to water is a long term process and is therefore to be achieved progressively.[10]

2.1 Content of the Human Right to Water

In order to assess the implementation of the right to water and to identify any deficits, it is important to first establish its normative content. In recent years, the content of the right to water has been determined, rather detailed, mainly as being derived from provisions of the Social Covenant. The General Comment No. 15[11] of the Committee on Economic, Social and Cultural Rights – the treaty body monitoring the implementation of the Social Covenant

---

[7] *Cf.* Committee on Economic, Social and Cultural Rights, General Comment 15: The Right to Water (Articles 11 and 12 of the International Covenant on Economic, Social and Cultural Rights), UN Doc. E/C.12/2002/11 (2002), para. 3. *Source:* http://www.ielrc.org/content/e0221.pdf [hereafter General Comment No. 15] and Eibe Riedel, 'The Human Right to Water', *in* Klaus Dicke ed., *Weltinnenrecht – Liber Amicorum Jost Delbrück* 585, 596 et seq. (Berlin: Duncker & Humblot, 2005).

[8] Francis, note 4 above at 156.

[9] Craig Scott and Philip Alston, 'Adjudicating Constitutional Priorities in a Transnational Context: A Comment on Soobramoney's Legacy and Grootboom's Promise', 16 *South African Journal on Human Rights* 206, 214 (2000) and Sage Russell, 'Minimum State Obligations: International Dimension', *in* Danie Brand and Sage Russell eds., *Exploring the Core Content of Socio-economic Rights: South African and International Perspectives* 11, 13 et seq. (Pretoria: Protea, 2002).

[10] For the discussion about core obligations that aim at guaranteeing an immediate minimum standard and arise regardless of the principle of progressive realisation, *cf.* below at 4.2.2.

[11] General Comment No. 15, note 7 above.

– is particularly relevant in this regard. It is not legally binding,[12] but an authoritative interpretation of the Social Covenant.[13] Not having ratified the Covenant, the General Comment is however of no direct relevance for South Africa. Yet, some provisions of the South African Bill of Rights are very similar to those of the Social Covenant and Section 39(1)(b) explicitly calls for the consideration of international law when interpreting the Bill of Rights. The right to water as contained in Section 27(1)(b) of the Constitution can thus be interpreted similarly to the right to water under the Social Covenant.[14]

Moreover, there are other instruments at the international level that are of direct relevance for countries that have not ratified the Social Covenant such as South Africa, as they are not treaty-based. The Special Rapporteur on the right to water appointed by the Sub-Commission on the Promotion and Protection of Human Rights, El Hadji Guissé, has issued a set of draft guidelines on the right to water.[15] Moreover, the Human Rights Council has requested the Office of the United Nations High Commissioner for Human Rights (OHCHR) to conduct a study on the scope and content of the relevant human rights obligations related to equitable

---

[12] Eckart Klein, 'General Comments: Zu Einem Eher Unbekannten Instrument des Menschenrechtsschutzes', *in* Jörn Ipsen and Edzard Schmidt-Jortzig eds., *Recht - Staat - Gemeinwohl, Festschrift für Dietrich Rauschning* 301, 307 et seq. (Cologne: Heymann, 2001) and Riedel, note 7 above at 592 et seq.

[13] Riedel, note 7 above at 592.

[14] Anton Kok and Malcolm Langford also interpret the right to water by referring to international law and in particular the interpretation provided by the Committee. *See* Anton Kok and Malcolm Langford, 'The Right to Water' *in* Stuart Woolman *et al.* eds., *Constitutional Law of South Africa* 56B–9 et seqq. (Lansdowne: Juta, 2nd ed., 2005). *Cf.* as well *Government of the Republic of South Africa and Others v Grootboom and Others*, Constitutional Court of South Africa, Judgment of 4 October 2000, 2000 (11) BCLR 1169 (CC) at 1185 [hereafter *Grootboom Judgment*].

[15] Report of the Special Rapporteur, El Hadji Guissé, Draft Guidelines for the Realization of the Right to Drinking Water and Sanitation, UN Doc. E/CN.4/Sub.2/2005/25, 11 July 2005. *Source:* http://www.ielrc.org/content/e0501.pdf [hereafter Draft Guidelines].

access to safe drinking water and sanitation.[16] After consulting with states and other stakeholders, the OHCHR submitted this study[17] in August 2007. In its most recent resolution on 'Human rights and access to safe drinking water and sanitation',[18] the Human Rights Council decided to appoint an independent expert on the issue. Catarina de Albuquerque has taken up her mandate in November 2008. As these instruments belong to the realm of the former Human Rights Commission and the new Human Rights Council respectively, they are not treaty- but Charter-based and are as such relevant for all member states of the United Nations including South Africa.

2.1.1 Sufficient Quantity

When determining the normative content of the right to water, the first question regards the *quantity* of water guaranteed. This amount has to be sufficient to meet drinking purposes and other basic human needs – personal and domestic uses such as washing, cooking, cleaning and personal hygiene.[19] It is difficult to determine the exact amount of water necessary to fulfill these needs as requirements vary, for example, due to climatic conditions, but also between different groups of people. For example, people living with HIV/AIDS require larger amounts of water.[20] Yet, several studies regard 20 to 25 litres per person per day as the absolutely

---

[16] Human Rights Council, Human Rights and Access to Water, UN Doc. A/HRC/2/L.3/Rev.1 (4 October 2006).

[17] Human Rights Council, Report of the United Nations High Commissioner for Human Rights on the scope and content of the relevant human rights obligations related to equitable access to safe drinking water and sanitation under international human rights instruments, UN Doc. A/HRC/6/3 (16 August 2007) [hereafter OHCHR Study].

[18] Human Rights Council, Human Rights and Access to Safe Drinking Water and Sanitation, UN Doc. A/HRC/7/L.16 (20 March 2008).

[19] General Comment No. 15, note 7 above at para. 2 and OHCHR Study, note 17 above at para. 13.

[20] *Cf. Lindiwe Mazibuko et al. v The City of Johannesburg et al.*, High Court (Witwatersrand Local Division) of South Africa, Judgment of 30 April 2008 [hereafter Mazibuko Judgment]. *Source:* http://www.cohre.org/store/attachments/Mazibuko%20Judgment.pdf.

necessary minimum amount[21] and it can hardly be assumed that basic needs can be met with a smaller amount. The provision of such an amount does not imply the full realisation of the right to water. To achieve this, a larger quantity has to be provided progressively.[22] The World Health Organisation (WHO) regards approximately 100 litres per day as sufficient to meet domestic needs. [23]

2.1.2 Other Features

Not only the quantity of water but its *quality* is also important. Water has to be safe and of such quality that it does not pose a threat to human health.[24] Furthermore, water has to be *physically accessible* in the household or its immediate vicinity.[25] The WHO assumes basic access when water is available at a distance of up to 1000 metres.[26] The South African Government aims to supply water at a distance of less than 200 metres to everyone.[27] Furthermore, water has to be *affordable*, that is economically accessible. People must be able to realise their right to water without

---

[21] Guy Howard and Jamie Bartram, Domestic Water Quantity, Service Level and Health 23 (Geneva: World Health Organization, WHO Doc. WHO/SDE/WSH/ 3 February 2003), United Nations Development Programme, note 1 above at 3; World Health Organisation and United Nations Children's Fund, *Global Water Supply and Sanitation Assessment 2000 Report* 77 (Geneva and New York, 2000); General Comment No. 15, note 7 above at para. 12 (a), OHCHR Study, note 17 above at para. 14 and World Bank *Technical Notes* 5, available at: http://siteresources.worldbank.org/INTPOVERTY/Resources/WDR/English-Full-Text-Report/ch12b.pdf. Others regard 50 litres as minimum. *See* Gleick, 'Basic Water Requirements for Human Activities: Meeting Basic Needs', 21 *Water International* 83, 88 (1996).

[22] On the distinction between core obligations and the obligation to progressive realisation see below 4.2.2.

[23] Howard and Bartram, note 21 above at 22.

[24] General Comment No. 15, note 7 above at para. 12 (b) and OHCHR Study, note 17 above at para. 17.

[25] General Comment No. 15, note 7 above at para. 12 (c); OHCHR Study, note 17 above at para. 25 and Draft Guidelines, note 15 above at para. 1.3 (a).

[26] Howard and Bartram, note 21 above at 22.

[27] Department of Water Affairs and Forestry, Strategic Framework for Water Services: Water is Life, Sanitation is Dignity 46 (2003) [hereafter DWAF Strategic Framework 2003]. *Source:* www.dwaf.gov.za/Documents/Policies/Strategic%20 Framework%20approved.pdf.

having to compromise other basic needs such as food and housing.²⁸ Affordability can be assessed by using the percentage of household income spent on water services as an indicator. It is difficult to determine the exact percentage which exceeds affordability, but international recommendations are in a certain range: the Human Development Report of the United Nations Development Programme (UNDP) regards three per cent of household income as an appropriate benchmark,²⁹ whereas the Camdessus Report assumes five per cent.³⁰

2.2 Concept of Obligations to Respect, to Protect and to Fulfill

Human rights correspond to obligations borne by the state, which can be categorised as duties to respect, to protect and to fulfill. This concept was first developed by Shue³¹ and has become widely used, for example, by the Committee on Economic, Social and Cultural Rights. Most importantly in the South African context, it is also laid down in Section 7(2) of the South African Constitution.³²

The *obligation to respect* requires states to refrain from interfering with the enjoyment of human rights thus aiming to prevent an infringement of rights that have already been realised.³³ States

---

²⁸ General Comment No. 15, note 7 above at para. 12 (c); OHCHR Study, note 17 above at para. 28 and Draft Guidelines, note 15 above at para. 1.3 (d).
²⁹ United Nations Development Programme, note 1 above at 97.
³⁰ Michel Camdessus and James Winpenny, Financing Water For All 19 (Report of the World Panel on Financing Infrastructure, 2003). *Source:* http://www.ielrc.org/content/e0315.pdf.
³¹ Henry Shue, *Basic Rights: Subsistence, Affluence and U.S. Foreign Policy* (Princeton: Princeton University Press, 1980).
³² Republic of South Africa, Constitution of 1996, Act No. 108 of 1996. *Source:* http://www.info.gov.za/documents/constitution/1996/a108-196.pdf.
³³ Matthew Craven, *The International Covenant on Economic, Social and Cultural Rights, A Perspective on its Development* 109 (Oxford: Oxford University Press, 1995); Asbjørn Eide, 'Economic, Social and Cultural Rights as Human Rights', *in* Asbjørn Eide, Catarina Krause & Allan Rosas eds. *Economic, Social and Cultural Rights – A Textbook* 9, 23 (Dordrecht, Boston, London: Martinus Nijhoff Publishers, 2nd ed., 2001) and Michael Kidd, 'Not a Drop to Drink: Disconnection of Water Services for Non-Payment and the Right of Access to Water', 20 *South African Journal on Human Rights* 119, 121 (2004).

have to avoid any law or conduct that would result in a deprivation of access to rights. As far as the right to water is concerned, this obligation requires states to respect existing access to water which is particularly relevant for the question whether water supplies may be disconnected in the case of non-payment. Fair procedures must be followed.[34] The same holds true for access to public standpipes on which many people rely particularly in rural areas. The state must not arbitrarily stop their operation and maintenance if people are left without an alternative access to water. In any case, it is indispensable that no one may be deprived of the minimum essential level of water under any circumstances.[35]

The *obligation to protect* refers to the duty of states to prevent third parties from interfering with the enjoyment of human rights.[36] This obligation places a duty on states to implement legislation and other measures that prevent (powerful) private parties from undermining the rights of others. Third parties must not interfere with access to safe drinking water, for example, by polluting water resources. Moreover, the obligation becomes particularly relevant in the case of water service privatisation. States have to ensure that private provision does not compromise access to water by establishing an effective regulatory framework.[37] The outcome, the state has to ensure, is the same as for the obligation to respect – the right to water must not be infringed, but the state

---

[34] For the South African context, *see* Kidd, note 33 above at 129 et seqq.

[35] General Comment No. 15, note 7 above at para. 56; Draft Guidelines, note 15 above at para. 2.3 (d), 6.4; OHCHR Study, note 17 above at para. 57 and Kidd, note 33 above at 120, 132.

[36] Craven, note 33 above at 109 and Eide, note 33 above at 24. A different question discussed under the term horizontal application is whether human rights obligations are extended to the private parties themselves. It has to be considered in light of Section 8(2) of the Constitution. *Cf.* Anton Kok, 'Privatisation and the Right to Access to Water', *in* Koen De Feyter and Felipe Gómez Isa eds., *Privatisation and Human Rights in the Age of Globalisation* 259, 269 et seq. (Antwerp: Intersentia, 2005) and Sean Flynn and Danwood Mzikenge Chirwa, 'The Constitutional Implications of Commercializing Water in South Africa', *in* David A. McDonald and Greg Ruiters eds., *The Age of Commodity, Water Privatization in Southern Africa* 59, 62 et seq. (London, Sterling: Earthscan, 2005).

[37] General Comment No. 15, note 7 above at para. 24; Draft Guidelines, note 15 above at para. 2.3 (e) and OHCHR Study, note 17 above at para. 38, 53.

has to adopt different measures as it does not itself act as the water supplier, but has to act as a regulator in order to prevent private suppliers from violating the human right to water.

The *obligation to fulfill*[38] requires states to adopt the necessary measures directed towards the full realisation of the human rights.[39] This obligation aims to ensure that those people who currently lack access gain access to these rights. States have to take positive measures to enable and assist people to enjoy the right to water, in particular by allocating sufficient water resources for personal and domestic needs as well as by maintaining and expanding infrastructure to currently un-served areas such as rural areas and informal settlements. As far as the actual provision of water is concerned it has to be kept in mind that every individual is generally expected to satisfy his or her basic needs through their own efforts and resources.[40] However, when people do not have sufficient means to provide for themselves, the state is required to adopt the necessary policies to ensure that water is affordable to everyone such as subsidisation mechanisms or even supplying water free of charge to ensure a minimum essential level of water.[41] In so far, South Africa's Free Basic Water (FBW) Policy becomes relevant.[42]

Legislation and Policies Towards the Implementation of the Right to Water

3.1 The Legislative Framework

This section lays down the legislative framework for the human right to water in South Africa. The National Water Act of 1998[43]

---

[38] The obligation to promote included in Section 7(2) of the South African Constitution can be regarded as part of the obligation to fulfill. Measures to fulfill this obligation could include awareness campaigns, educational programmes, etc. *Cf.* de Visser, Cottle and Mettler, note 5 above at 29.

[39] Craven, note 33 above at 109 and Eide, note 33 above at 24.

[40] Eide, note 33 above at 23.

[41] General Comment No. 15, note 7 above at para. 25, 27; Draft Guidelines, note 15 above at para. 6.1, 6.2 and OHCHR Study, note 17 above at para. 41.

[42] See below section 3.2.

[43] Republic of South Africa, National Water Act, Act No. 36 of 1998. *Source:* www.dwaf.gov.za/Documents/Legislature/nw_act/NWA.pdf.

and the Water Services Act of 1997[44] are the most relevant legislative acts in the water sector. The National Water Act is mainly concerned with water resources and their management and protection, whereas the Water Services Act deals with the regulatory framework for water supply.[45] Important specifications are provided by a number of Ministerial Regulations. Some provisions relate specifically to the obligation to respect, to protect or to fulfill respectively, which will be pointed out.

3.1.2 The National Water Act

Section 2 of the National Water Act outlines the purposes of the Act stressing the factors of meeting basic needs, promoting equitable access and redressing historical discrimination.[46] To this end, the Act establishes the 'Reserve', a certain quantity of every single water resource reserved for basic human needs. It is defined in Section 1(1)(xviii)(a) as referring to the quantity and quality of water required to satisfy basic human needs by securing basic water supply as prescribed in the Water Services Act. It thus provides for the essential needs of individuals relying upon the water resource in question by setting aside the necessary amount. As such, the human needs reserve is an instrument to ensure that basic human needs enjoy priority in the allocation of water resources without being subject to competition with other water demands.[47] It can be regarded as a unique concept which reflects the human right to water.[48]

3.1.3 The Water Services Act

In order to fulfill basic human needs, it is not sufficient to set aside a specified amount of water; rather, it also has to be supplied. As

---

[44] Republic of South Africa, Water Services Act, Act No. 108 of 1997. *Source:* http://www.ielrc.org/content/e9705.pdf.

[45] Republic of South Africa, Department of Water Affairs and Forestry, Report of the Department of Water Affairs and Forestry 7 (Pretoria, 2007). *Source:* http://www.dwaf.gov.za/Documents/AnnualReports/2007/AnRep07full.pdf.

[46] Section 2(a), (b) and (c) of the Water Services Act, note 44 above. *See also* Conca, note 4 above at 342.

[47] Sandy Liebenberg, 'The National Water Bill – Breathing Life into the Right to Water', 1 *Economic and Social Rights Review* 1 (1998).

[48] Conca, note 4 above at 346.

its title suggests, the Water Services Act is concerned with provision of water services. Section 2(a) lists as the first of the main objectives of the Act to provide for 'the right of access to basic water supply and the right to basic sanitation'. This significance conferred to the fulfillment of basic human needs is reinforced by Section 3 which guarantees the 'right of access to basic water supply and basic sanitation' and stipulates that every 'water services institution must take reasonable measures to realise these rights'. Moreover, Section 5 gives preference to the provision of basic water supply over other uses of water. The term basic water supply is defined in Section 1(iii) as 'the prescribed minimum standard of water supply services necessary for the reliable supply of a sufficient quantity and quality of water to households, including informal households, to support life and personal hygiene'.

In the context of the *obligation to respect* an existing water supply, Section 4(3)(c) of the Water Services Act is of particular relevance. It is part of the procedures for the limitation of discontinuation of water services and provides that procedures must 'not result in a person being denied access to basic water services for non-payment, where that person proves, to the satisfaction of the relevant water services authority, that he or she is unable to pay for basic services'.[49] However, Section 11 which determines the duty of water service authorities to provide access states in its paragraph 2(g) that this duty is subject to 'the right of the relevant water services authority to limit or discontinue the provision of water services if there is a failure to comply with reasonable conditions set for the provision of such services'. These two provisions seem contradictory: Section 4(3)(c) prohibits disconnections for non-payment when people are unable to pay for services, whereas Section 11(2)(g) allows disconnections under certain circumstances. The very broad term 'reasonable conditions' in Section 11(2)(g) needs to be interpreted in terms of the more specific Section 4(3)(c). It has to be understood in a way that the disconnection must not lead to the denial of basic services for indigent people.

Section 19 of the Water Services Act is relevant for the privatisation of water services and thus relates specifically to the

---

[49] *Cf.* the analysis by Kidd, note 33 above at 131 et seq.

obligation to protect.[50] The Water Services Act determines the local governments as 'default service provider', but allows them to subcontract that task to private service providers under certain circumstances.[51] However, in this case, the obligation to protect remains with the state. It thus has to ensure through its regulatory framework that the private provider acts in accordance with the human right to water.[52]

Most important for the *obligation to fulfill* is the duty of water services authorities to provide access to water services. It is laid down in Section 11 of the Water Services Act. However, paragraph 2 qualifies this duty and states that it is inter alia subject to the availability of resources and to the duty of consumers to pay reasonable charges in accordance with any prescribed norms and standards for tariffs for water services.

3.1.4 Ministerial Regulations

In line with the authorisation in Sections 9 and 10 of the Water Services Act the Minister has issued 'Regulations relating to compulsory national standards and measures to conserve water' in April 2001.[53] Regulation 3 is especially important, as it further specifies the term basic water supply as a 'minimum quantity of potable water of 25 litres per person per day or 6 kilolitres per household per month'.

3.2 Policies Aiming at Implementation

It is not sufficient to put into place a legislative framework for the right to water; rather, its implementation through policies is essential. The Government's overall aim is to provide more people

---

[50] *Cf.* as well Sections 76 et seqq. of the Municipal Systems Act. *See* Republic of South Africa, Local Government: Municipal Systems Act, Act No. 32 of 2000, 425 Government Gazette No. 21776, 20 November 2000. *Source:* www.info.gov.za/gazette/acts/2000/a32-00.pdf.

[51] Conca, note 4 above at 353.

[52] Kok, note 36 above at 280 et seq.

[53] Republic of South Africa, Department of Water Affairs and Forestry, Regulations Relating to Compulsory National Standards and Measures to Conserve Water, 2001. *Source:* www.dwaf.gov.za/Documents/Notices/Water%20Services% 20Act/SEC9DREG-20%20April%202001.doc.

with access to water supply. Large parts of the black population did not have access when the African National Congress (ANC) came to power in 1994, an estimated 12 to 14 million people.[54]

According to government statements, ten million people had gained access in 2004 since the end of the apartheid era.[55] In its 2007/2008 Annual Report, the Department of Water Affairs and Forestry (DWAF) states that access has improved from 59 per cent in 1994 to 95 per cent of the population.[56] However, not all of these water connections meet the level of basic services, in particular because services are only provided at a distance of more than 200 m from the households.[57]

However, these figures refer to access to infrastructure which does not necessarily imply that people can afford these services. The government aims to overcome this deficit with its FBW Policy seeking to provide everyone with a minimum amount of water for free. The idea of FBW dates back to 1994 when the ANC won the first democratic elections. However, the Policy was only introduced in 2000/2001 in the wake of the rising community struggle and controversy over water cut-offs, the introduction of prepayment

---

[54] Republic of South Africa, Department of Water Affairs and Forestry, White Paper on a National Water Policy for South Africa 15 (1997). *Source:* www.dwaf.gov.za/Documents/Policies/nwpwp.pdf; Conca, note 4 above at 319; Eddie Cottle, The Class Nature of Free Water in South Africa: From Past to Present 19 (Durban: University of KwaZulu-Natal, Paper for the Centre for Civil Society, 2004). *Source:* www.nu.ac.za/ccs/default.asp?3,28,10,1186, and David A. McDonald, 'The Bell Tolls for Thee: Cost Recovery, Cutoffs, and the Affordability of Municipal Services in South Africa', *in* David A. McDonald and John Pape eds. *Cost Recovery and the Crisis of Service Delivery* 161, 162 (Cape Town: Human Sciences Research Council Publishers, 2002).

[55] Republic of South Africa, Department of Water Affairs and Forestry, Parliamentary Media Briefing 1, 10 February 2004. *Source:* www.dwaf.gov.za/Communications/PressReleases/2004 Parliamentary%20Media%20Briefing%20Release%202004.doc.

[56] Republic of South Africa, Department of Water Affairs and Forestry, Annual Report 2007/2008 at 54 (Pretoria, 2008). *Source:* www.dwaf.gov.za/documents/AnnualReports/ANNUALREPORT2007-2008.pdf [hereafter DWAF Annual Report 2008].

[57] DWAF Annual Report 2008, note 56 above at 54.

water metres and a cholera outbreak.[58] In the 2000 local elections, the ANC announced that it would provide all residents with a free basic amount of water.[59] In February 2001, the policy was officially announced by the Minister of Water Affairs and Forestry.[60] It means to provide each household with 6000 litres of water every month free of charge which amounts to 25 litres per day per person in a household of eight. FBW is financed via cross-subsidisation through a rising block tariff system. Users who consume more than the basic supply have to pay higher tariffs for additional units which results in a cross-subsidisation from high volume to low volume users.[61] Moreover, financial support is provided to municipalities through the 'equitable share', a portion of the national annual budget transferred to local governments that is calculated on the basis of the percentage of poor people living in a municipality.[62] According to the 2007/'08 DWAF Annual Report, 90 per cent of the population with access to water infrastructure benefit from FBW which translates to 84 per cent of the South African population.[63]

---

[58] Patrick Bond, De-Commodification in Theory and Practice: Fighting Human Insecurity in Post-Apartheid South Africa's Water Wars 15 (Paper Presented to the International Sociological Association, 28 July 2006). *Source:* www.ukzn.ac.za/ccs/default.asp?3,28,10,2650 and Julie A. Smith and J. Maryann Green, 'Free Basic Water in Msunduzi, KwaZulu-Natal: Is It Making a Difference to the Lives of Low-Income Households?', 7 *Water Policy* 443, 445 (2005).

[59] *Cf.* African National Congress, Local Government Elections 2000 Manifesto – Together Speeding Up Change, 2000. *Source:* www.anc.org.za/elections/local00/manifesto/manifesto.html.

[60] Republic of South Africa, Department of Water Affairs and Forestry, Media Statement by the Minister of Water Affairs and Forestry, Mr. Ronnie Kasrils, 14 February 2001. *Source:* http://www.dwaf.gov.za/Communications/Press Releases/2001/Free%20water%206000%20litres%2014%20February%202001.doc.

[61] Francis, note 4 above at 180 and United Nations Development Programme, note 1 above at 64.

[62] Francis, note 4 above at 180; de Visser, Cottle and Mettler, note 5 above at 37 and Arnold M. Muller, Sustaining the Right to Water in South Africa 5 (New York: UNDP Human Development Report Office, Occasional Paper 2006/29, 2006). *Source:* http://hdr.undp.org/en/reports/global/hdr2006/papers/muller_arnold.pdf.

[63] DWAF Annual Report 2008, note 56 above at 58.

Challenges to the Implementation of the Right to Water

In spite of the progress made by the increasing access to water services and the expansion of the FBW Policy, there are certain areas of concern in the implementation of the right to water. These will be analysed under the framework of obligations to respect, to protect and to fulfill.

4.1 Obligation to Respect and to Protect

The obligations to respect and to protect are two separate obligations as outlined above. However, they are associated with the same challenges in the implementation of the human right to water:[64] Increases in water tariffs can be implemented either by public or private water service providers. Similarly, disconnections of water services can be carried out and prepayment metres be installed by both types of service providers. Depending on their type, the role of the state changes and it bears different obligations: the obligation to respect when measures are carried out by public water providers and the obligation to protect in cases of water service privatisations with the state acting as regulator.

4.1.1 Policy of Cost Recovery

Increases in tariffs, water service disconnections and the installation of prepayment water metres cannot be understood without reference to the principle of cost-recovery. It signifies that consumers are charged the full (or nearly full) cost of providing water services[65] and is the basis for privatisation as municipalities try to attract private (foreign) investment.[66] In 1996, the principle

---

[64] *Cf.* David A. McDonald and Greg Ruiters, 'Theorizing Water Privatization in Southern Africa', *in* David A. McDonald and Greg Ruiters eds., *The Age of Commodity, Water Privatization in Southern Africa* 13, 18, 28 et seq. (London, Sterling: Earthscan, 2005). It is pointed out that commercialised public utilities are very similar to privatised water services.

[65] David A. McDonald, 'The Theory and Practice of Cost Recovery in South Africa', *in* David A. McDonald and John Pape eds., *Cost Recovery and the Crisis of Service Delivery* 17 (Cape Town: Human Sciences Research Council Publishers, 2002).

[66] Francis, note 4 above at 157 et seq. and McDonald and Ruiters, note 64 above at 18 et seq.

of cost recovery became official policy with the adoption of the 'Growth, Employment and Redistribution' (GEAR) policy. It includes the government's commitment 'to the application of public-private sector partnerships based on cost recovery pricing where this can practically and fairly be effected'.[67]

4.1.2 Increases in Water Tariffs

While many water activists in South Africa demand water to be provided free of charge, this is not necessarily required from a human rights perspective.[68] Only if people live in extreme poverty and have hardly any income at all, does water have to be provided for free. As long as people can fulfill their basic human needs, there is no violation of the human right to water, even if people have to pay in order to do so. The decisive criterion is that of affordability. If people spend a large percentage of their income on water supply, services have to be regarded as unaffordable.

As a consequence of the introduction of the principle of cost recovery and the privatisation of water supply, there have been steep price increases in many areas. In Johannesburg, for example, tariffs have doubled while they have even tripled in Queenstown, Eastern Cape. People spend an average one-fifth of their income on their water bill.[69] In many cases, residents of poor black communities pay higher tariffs than residents of more affluent, historically white communities.[70] Such percentages are far beyond

---

[67] Republic of South Africa, *Growth, Employment and Redistribution, A Macroeconomic Strategy*, 1996, para. 7.1. *Source:* www.info.gov.za/otherdocs/1996/gear.pdf. For more details on cost recovery and its underlying rationale *cf.* McDonald, note 65 above.

[68] Kok, note 36 above at 274 and Kok and Langford, note 14 above at 56B–13.

[69] Lena Partzsch, 'Wasser in der Krise – Das Beispiel Südafrika', 196 *Solidarische Welt* 4 (2007).

[70] *Cf.* the example at Cottle, note 54 above at 31. Some authors point out that full-cost recovery may include the initial costs of infrastructure thus leading to higher prices for historically disadvantaged areas with no water services infrastructure. *See* Hameda Deedat and Eddie Cottle, 'Cost Recovery and Prepaid Meters and the Cholera Outbreak in KwaZulu-Natal, A Case Study in Madlebe', *in* David A. McDonald and John Pape eds., *Cost Recovery and the Crisis of Service Delivery* 81, 94 (Cape Town: Human Sciences Research Council Publishers, 2002); McDonald, note 65 above at 27 and Flynn and Chirwa, note 36 above at 65.

the international recommendations of three to five per cent of household income. In these cases, water services can no longer be regarded as affordable and thus fail to meet this criterion of the human right to water.

Price increases can be a problem even with the provision of FBW when prices increase very steeply after the basic amount of six kilolitres. For example, Johannesburg sets a high price increase for the second block of consumption of seven to ten kilolitres. After this initial increase, prices level off even resulting in a flat tariff after 40 kilolitres per month.[71] Such a tariff system does not help to decrease luxury consumption and encourage water conservation. Rather, it puts a high burden on poor households that use a little more water than the six kilolitres provided for free.[72] However, tariffs are not necessarily structured in this way in all municipalities. A different trend can also be observed. In its 2007/'08 Annual Report, DWAF states that the highest price increases were found in the upper blocks.[73]

In order to determine whether tariffs meet the standards set by the human right to water, it is critical to look at affordability. It has to be answered in the negative whenever people have to spend a large percentage of their income on water.

4.1.3 Disconnections

To implement cost recovery, it seems a logical consequence to disconnect water supplies of people who do not pay their water bills which is often the result of increasing tariffs. The main controversy in South Africa is concerned with such disconnections due to non-payment. However, interruptions of service delivery that are caused by management problems and nonexistent or dysfunctional infrastructure also pose a huge problem.[74]

---

[71] Johannesburg Water, Schedule of Water Tariffs 2006/2007. *Source:* www.johannesburgwater.co.za/uploads/documents/TARIFFS-WS-%202006-2007.xls. *See also* South African Human Rights Commission, note 6 above at 26, 53; Bond, note 58 above at 17 and McDonald, note 65 above at 28.

[72] Bond, note 58 above at 28.

[73] DWAF Annual Report 2008, note 56 above at 62.

[74] South African Human Rights Commission, note 6 above at 19, 37, 43 et seq.

There has been a significant number of disconnections in South Africa,[75] particularly between 1994 and 2000.[76] The exact number is subject to extensive debate and controversy. According to a survey of the *Municipal Services Project*, whose estimates were widely spread, it is assumed that as many as ten million people in South Africa had experienced water cut-offs since 1994.[77] The DWAF refuted these figures, but admitted that two per cent of connected households may have suffered from the discontinuation of services and that disconnections by local authorities are therefore a matter of concern.[78] Furthermore, a survey conducted by the Department in 2004 found that 30,000 households reported to have water services cut off due to non-payment in the year before.[79]

Until 2003, the government did not take a clear stand on disconnections.[80] This ambivalence corresponds to the Water Services Act's seemingly contradictory provisions on the discontinuation of water services. Only in 2003, after reports in high-profile media such as the *New York Times*,[81] DWAF adopted the position that municipalities should refrain from complete disconnection.[82] The DWAF 2003 Strategic Framework stresses the importance of fair procedures and establishes that domestic water supply connections must be restricted in the first instance, and not disconnected in order to ensure that at least a basic supply of

---

[75] Ashfaq Khalfan and Anna Russell, 'The Recognition of the Right to Water in South Africa's Legal Order', *in* Henri Smets ed., *Le Droit à l'Eau dans les Législations Nationales* 121, 128 (Nanterre: Académie de l'Eau, 2005); Francis, note 4 above at 174.

[76] Partzsch, note 69 above at 4.

[77] McDonald, note 54 above at 170; Ginger Thompson, 'Water Tap Often Shut to South Africa Poor', *New York Times*, 29 May 2003; Conca, note 4 above at 353; Partzsch, note 69 above at 4; Cottle, note 54 above at 26 and Francis, note 4 above at 174.

[78] Republic of South Africa, Department of Water Affairs and Forestry, Minister Kasrils Responds to False Claim of 10 Million Cut-Offs, 8 June 2003. *Source:*www.dwaf.gov.za/Communications/Articles/Kasrils/2003/cutoffs%20article%20WEBSITE.doc.

[79] Mike Muller, 'Keeping the taps open', *Mail and Guardian*, 30 June 2004.

[80] Bond, note 59 above at 19.

[81] Thompson, note 77 above.

[82] Bond, note 59 above at 19.

water is available. Disconnection is only regarded as appropriate in the case of tampering with the service equipment.[83]

Some of these cases of water disconnections have been heard before South African courts. *Residents of Bon Vista Mansions*[84] can be regarded as the most important one. The residents launched an urgent application for interim relief as their water supply had been disconnected which they regarded as unlawful. In order to interpret Section 27(1)(b), the judge considered international law as stipulated by Section 39(1)(b) of the Constitution. He held that the matter relates to the duty to respect access to water and that the state has to refrain from actions that deprive individuals of their rights.[85] The discontinuation of water services is *prima facie* in breach of this obligation and requires constitutional justification.[86] The onus rests on the respondent who has to show that the disconnection was legal.[87] At the time of the interim order, the respondent had not yet discharged that onus. Thus, the Court ordered the restoration of water supply to the residents pending the final determination of the application.[88]

Irrespective of the exact figures, water cut-offs have been widespread in South Africa. In cases where they left people without access to basic water supply, they constitute a violation of the human right to water. A public water supplier violates its obligation to respect existing water supply, while the state fails to meet its obligation to protect when disconnections are carried out by private providers. An indigent person may not have access to basic water services denied for reasons of non-payment. Water services may

---

[83] DWAF Strategic Framework 2003, note 27 above at 37.

[84] *Residents of Bon Vista Mansions v Southern Metropolitan Local Council*, High Court (Witwatersrand Local Division) of South Africa, Judgment of 5 September 2001, 2002 (6) BCLR 625 (W). *Cf.* as well *Manqele v Durban Transitional Metropolitan Council*, High Court (Durban and Coast Local Division) of South Africa, Judgment of 7 February 2001, (2002) 2 All SA 39 (D), and *Highveldrige Residents Concerned Party v Highveldridge TLC and Others*, High Court (Transvaal Provincial Division) of South Africa, Judgment of 17 May 2002, 2003 (1) BCLR 72 (T).

[85] *Residents of Bon Vista Mansions v Southern Metropolitan Local Council*, note 84 at 629.

[86] Ibid., 630.

[87] Ibid., 632.

[88] Ibid., 633.

be limited to the basic amount as stipulated in Section 4(3)(c) of the Water Services Act in conjunction with Regulation 3, but not be completely disconnected.[89] It has to be ensured that no one is 'deprived of the minimum essential level of water'[90] under any circumstances.

4.1.2 Installation of Prepayment Water Metres

The installation of prepayment metres is also a measure to implement the principle of cost recovery very similar to cut-offs of conventional connections. The first prepayment water metres were installed in rural communities in 1997.[91] To use the metres people are required to obtain water cards which work like prepaid phone cards. Usually, these cards are charged with six kilolitres of water per household per month (the FBW amount). Once this amount is exhausted, people are required to purchase water units. When they cannot afford to do so, people are no longer able to obtain water from the metre which has the same effect as the disconnection of services.[92]

The installation of prepayment metres can result in people not having the possibility to access safe water when they cannot pay for it and then turning to unsafe water,[93] which can have serious consequences. In 2000, South Africa experienced one of the worst cholera epidemics. The reasons were traced back to the installation of prepayment water metres in Kwazulu. As thousands of people were unable to pay for water, they turned to the use of polluted river water,[94] which resulted in a cholera outbreak that affected about 120,000 people and caused at least 265 deaths.[95]

Prepayment metres raise further concerns. In the case of malfunctioning – which is not uncommon – repairing the metres

---

[89] Kidd, note 33 above at 133.
[90] General Comment No. 15, note 7 above at para. 56.
[91] Cottle, note 54 above at 20 et seq.
[92] *Cf. Mazibuko Judgment*, note 20 above at para. 84.
[93] South African Human Rights Commission, note 6 above at 45.
[94] For details see Deedat and Cottle, note 70 above.
[95] Conca, above note 4 at 353; Cottle, note 54 above at 22 and Smith and Green, note 59 above at 445. *See also* Bond, note 59 above at 13 and Francis, note 4 above at 174 et seq. These figures are confirmed by the Cholera statistics of the South African Department of Health. *Source:* www.doh.gov.za/facts/index.html.

often takes a long time and people are without access to water in the meantime.[96] The High Court judgment in the *Mazibuko* case, in which the issue of prepayment metres has been raised, states that the installation of prepayment metres is unlawful and unreasonable as it amounts to a violation of Section 33 of the Constitution providing for lawful, reasonable and procedurally fair administrative action.[97] In the case of prepayment metres, cut-offs occur without reasonable notice and do not allow for making representations and explaining financial difficulties.[98] This does not agree with fair and equitable procedures for the discontinuation of water services as stipulated in Section 4(3) of the Water Services Act.[99]

Furthermore, the installation of prepayment metres is regarded as being discriminatory on the basis of colour. While historically 'white' areas receive water services on credit with the opportunity to settle arrears and make arrangements in the case of financial difficulties, prepayment metres are installed in poor and predominantly 'black' areas such as in Phiri, the township in question in the *Mazibuko* case.[100] The judgment therefore declared the prepayment water system to be discriminatory and unconstitutional and ordered that the residents of Phiri must be provided with the option of a metered supply.[101]

---

[96] Deedat and Cottle, note 70 above at 89.

[97] *Mazibuko Judgment*, note 20 above at para. 92. The Supreme Court of Appeal delivered its judgment in this case on 25 March 2009. *See City of Johannesburg v Lindiwe Mazibuko*, Supreme Court of Appeal of South Africa, Judgment of 25 March 2009, (489/08)[2009] ZASCA 20. *Source:* http://web.wits.ac.za/Academic/Centres/CALS/Basic Services/Litigation.htm. The Court held that the installation of prepayment metres has been unlawful, yet on much narrower grounds than the High Court stating that the City's bylaws do not authorise their installation. The Court supended its order allowing the City to legalise the use of prepayment metres. Concerning the amount of water to be provided (further discussed below), it declared that 42 litres of water per person per day would constitute sufficient water in terms of Section 27(1) of the Constitution under the circumstances in Phiri and ordered the City of Johannesburg to provide such an amount to each Phiri resident who cannot afford to pay having regard to its available resources and other relevant considerations determining reasonableness.The case is now on appeal to the Constitutional Court and is expected to be heard in September 2009.

[98] Ibid., para. 93.

[99] *Cf.* Flynn and Chirwa, note 36 above at 71.

[100] *Mazibuko Judgment*, note 20 above at para. 94, 151, 153.

[101] Ibid., para. 183.

4.2 Obligation to Fulfill

The analysis now turns to challenges in the implementation of the right to water regarding the obligation to fulfill. It raises some points of critique regarding the FBW Policy and then concentrates on the more fundamental issue of the complete lack of access to water supply. Apart from the issue of prepayment metres, the *Mazibuko* judgment is the first judgment that also addresses a number of questions related to the obligation to fulfill.

4.2.1 Critique of the FBW Policy

The FBW Policy is an instrument to meet the obligation to fulfill the right to water. Since its adoption in 2001, enormous progress has been made and many people benefit from the policy, but there are also some concerns.

FBW is still not provided to a great number of people: As of 2008, 16 per cent of the entire population and 27 per cent of those defined as poor do not receive FBW[102] signifying that the affordability of the minimum amount of water remains critical for several million people in South Africa. Originally, FBW was intended as an instrument to provide the poor with free water. But due to management reasons it is served to everyone in many communities.[103] This leads to the peculiar result that a greater percentage of the entire population than of the poor population is served by FBW. In particular, the poorest in society are excluded from its implementation[104] as they often live in areas that lack the necessary infrastructure by which FBW could be provided. Moreover, many poor municipalities do not have sufficient financial resources to implement the FBW Policy, as cross-subsidisation is difficult to realise in communities with only a small number of affluent high-volume users.[105]

Another concern refers to the calculation of the basic water supply on a per-household basis which seems unsatisfactory as it

---

[102] DWAF Annual Report 2008, note 56 above at 58.
[103] South African Human Rights Commission, 6th Economic and Social Rights Report 111 (Johannesburg, 2006).
[104] South African Human Rights Commission, note 6 above at 44.
[105] South African Human Rights Commission, note 6 above at 47; Francis, note 4 at 180 and de Visser, Cottle and Mettler, note 5 above at 43, 50.

does not take into account the number of people living in one household.[106] Depending on the household's size 200 litres per household per day may or may not result in the minimum amount of 25 litres per person per day.[107] Households of more than eight people are not unusual, especially in poor black communities. According to the Census 2001, approximately 620,000 households had nine or more members.[108] Thus, the right to a sufficient amount of water of several million people is infringed due to this form of calculation. It is indispensable to supply large households with an increased amount of FBW that adequately reflects the number of people and guarantees a minimum of 25 litres to everyone. In so far, it has to be noted that the *Mazibuko* judgment requires the calculation on a per capita basis.[109]

A further point of critique is that only 25 litres per day are provided (assuming the government maximum of eight people per household). This was the second main issue addressed in the *Mazibuko* case apart from the question of prepayment metres. The judgment concluded that an amount of 25 litres is insufficient, particularly where water is also used for water-borne sanitation systems[110] as well as for people living with HIV/AIDS.[111] The City of Johannesburg was ordered to provide 50 litres per capita per day of free basic water. In this regard, the decisive question was whether the City had the resources to do so, which it did not contest.

Moreover, DWAF considers generally increasing the free basic amount to at least 50 litres per person per day for poor households in its 2003 Strategic Framework.[112] However, it remains to be seen whether this will be turned into practice. The *Mazibuko* judgment can provide an important impetus in this regard.

---

[106] South African Human Rights Commission, note 6 above at 37; Francis, note 4 above at 182 and McDonald, note 65 above at 29.
[107] Smith and Green, note 58 above at 449.
[108] South African Human Rights Commission, note 6 above at 58 endnote 8.
[109] *Mazibuko Judgment*, note 20 above at para. 168, 183.
[110] *Mazibuko Judgment*, note 20 above at para. 179. *See also* Francis, note 4 above at 181; Bond, note 58 above at 17; Cottle, note 54 above at 30 and de Visser, Cottle and Mettler, note 5 above at 43.
[111] *Mazibuko Judgment*, note 20 above at para. 179.
[112] DWAF Strategic Framework 2003, note 27 above at 29.

4.2.2 Lack of Access to Water Supply

An even more fundamental question related to the obligation to fulfill refers to the situation of those who still completely lack access to water supply. According to government estimates in the 2007/'08 DWAF Annual Report, this affects 2.4 million people.[113] Can they claim to be connected to water services and receive a minimum quantity of water necessary to satisfy their basic needs? Does the state have a positive duty to fulfill this obligation?

The question of positive duties to fulfill the right to water is related to the discussion about the minimum core content of the right. As outlined above, Section 27(2) of the Constitution stipulates that the state must take reasonable measures within its available resources to achieve the progressive realisation of the right to water. Article 2(1) of the Social Covenant contains a similar clause. An extensive debate has evolved around the interpretation of this clause, in particular around the question whether states have minimum core obligations that are to be fulfilled immediately. In this context, the landmark *Grootboom judgment*[114] of the South African Constitutional Court has to be taken into account. It is primarily concerned with the right to housing, but the Court also refers to the right to water[115] and emphasises that all socio-economic rights have to be interpreted together.[116] The notion of reasonableness developed in the *Grootboom judgment* has become the litmus test against which the realisation of socio-economic rights is tested.[117]

According to the Constitutional Court, Section 27(2) of the Constitution obliges the state to establish a coherent programme directed towards the progressive realisation of socio-economic

---

[113] DWAF Annual Report 2008, note 56 above at 55.
[114] *Grootboom Judgment*, note 14 above.
[115] Ibid., 1204, 1208.
[116] Ibid., 1181, 1184.
[117] *Cf.* Kok, note 36 above at 274; South African Human Rights Commission, note 6 above at ix and DWAF Strategic Framework 2003, note 27 above at 51. The Constitutional Court itself used the concept again in the *TAC Judgment-Minister of Health and Others v Treatment Action Campaign and Others*, Constitutional Court of South Africa, Judgment of 5 July 2002, 2002 (10) BCLR 1033 (CC) [hereafter TAC Judgment].

rights.[118] It has to be ensured that measures are reasonable in their conception and their implementation. Programmes must be balanced and flexible and take account of short, medium and long term needs.[119] Moreover, the Court explicitly states that it must be guaranteed that a significant number of people in desperate need are afforded relief.[120] Following the Court's view, water legislation and policy have to be tested against the concept of reasonableness. Access to infrastructure and the implementation of the FBW policy both show progress in their extension and even though the FBW policy does not only aim at the indigent population, it reaches a significant number of poor people. The DWAF reports specifically on the extension of FBW to the indigent population which underlines the special consideration of this population group. Therefore, it can be assumed that the state's policy meets the requirements set out by the Constitutional Court in *Grootboom*.[121]

However, the Court does not demand that all people in desperate need are afforded immediate relief and has rejected to apply the minimum core approach.[122] In this regard, its approach has been criticised for not being far-reaching enough.[123] The minimum core approach developed by the Committee on Economic,

---

[118] *Grootboom Judgment*, note 14 above at 1190.
[119] Ibid., 1191.
[120] Ibid., 1202.
[121] Likewise Francis, note 4 above at 194 et seq.
[122] *TAC Judgment*, note 117 above at 1046 and *Grootboom Judgment*, note 14 above at 1188. *See also* Sandra Liebenberg, 'The Interpretation of Socio-Economic Rights', *in* Stuart Woolman *et al.* eds., *Constitutional Law of South Africa* 33-30 (Lansdowne: Juta, 2nd ed., 2005).
[123] David Bilchitz, 'Giving Socio-Economic Rights Teeth: The Minimum Core and its Importance', 119 *South African Law Journal* 484 (2002). *See also* Scott and Alston, note 9 above at 262 et seqq.; Kok and Langford, note 14 above at 56B–19 and Liebenberg, note 122 above at 33–27 et seqq., 41; *but also* Elisabeth Wickeri, Grootboom's Legacy: Securing the Right to Access to Adequate Housing in South Africa 19 (New York: Center for Human Rights and Global Justice, Working Paper No. 5/2004, 2004). *Source:* www.chrgj.org/publications/docs/wp/Wickeri%20Grootboom%27s%20 Legacy.pdf, and Murray Wesson, 'Grootboom and Beyond: Reassessing the Socio-Economic Jurisprudence of the South African Constitutional Court', 20 *South African Journal on Human Rights* 284, 300 et seqq. (2004).

Social and Cultural Rights[124] would be more far-reaching. It aims at guaranteeing a minimum essential level of each right,[125] which is indispensable for human survival and dignity. These minimum needs, such as the need for a basic amount of water, are more urgent than others, have to be accorded priority[126] and are therefore not subject to progressive realisation, but are to be fulfilled immediately.[127] The minimum core content is the baseline from which progressive realisation of the right to water has to start improving the level of realisation over time.[128]

The approach is based on the assumption that the rights would be largely deprived of their *raison d'être* without such minimum core obligations.[129] It acknowledges that there are fundamental obligations appertaining to each right whose immediate fulfillment is of such central importance for the realisation of the right that it would otherwise lose its significance as human right.[130] Moreover, without at least protecting people's survival interests all other human rights become meaningless.[131] Thus, the state is obliged to immediately guarantee the minimum core content of the right to water to everyone.[132]

---

[124] General Comment No. 3: The Nature of States' Parties Obligations, para. 10, UN Doc. E/1991/23 Annex III (14 December 1990) [hereafter General Comment No. 3]. *See also* Philip Alston, 'Out of the Abyss: The Challenges Confronting the New U.N. Committee on Economic, Social and Cultural Rights', 9 *Human Rights Quarterly* 332, 352 et seq. (1987).

[125] General Comment No. 3, note 124 above at para. 10; General Comment No. 15, note 7 above at para. 37; David Bilchitz, 'Towards a Reasonable Approach to the Minimum Core: Laying the Foundations for Future Socio-Economic Rights Jurisprudence', 19 *South African Journal on Human Rights* 1, 12 (2003) and Russell, note 9 at 15.

[126] Bilchitz, note 125 above at 11 and Bilchitz, note 123 above at 485.

[127] General Comment No. 15, note 7 above; Bilchitz, note 125 above at 11 et seq. and *Maastricht Guidelines on Violations of Economic, Social and Cultural Rights* para. 9 (1997), reprinted in 20 *Human Rights Quarterly* 691 (1998).

[128] Bilchitz, note 123 above at 493; Bilchitz, note 125 above at 11 et seq.; Liebenberg, note 122 above at 33–41 and Scott and Alston, note 9 above at 250.

[129] General Comment No. 3, note 124 above at para. 10.

[130] *Cf.* Russell, note 9 above at 15.

[131] Bilchitz, note 125 above at 12.

[132] General Comment No. 15, note 7 above at para. 37 (a).

However, the minimum core approach does not prescribe the impossible. It has to be recognised that it is not only impossible to achieve the full realisation of socio-economic rights in a short period of time,[133] but not even always possible to guarantee the minimum core of every right to everyone immediately. Yet, the minimum core approach requires that the minimum essential level is realised whenever and as soon as this is possible.[134] Moreover, it signifies a significant change – the onus then rests on the state. The state *prima facie* fails to meet its obligations and has to demonstrate that every effort has been made and that all available resources have been used to satisfy these minimum needs as a matter of priority.[135]

The *Mazibuko* judgment deals with the minimum core approach in depth and provides an interpretation to combine it with the reasonableness review of the Constitutional Court. Admitting that the Constitutional Court has not applied the minimum core approach in the *Grootboom judgment*, the High Court held in *Mazibuko* that it has nevertheless left a caveat to potentially consider it.[136] While the Constitutional Court had argued that the determination of the minimum core presents difficult questions,[137] the High Court held that it is possible to determine the minimum core if the Court is provided with sufficient information.[138]

According to this reasoning, South Africa is obliged to immediately realise the core content of the right to water for everyone. Yet, it has to be acknowledged that water infrastructure cannot be built overnight and that it is thus impossible to immediately supply all people with access to water. However, the state has to make use of all possible means. Moreover, the extension of FBW to all indigent people in need with access to infrastructure would be possible in relatively little time. The DWAF itself admits that '[t]he cost associated with providing free basic water to poor

---

[133] General Comment No. 3, note 124 above at para. 9.
[134] Bilchitz, note 125 above at 18.
[135] General Comment No. 3, note 124 above at para. 10; Scott and Alston, note 9 above at 250; Bilchitz, note 125 above at 16; Liebenberg, note 122 above at 33–31 and Russell, note 9 above at 16.
[136] *Mazibuko Judgment*, note 20 above at para. 131.
[137] *Grootboom Judgment*, note 14 above at 1188.
[138] *Mazibuko Judgment*, note 20 above at para. 131 et seqq.

households is not large for a country of our economic size and strength'.[139]

Conclusion

South Africa's commitment to the human right to water in its Constitution and legislation is outstanding and hardly found in any other country. Its acts, regulations and policies set up a very detailed and precise framework for the implementation of the right to water. Moreover, the country has also made significant progress in turning these into practice by expanding access to infrastructure as well as to FBW.

Yet, there remain a number of concerns in implementing the right to water. The overall policy of cost recovery entails increases in water tariffs, disconnection of services and the installation of prepayment metres. Price increases have often occurred after water services were privatised thus relating to the obligation to protect, but public water providers have also increased tariffs. When water services become unaffordable the human right to water is violated.

Disconnection of water services has been a widespread concern in South Africa. When people are left without access to basic water supply, they constitute a violation of the obligation to respect or to protect the right to water depending on the type of service provider. In recent years, disconnections seem to have become less common. Court decisions that judge them illegal such as in the *Bon Vista Mansions* case are an important signal in so far. Also, the DWAF adopted the position that water service providers have to refrain from complete disconnections in 2003. It is critical to ensure that no one is deprived of minimum essential water services even without payment. The increasing access to water infrastructure and expansion of FBW would turn meaningless if a significant number of people lose access at the same time.

Currently, the installation of prepayment metres poses a greater concern than cut-offs of conventional connections, but they can have the same effect. Whenever they leave people without access to basic water supply they violate human right to water. The *Mazibuko* judgment has declared the installation of such metres to

---

[139] DWAF Strategic Framework 2003, note 27 above at 29.

be unconstitutional, particularly because they do not allow for fair procedures and have been installed in a discriminatory manner.

The FBW Policy relates to the obligation to fulfill and addresses the issue of inability to pay for water services by securing a minimum amount of free water for a great number of people. Yet, it remains inadequate as millions of people are still not supplied with FBW and affordability remains critical. Moreover, it has to be ensured that all indigent people in need receive the minimum of services without putting people living in large households at a disadvantage.

The complete lack of access to water supply due to missing water infrastructure is the most fundamental concern. Under its minimum core obligations the state has the duty to provide everyone with minimum services. This can be realised by using all possible means to develop infrastructure. The FBW policy allows ensuring that services are affordable to everyone and therefore needs to be extended to all indigent people. To this end, increasing financial resources available to poor communities is crucial. This could be achieved by augmenting allocations from the national budget or via a cross-subsidisation mechanism between municipalities. It is to hope that the Government reaches its goal of universal access as soon as possible. Even then, it is important to keep in mind that the human right to water is not completely fulfilled as soon as everyone has access to minimum services. Rather, it is an ongoing obligation of the state to progressively realise the right to water until everyone has access to sufficient water for an adequate standard of living.

UNIT V
Comparative Perspectives on Reforms

16

Learning from Water Law Reform in Australia*

POH-LING TAN

1. Introduction

Parts of Australia have recently experienced the lowest rainfall on record since 1900.[1] For a period of six years from 2001 to 2007 drought affected southern and eastern Australia in a broad arc extending across southern South Australia, most of Victoria and New South Wales (NSW), and a large part of southeast Queensland.[2] Storages across the country, particularly in the Murray and in South East Queensland were at record lows. In the former, storages were at 13 per cent capacity in mid 2007.[3] An established La Niña event in December 2007 and January 2008 brought heavy rainfall across the eastern states, especially in

---

* This paper was written in April 2007 and updated in March 2008.
[1] These record low falls occurred in a strip from Melbourne, Victoria to central New South Wales, Southern Queensland and much of Western Australia's west coast. *See* Australian National Climate Centre, Statement on Drought for the 12-month period ending 28th February 2007, 6 March 2007. *Source:* http://www.bom.gov.au/climate/drought/drought.shtml.
[2] *See* Australian Government, Bureau of Meteorology, National Climate Centre, Special Climate Statement 14, 1 November 2007. *Source:* http://www.bom.gov.au/climate/current/statements/scs14.pdf.
[3] Asa Wahlquist, 'A Big Drought Should Bring Big Ideas with it' *The Australian*, 24 February, 2007. *Source:* http://www.theaustralian.news.com.au/story/0,20867, 21276403-28737,00.html.

Queensland. However long-term deficiencies remain, especially in Tasmania, southeast Queensland, the southwest coast of Western Australia (WA) and central Victoria into southern NSW.

Most of the cities in Australia are subject to water restrictions. The most severe is in southeast Queensland. In September 2007 householders in the fastest growing region in Australia were imposed with Level Six water restrictions. Households which use more than 800 litres a day of water were asked to file a water usage audit report to provide information on their household, water use and water efficiency. Certain types of water use will be banned. The target is for individuals to use no more than 140 litres a day. Tough penalties apply for households using excessive amounts of water. This includes an outdoor watering ban, fines up to $1050 and restricted flows.[4]

For the rural sector, the drought has brought social and family disruption far beyond that felt by city dwellers, with rural suicide rates receiving widespread attention in the media. Some farmers in New South Wales will not get one drop of water from their water entitlements.[5]

Water reform has been high on the political agenda for over ten years in this country. A broad summary of the factors that gave rise to that policy and law reform, and a description of the features of the new legal framework for water have been given in this paper. The paper outlines where early reform has succeeded and gives a concise analysis of the problems that have been encountered in the implementation of the reform. As the reform framework relies on water resource planning, the paper discusses issues relating to participatory planning in water and considers suggestions as to how this may be addressed particularly in remote communities.

2. A Snapshot of Australia's Water

Australia is a large and old continent, with a wide range of environments. It has tropical savannahs in the North, deserts in

---

[4] *See* Queensland Water Commission Webpage, Household Restrictions Section. *Source:* http://www.qwc.qld.gov.au/Household+restrictions.

[5] Asa Wahlquist, 'A Big Drought Should Bring Big Ideas with it' *The Australian*, 24 February, 2007. *Source:* http://www.theaustralian.news.com.au/story/0,20867, 21276403-28737,00.html.

the Centre and many areas that are prone to salinity problems. Rainfall patterns are highly variable. It has some of the wettest areas on earth, but there are regions that experience prolonged droughts, broken by sweeping floods. Ground water is an important part of the resource. Due to the relative dryness of the continent, Australia has the highest per capita storage capacity of all countries in the world, over 4 million litres per person.[6]

Much of the water is supplied by organised service providers or public dams. But in many catchments in Northern New South Wales, Queensland and Western Australia water users directly access their water either through pumping ground water or dams. 85 per cent of water used within the Australian economy was extracted directly by water users.[7] Approximately 80 per cent of water used was supplied from surface water (rivers and overland flows) with ground water resources accounting for the remaining 20 per cent.[8]

Prior to colonisation, indigenous people exercised communal property rights over water. Under British colonial rule, the English common law became the basis of management of water resources. It soon became apparent that common law rules were inadequate. There were concerns that efforts at water supply and irrigation would be undermined by common law claims. In 1886, after worldwide study of approaches to water supply and allocation, Victorian legislation vested the right to use water in rivers, lakes, swamps, etc., in the Crown. A royal commission in New South Wales made similar recommendations.

Reform occurred in the late nineteenth century imposing an administrative (licensing) regime in all states. Water is vested in

---

[6] Dennis Trewin ed., 2007 Year Book Australia (Canberra, Australian Bureau of Statistics, Yearbook No. 89, ABS Catalogue No. 1301.0, 2007). *Source:* http://www.ausstats.abs.gov.au/Ausstats/subscriber.nsf/0/D6C6B02D31617 DA4 CA25726D000467A6/$File/13010_2007.pdf.

[7] Australian Bureau of Statistics, Water Account, Australia, 2004–05, 28 November 2006. *Source:* http://www.abs.gov.au/AUSSTATS/abs@.nsf/ Productsby ReleaseDate/9F319397D7A98DB9CA256F4D007095D7? Open Document [hereafter Water Account 2005].

[8] Australian Government, Department of Agriculture, Fisheries & Forestry, Bureau of Rural Sciences, Australia – Our Natural Resources At A Glance 2007. *Source:* http://affashop.gov.au/product.asp?prodid=13637.

the State. Water for urban supply and mining was generally granted directly by specific acts of legislation. But the majority of grants of water for irrigation took place by the licensing system established under water statutes. In that time public policies promoted increased water use, and farmers were charged for water they were entitled to take under their grants regardless of whether all of it was used. The price of water was also heavily subsidised. Storages and supply systems were mainly publicly funded.

A great variety of administrative grants with different tenure, rights and obligations (collectively referred to in this chapter as 'administrative grants') developed over the years. In each of the States, and regions within those states, water managers had a great deal of autonomy over how they expended the supply of water. Many of the differences in these administrative grants relate to the subtleties associated with local land use and development.[9]

The type of crops grown in a region/state also determined how dams were managed. Where horticultural crops were grown, dams were managed conservatively. This meant that water was held for several years in large storages to ensure there was enough for supply during dry years. In regions where rice or cotton, both annual crops, were grown, then the policy was to let as much water be used in a wet year. In a dry year these crops were simply not grown at all. As we will observe later in this chapter, these patterns of management became institutionalised as part of a post-reform water entitlement.

The main period of expansion took place within a 50-year period of the twentieth century that was a relatively wet period. During these years rivers were treated as supply channels for irrigation. This disrupted their natural flow patterns. Wetlands were drained for agricultural use, and so much water was pumped that rivers are running dry. Floodplains and ecosystems have been affected. Over-allocation of water is now a significant issue.

---

[9] See T. Shi, *Simplifying Complexity: A Framework for the Rationalisation of Water Entitlements in the Southern Connected River Murray System* (Adelaide: CSIRO, 2005).

Consumptive use of water is shown by the following table.

Table 1: Water consumption in Australia 2004–2005.[10]

Sector	Use
Agriculture	65%
Household	11%
Water Supply Industry	11%
Manufacturing, Mining, Electricity & Gas	3%
Total	**100%**

Water consumption patterns in Australia have changed significantly since 2001–2002. Overall water consumption has decreased by 14 per cent with the greatest reduction (19 per cent) being within the agricultural sector. Household water use has decreased by an average of 8 per cent.[11]

3. Co-operation in Water Law Reform

When Australia became a federation in 1901, inland rivers had already been used for decades as highways for getting produce to markets. They were also very important to the states for irrigation. To reflect the states' concerns, the Constitution was silent on the issue of water resources therefore according to common law principles about sovereign legislative power, this power remained with the states.[12] The only explicit reference to water resources is found in Section 100 of the Constitution which reads:

> The Commonwealth shall not, by any law or regulation of trade or commerce, abridge the right of a State or of the residents therein to the reasonable use of waters of rivers for conservation or irrigation.

---

[10] *See* Water Account 2005, note 7.
[11] Ibid.
[12] *See* S.D. Clark and I.A. Renard, 'Constitutional, Legal and Administrative Problems', *in* HJ Firth and G. Sawer, *The Murray Waters: Man, Nature and a River System* (Sydney: Angus Robertson, 1974), 265–266.

Thus the Commonwealth's powers over water came from its power to legislate for defence, trade and commerce, and external matters. In practice the Commonwealth has had an important role in the management of the internal waters of Australia through policy formulation and the provision of financial assistance for schemes related to water resources.

By the late 1970s it became apparent in the heavily irrigated states of Victoria and NSW that water was over-allocated. The main problems were:

- The lack of security for entitlement holders. They were like members of a club that kept taking in new members, resulting in the erosion of the enjoyment of the club facilities. In addition membership rights were not clearly specified. Managers determined the rules of water access. In parts of NSW there were a few examples where managers had seriously underestimated losses in the 'system' and later that year had to cancel water that had been promised to farmers.
- Governments were authorised to revoke the entitlements for 'bad behaviour' but this did not happen.
- Entitlements were often limited to a definite period (this period varied from state to state) but again the rights were simply rolled over to a new period.
- The legal regime did not recognise that the environment was a legitimate user of water.[13]

Early measures were put in place in the form of volumetric allocation schemes, embargoes on new water licences, temporary and permanent trade of licences particularly in Victoria and NSW.[14] These did not adequately address the two main problems of water use – the opposing demands of security for consumptive users and the growing awareness that water needed to be allocated to ecosystem needs.

---

[13] Poh-Ling Tan, Legal Issues Relating to Water Use (Murray-Darling Commission Project MP2002 Issues Paper No 1, Report to the Murray-Darling Basin Commission, 2002) reprinted in *Property: Rights and Responsibilities, Current Australian Thinking* 13 (Canberra: Land and Water Australia, Report of the Social and Institutional Research Program, 2002). *Source:* http://products.lwa.gov.au/files/PR020440.pdf.

[14] For details see note 13 above.

Wide scale reform was sparked by Industry Commission findings 1992 that the water 'industry' was inefficient and unsustainable. The enquiry recommended:

- Payment of full cost of use by users
- Integrated approach to water management
- Institutional change – separating water provision from resource management, standard setting and regulatory enforcement
- Tradeable water rights to allow reallocation of water from unproductive uses to efficient uses.

This report was handed down in a political climate that augured well for its adoption. International treaties and conventions placed obligations on the Commonwealth government to manage natural resources in a sustainable manner and to protect biodiversity. Domestically, Commonwealth, State and local governments adopted a new cooperative approach to environmental matters. This approach is exemplified by the Intergovernmental Agreement on the Environment concluded in May 1992.[15]

3.1 National Water Reform Framework, 1994

The Council of Australian Governments (COAG) comprising federal, state and local governments in Australia embarked on an ambitious agenda for reform in 1994. They recognised that better management of Australia's water resources was a national issue and agreed on a Water Reform Framework. The reforms were based on two main objectives – to introduce a water market while at the same time protecting environmental uses of water. A raft of actions were required including:

- separating water access entitlements from land titles
- allowing for trade in water entitlements, both intra and interstate.

---

[15] Australian Government, Department of Environment and Water Resources, Intergovernmental Agreement on the Environment, 1 May 1992. *Source:* http://www.environment.gov.au/esd/national/igae/index.html. *See* Jacqueline Peel and Lee Godden, 'Australian Environmental Management: A 'Dams' Story', 28 *UNSW Law Journal* 668, 676 (2005).

- consulting the public where new initiatives are proposed especially in relation to pricing, specification of water entitlements and trading in those entitlements
- making explicit provision for environmental water and where river systems were over-allocated to provide for a better balance in water use, and
- separating the functions of water delivery from that of regulation.

As a result, over the last decade states and territories have a significant program of reforms to their water management regime. All jurisdictions have introduced new legislation and statutory water plans.

3.2 The National Water Initiative (2004)

Australian governments revisited policy-making in 2003/4 and agreed on the National Water Initiative (NWI).[16] This policy document, agreed on after months of negotiations between states confirms the direction of earlier reform, recognises that more work needs to be done and sets new and specific targets.

It seeks to achieve nationally-compatible trading system and to that end states that entitlements are to be exclusive, tradeable, enforceable, defined separately from land and as a perpetual share of a specified water resource.[17] Registers of entitlements are to be set up. Among an extensive list of measures required, the NWI seeks to remove institutional barriers to permanent trades by 2014.

The NWI sets a detailed implementation schedule. Water planning is critical to the NWI outcomes and is the instrument which delivers both a highly specified 'product' for trade while at the same time providing for sustainable management. Achievement of the central objectives of the reform depends on comprehensive planning systems based on hydrological assessment of the resource. A cardinal principle that underpins planning is that water users, interest groups and the general community are to be involved as

---

[16] Council of Australian Governments, Intergovernmental Agreement on a National Water Initiative, 25 June 2004. *Source:* http://www.ielrc.org/content/e0403.pdf [hereafter NWI].

[17] Ibid., Section 31.

partners in catchment planning processes. The baseline for planning is sustainable management.

> ... all water bodies, no matter what level of modification is accepted as the appropriate balance between production and the environment, must be maintained in or restored to an environmentally sustainable condition as the first priority of management.[18]

4. An Assessment of Water Reform

Reform has been successful in three areas.

First, an independent 'watch-dog' has been set up to assess compliance with policy. The National Water Commission (NWC), a statutory body was set up in 2004 to drive the national water reform agenda.[19] It provides advice to COAG and the Commonwealth Government on national water issues. Commissioners are appointed in recognition of their skills for example expertise in water resource policies and relevant scientific disciplines. One of its main functions is to assess governments' progress in implementing the NWI which sets national standards for water reform. This is done through biennial assessments of progress commencing in 2006–07.

Secondly, a water market has been established. Trade in surface administrative grants were introduced in 1983 in NSW and when introduced were restricted to temporary trade between irrigators within the same irrigation district. Over time permanent trades were allowed, but markets worked in a limited fashion. The term 'permanent' refers to sale (or transfer) of access entitlements. The transfer of the seasonal allocation for a year or period of years is often referred to as temporary trade or renting or leasing of the entitlement. Since the 1994 reforms, the legislative barriers for trade have been minimised. For example in Queensland within a

---

[18] Daniel Connell and Stephan Dovers, 'Tail Wags Dog – Water Markets and the National Water Initiative', July–September *Public Administration Today* 18 (2006).

[19] Commonwealth of Australia, National Water Commission Act 2004, Act No. 156 of 2004, 17 December 2004.

catchment, there is what can be considered a trading zone. If the proposed trade, whether temporary or permanent complies with the Resource Operations Plan (ROP) for that catchment, then there are very limited procedures to comply with. Only if the trade falls outside of the ROP then procedures such as public notice, objections, scientific reports supporting that externalities are not significant, etc., are required.

In 1998 the Murray-Darling Basin Commission (MDBC) put in place a pilot interstate water trading project between NSW, Victoria and South Australia (SA). Trade within the southern Murray-Darling Basin (MDB) has been most active. However water trade outside the MDB is limited mainly because of the lack of hydrological connectivity of and a lack of demand for water relative to supply and availability.[20]

Open and robust markets are intended to provide more efficient use of water by allowing water to be 'retired' from less productive use.[21] Some of the economic, environmental and social benefits that may accrue from water trading include:

- reducing environmental effects of irrigation by transferring water away from highly saline areas or areas of high leakage to areas of more sustainable use
- allowing landholders to retire, restructure or leave agriculture
- allowing flexible use of water for annual crops such as vegetables and cotton in times where permanent plantings may have surplus water
- providing regional development with water made available to new industries and investments.[22]

---

[20] Productivity Commission, *Rural Water Use and the Environment: The Role of Market Mechanisms* (Melbourne: Productivity Commission, Final Research Report, 25 August 2006) [hereafter Productivity Commission].

[21] Henning Bjornlund and J. McKay, 'Do Permanent Water Markets Facilitate Farm Adjustment and Structural Change within Irrigation Communities?', 9 *Rural Society* 3 (1999).

[22] Frontier Economics *et al.*, *The Economic and Social Impacts of Water Trading: Case Studies in the Victorian Murray Valley* (Canberra: Rural Industries Research and Development Corporation, Publication No. 07/121, 2007).

At the same time number of concerns have been expressed about the economic and social impacts of water trading. These include (i) that sale of water out of a region may leave the costs of maintaining existing water infrastructure on fewer remaining users, referred to as 'stranded assets'; (ii) that contraction of irrigated agriculture because of water being sold to metropolitan areas will have a flow-on effect to various support industries in the region.

A recent report on this subject examines the evidence over a ten year period from irrigation areas in Victoria where water trading has been the most prevalent.[23] The report notes the difficulties of untangling the effects of water trading from other events which have occurred in rural communities such as drought, and changing commodity markets, and finds that it is necessary to consider permanent and temporary trading together. There was net increase of water for irrigation use in four out of five areas studied. The report has found that the impacts of trade are generally consistent with rural sociology theories. Trading was found to have affected third parties, with trade into the regions leading to increased competition in production, higher water tables; and trade out of regions leading to 'stranded assets', and depopulation. Rapid structural and social change placed a strain on regional infrastructure, and affected the social cohesion of communities.

Advantages of trading were also evident, and again were consistent with economic theories of trade and investment. It allowed horticulturalists to secure access to water for new plantings of land not previously irrigated, and led to increased regional development. Trading also improved the capacity of parties to manage risk and farm decision-making. It allowed farmers in sectors such as the dairy industry to clear debt. Permanent trading tended to occur only when a farm was sold as a 'going concern'.

Thirdly, public and political attitudes are changing with options of recycling and desalination taken seriously. In 2006 Toowoomba – a country town in Queensland which is fast running out of water – planned to spend $68 million on recycling water to add to drinking supplies. When the council's plan was put to a referendum, two-thirds of residents voted against it. However this is not indicative of wider community views. An overwhelming 70

---

[23] Ibid.

per cent of Australians are now prepared for recycled water to be added to potable supplies.[24]

Political commitment to water reform means that money is available for the reform agenda. Over thirteen billion dollars will be allocated in the period 2006–2011 for national water projects. Infrastructure solutions still dominate the water reform agenda, for example a desalination plant in Western Australia costing $387 million, a water pipeline in Queensland costing $500 million, and a water recycling plant in NSW of $500 million. The allocation of money is still very much subject to political pressures.[25]

The following parts of this paper refer to problems encountered in the implementation of water reform.

4.1 Constitutional Barriers and the 2007 'Water Security Plan'

Rivers and aquifers are transboundary resources, and the most prominent example is the MDB. The major river system running through the basin, the Murray-Darling, is Australia's largest and one of the world's major river systems. The Basin comprises 14 per cent of Australia's total land mass across four states. It is an important bio-region – within it is found over 30,000 wetlands some of which are listed under the Ramsar Convention of Wetlands of International importance. Irrigated agriculture in the Basin has had a history of over 140 years and accounts for an estimated 70–72 per cent of all irrigation water use in the country.[26]

The MDB is governed by an inter-jurisdictional pact which has its historical roots in the early 1900s. Presently an agreement exists between the governments of NSW, Victoria, SA, Queensland, the Australian Capital Territory (ACT) and the Commonwealth.

---

[24] Selina Mitchell, '70pc Would Drink Recycled Sewage', *The Australian*, 26 December 2006, 1. 1200 people participated in a poll conducted for the newspaper. 69 percent of respondents were prepared to accept treated effluent for household use including drinking if treated to the same quality as current supplies. 29 percent were prepared to accept it for non-drinking purposes. Only two percent opposed any use of recycled effluent.

[25] Fiona Carruthers, 'Politics Muddies $13bn Water Reform Plan', *The Australian Financial Review*, 19 May 2006, 1, 80–81.

[26] *See* Productivity Commission, note 20 above.

Figure 1: Murray-Darling Basin (Murray Darling Association, 2007).

In 1996 the Murray-Darling Basin Ministerial Council imposed a 'Cap' on extraction of water for consumptive use. It was set at a level of extraction under management rules and infrastructure development as of 1993/94.[27] Connell observes that this reference point has almost no connection with environmental perspectives and provides first for the need for production.[28] States were responsible for implementing the Cap. Annual Independent audits showed that NSW regularly breached the Cap and Queensland had not set a state target as was required.

---

[27] The Cap Formula 'is the volume of water that would have been used with the infrastructure (pumps, dams, channels, areas developed for irrigation, management rules, etc.) that existed in 1993/94, assuming similar climatic and hydrological conditions to those experienced in the year in question'. *See* Murray-Darling Basin Commission, Annual Report 2000–2001 66 (Canberra: Murray-Darling Basin Commission, 2000). *Source:* http://www.mdbc.gov.au/__data/page/46/annual_report_00-01.pdf.

[28] D. Connell, *Water Politics in the Murray-Darling* (Annadale: Federation Press, 2007).

A review of the Cap in 2000 concluded that although the Cap provided some measure of control, the activation of 'sleeper' and 'dozer' licences by trade meant that overall security of access had been reduced. Critically, because the agreed level was meant to be an interim measure from the start, the Cap level needed to be lowered in some places within the Basin. Further, as surface water allocation was reduced, groundwater was being 'mined' and the Cap needed to extend to this sector of the resource. It was becoming apparent that the Cap was not delivering on environmental benefits for the MDB.

After lobbying by the environmental sector and South Australians, in 2002 the Murray-Darling Basin Commission (MDBC) started exploring options of further reducing water extraction. The options considered included reductions of (i) 750 GL, (ii) 1,630 GL and (iii) 3,350 GL. A scientific panel was set up to assess the options. It gave the first option a 'low to moderate' probability of returning the River Murray to healthy working state. Only the third option was given a high probability of doing so.[29] Despite these findings the 'Living Murray Initiative', announced in 2004 with much fanfare, adopted a reduction level of 500 GL for use on six iconic sites.

Decision-making under the Murray-Darling Agreement is slow and laborious. All signatories need to agree on measures taken and members on the governing Ministerial council reflect parochial interests. Often difficult decisions are side-stepped. The decision to Cap use in the MDB was bold, but standards adopted were low, and although this was meant to be adjusted upwards, it did not happen. Implementation of the Cap was also uneven. With the Living Murray Initiative, environmentalists say that very little water has actually been returned to the river.[30]

These criticisms have recently been validated. Attempting to break the stalemate, the Commonwealth government announced

---

[29] Gary Jones *et al*, *Ecological Assessment of Environmental Flow Reference Points for the River Murray System: Interim Report Prepared by the Scientific Reference Panel for the Murray-Darling Basin Commission, Living Murray Initiative* 18 (Canberra: Cooperative Research Centre for Freshwater Ecology, 2003), as cited by Connell in note 28 above.

[30] R. Keating, 'Living Murray Plan 'Drains Faith in Govts'', *Australian Broadcasting Commission* 22 February 2006. *Source:* http://www.abc.net.au/water/stories/s1576062.htm.

in January 2007 that 10 million dollars will be invested over ten years to improve water management across the nation.[31] Citing the severity of the drought in south eastern Australia, various factors such as climate change and changes in land use reducing inflows in the catchments, the Commonwealth proposed new governance arrangements 'to ensure decisions affecting it are made promptly and with a Basin-wide perspective'. The Plan focused primarily on issues in the south of Australia, specifically the Murray-Darling Basin.

The investment plan for water security was aimed at improving water efficiency and addressing over-allocation of water in rural Australia through targeted works in key areas namely:

- modernising irrigation by reducing the loss of water mainly in the Murray-Darling Basin through leakage, evaporation and obtaining efficiencies through updating outmoded irrigation methods. The funds will also improve metering, monitoring and accounting of water and improve the efficiency of river flow and water storage operations. The target is to achieve 25 per cent savings in total irrigation water use ($6 billion over 10 years)
- addressing over-allocation in the Murray-Darling Basin including reconfiguring irrigation systems, retiring non viable areas, providing structural adjustment assistance, and using water saved to restore the health of rivers and wetlands ($3 billion)
- reforming governance in the Murray-Darling Basin ($600 million)
- developing capacity to enable management, interpretation and free public access to water information ($480 million)
- examining future land and water development in Northern Australia
- ensuring a sustainable future in the Great Artesian Basin through further investment in works ($85 million).

Cautiously welcomed by conservation and farming groups, the Plan was subject to intense negotiations between the Common-

---

[31] Australian Commonwealth Government, A National Plan for Water Security, 25 January 2007. *Source:* http://www.environment.gov.au/water/publications/action/npws-plan.html.

wealth, State and related parties. The overwhelming focus of spending under the Water Security Plan is on irrigation works in the Murray-Darling Basin. Over 60 per cent of total funds will be spent on improving irrigation efficiency, and 30 per cent of total funds addresses over-allocation in the Murray-Darling Basin.

On 26 March, 2008 a new political agreement renamed the National Water Plan was made between the Commonwealth and Victorian governments.[32] This means that all of the Murray-Darling jurisdictions have now agreed to refer their power over the rivers in the Murray-Darling Basin to the Commonwealth. The key additional terms include the continuation of the State's role in setting annual water allocations; and an in-principle commitment from the Commonwealth to invest up to $1 billion in Stage 2 of the Food bowl Project in Victoria to return 100 billion litres of water to the Murray River and an equivalent volume of water to farmers in the Goulburn region.[33]

The scope of new arrangements is as yet unclear. However it is likely that the Commonwealth will exercise these referred powers to, amongst other things, prepare a Basin wide strategic plan setting a sustainable Cap on surface and groundwater use at the Basin and individual catchment level; set Basin-wise water quality objectives direct operation on supply systems and manage water for the environment. A new MDB authority will be set up, with independent experts and not representatives of jurisdictions. That Authority will report to a single Minister and will be charged with setting a new 'Cap'. It appears that the new Murray-Darling Basin authority will not produce its first plan for the Basin until 2011.

The federal Minister for Water intends to use up to $ 3 million to buy-back access entitlements from non-viable or inefficient irrigators. This aspect of the Plan has met with criticism from within the Commonwealth government itself.[34] It is feared that irrigators who have been allocated entitlements without paying

---

[32] Senator Penny Wong, Minister for Climate Change and Water, Murray Darling Deal Delivered, (Media Release PW 41/08, 26 March 2008). *Source:* http://www.environment.gov.au/minister/wong/2008/pubs/mr20080326.pdf.

[33] Asa Wahlquist, 'Filling the Begging Bowl', *Weekend Australian* (29–30 March 2008), 23.

[34] Angus Griff and David Crowe, 'NATS Threaten PM's $10bn Water Plan', *The Australian Financial Review,* 9 March 2007, 1.

for them will now be unjustly enriched by selling to government. The use of public funds to buy back entitlements for stemming overuse of water can be justified only if there is sufficient public confidence in the market. There are concerns that a number of factors erode the long-term availability of water in rivers. The most significant of these are climate change, the proliferation of small farm dams, de-forestation, ground water extractions, bushfires, management of surplus flows, and changes to irrigation water management and return flows.[35] Buying back of access entitlements need to be accompanied by a resolution of these issues. This issue of buying back water to address over allocation is also very much related to continuing reform of the specification of entitlements.

The Water Act 2007 gives effect to a number of key elements of the Commonwealth Government's National Plan for Water Security. As yet in early 2008, the Water Act 2007 has yet to commence; with the renegotiation water pact resulting from the change of Commonwealth government in November 2007 significant amendments to the Act are expected. Accompanying changes will need to be made by each of the participating states.

It is significant that the two versions of the national water plans under the former Howard and the present Rudd government focus primarily on issues in south eastern Australia, specifically the Murray-Darling Basin.

4.2 The Specification of Private Entitlements is Incomplete

The COAG water reform framework required that access to water be specified as a tradeable entitlement. Specification has occurred in varying ways in states and is still subject to a high degree of difference.[36] The NWI requires that these entitlements must be subject to adaptive management. It also requires that entitlements are a right to a share of the consumptive pool of water available in a resource from year to year and not a fixed maximum volume.

---

[35] *See* Productivity Commission, note 20 above at 16.
[36] In perhaps overstating the problem Shi finds over 400 types of entitlements that are still being used in the River Murray to define opportunities to access and use water.

Even within the water sector there is confusion over terms.[37] For the purposes of clarity it is suggested that the term 'access entitlement' refers to the specified right with the following characteristics:

- usually perpetual but capable of being adjusted in a planning cycle of 10 years
- relates to a share of a pool of water.

The periodic opportunity provided by the entitlement is commonly referred to as 'seasonal allocation'. This is the volume of water that can be extracted, used, or sold within an irrigation season in a year. A sale (or transfer) of access entitlements is often referred to as permanent trade. The transfer of the seasonal allocation for a year or period of years is often referred to as temporary trade or renting or leasing of the entitlement.

Two examples of present entitlements illustrate the degree of variability between states. At present NSW's form of an entitlement is called an 'access licence'. It has two components. The first, the *share component*, is defined as a specified share in the available water within a specified water management area or water source. The second, the *extraction component*, is the right to take water at specified times, or rates or circumstances or a combination of these; and in specified areas or from specified locations.[38] In addition, the share component may be expressed as a specified maximum volume over a period, or a specified proportion of the available water, or a specified proportion of the storage capacity of a particular storage.[39] There are over twelve categories of access licences with differing priorities. The priorities became relevant in times of drought when allocations have to be diminished. Those of higher priority e.g., local water utility licences, regulated (high security) access licences will be least subject to any cutting back of their access to water.[40]

In Queensland, an entitlement is called a 'water allocation'. It is specified by a reference to a number of matters – a nominal

---

[37] *See* Productivity Commission, note 20 above at 45.
[38] New South Wales, Water Management Act 2000, Act 92 of 2000, 12 August 2000, Section 56(1).
[39] Ibid., Section 56(2).
[40] Ibid., Sections 57 and 58.

volume, the location from which the water may be taken, the purpose for which it is taken, any conditions attached to its use, the resource operations plan under which the entitlement is managed, and other matters prescribed under a regulation.[41] If the water is delivered in an irrigation scheme from a storage, then further particulars are required to be registered – the resource operations licence under which the entitlement is managed, and the priority group to which the entitlement belongs.[42] Alternatively if the water is pumped by the entitlement holder, then the entitlement should also state the maximum rate of taking the water, the flow conditions under which the water may be taken, the volumetric limit and the group to which the entitlement belongs.[43]

Under the NSW and Queensland models, entitlements are perpetual and capable of trade. They are reviewed under a ten year planning framework. Queensland and Victorian entitlements come with a 'security' of supply levels. This security level is embedded in Queensland within a Resource Operations Plan, and in Victoria within a Bulk Entitlement. 'Security' is also referred to as a reliability factor. It relates to the frequency and severity of shortfalls between the quantity of water desired and the quantity of water that can be supplied. It is often indicated as a statistical probability, for example urban users in the Goulburn catchment, Victoria will have 99 per cent security, meaning in 99 years out of a hundred, those users will get a full supply of water.[44] NSW entitlements do not refer to the notion of a statistical probability.

4.3 Interstate Markets not Active

It is an objective of the NWI that states put in place legal arrangements to allow interstate trade of entitlements. Legislative impediments that disallow interstate trade have been mainly dismantled. As a broad generalisation, interstate markets are still not active.

---

[41] Queensland Water Act 2000, Act No. 34 of 2000, 13 September 2000, Section 127(1).
[42] Ibid., Section 127(2).
[43] Ibid., Section 127(3).
[44] Bulk Entitlement Order (Kyabram) Conversion Order 1995 (Vic) Clause 7.

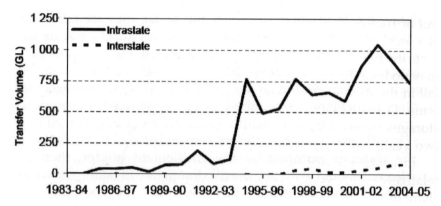

Figure 2: Seasonal allocation trade in the southern Murray-Darling Basin.
Source: See note 20 (Productivity Commission, 2006)

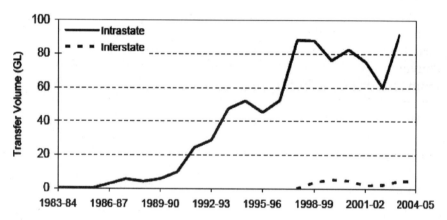

Figure 3: Water entitlement trade in the southern Murray-Darling Basin.
Source: See note 20 (Productivity Commission, 2006)

It is accepted that markets work best when the product is well-specified. Because entitlements are defined in a variable manner, buyers are uncertain of the product on offer when that product comes from a different state. This uncertainty results in a reluctance to enter the interstate market and affects price.

For example in Victoria holders of high security entitlements know that even in very dry years they can expect to receive almost all of the volume of water they are promised. A security factor is attached to the entitlement, as much as 97 per cent for rural users meaning that holders can expect to receive all of the water in 97 out of 100 years.⁴⁵ In the context of climate change, whether there is a level of security for entitlements is an important issue. These security factors are based on historical records of rainfall which may no longer be relied on for predicting how and where rain will fall in the future. Even without climate change, security is still an issue. Dams in Victoria have been conservatively managed. The storages are large, and as far as possible, water is held to ensure two years worth of supply of high security water in a dam.

In contrast, historically NSW entitlements are managed in a relatively less secure manner. The entitlements are framed as a share of the resource available for consumptive use in a particular catchment. These shares are not defined with a level of security that entitlement holders expect in Victoria. Therefore Victorian farmers are wary of buying water from NSW.⁴⁶

A high level of co-operation between SA and Victoria has resulted in an interstate trading arrangement for the River Murray based on an 'exchange-rate' system. The term refers to a rate of conversion calculated and agreed to be applied to water traded from one state (or trading zone) to another.⁴⁷ The federal government wants a collective agreement on trade between Victoria, NSW and SA, not bilateral agreements. All three states were penalised in 2006 by the National Water Commission for failing to reach agreement on interstate water trading.⁴⁸

A seven year study into the market of permanent water trades in an irrigation district in Victoria shows that prices and volumes of water traded have steadily increased but prices for permanent water have half the rate of increase of temporary water.⁴⁹ What is most telling is that demand for permanent water stagnates when

---

⁴⁵ Poh-Ling Tan, 'Irrigators Come First: A Study of the Conversion of Existing Allocations to Bulk Entitlements in the Goulburn and Murray Catchments, Victoria', 18 *Environmental and Planning Law Journal,* 154, 171 (2001).

⁴⁶ Duncan Hughes, 'States Fined for Failing to Agree on Water Deal' *The Australian Financial Review* 21 (21 April 2006).

⁴⁷ *See* Productivity Commission, note 20 above at 18.

⁴⁸ Ibid. Victoria was fined $10 million, NSW $13 million and SA $3 million.

⁴⁹ Henning Bjornlund and Peter Rossini, An Empirical Analysis of Factors Driving Outcomes in Markets for Permanent Water – An Australian Case-Study (Paper submitted to the Twelfth Pacific Rim Real Estate Society Conference, Auckland, January 2006).

supply is very low and irrigators struggle to cover their short term needs. In a drought it appears that there is little water to sell and few have money to buy.

4.4 Ecological Considerations Still Come Last

National principles developed under the Water Reform Framework and NWI accept that:

- environmental flows should be given statutory recognition and have at least the same degree of security as consumptive entitlements
- environmental flows should be determined under statutory water plans based on the best available scientific information available to sustain the ecological values of water dependent ecosystems
- water plans should incorporate ecological outcomes and define management arrangements to achieve those outcomes
- a better balance for over-allocated river systems and water to be reallocated from consumptive to ecological use
- environmental water managers be established to audit, review and publicly report on the achievement of environmental and other public benefit outcomes.[50]

In 2005 the first national audit of water resources found that ecosystems are still being short charged.[51] Only a handful of high-conservation value rivers are protected, there is inadequate knowledge of connections between surface and groundwater systems and poor understanding of over allocation of water licences.

NSW arguably had the firmest statutory commitment to providing for environmental flows. A statutory duty was imposed on the Minister to set environmental water allocations that maintain

---

[50] Alex Gardner, 'Environmental Water Allocations in Australia', 23 *Environmental and Planning Law Journal* 208, 214 (2006).

[51] Dennis Shanahan and Matthew Warren, 'States 'Fail' on Water Crisis, *The Australian* 1 (13 October 2006) and Australian National Water Commission/ WRON Alliance, Australian Water Resources 2005 – Discovery Phase Integrated Theme Discussion Paper (2006). *Source:* http://www.nwc.gov.au/nwi/docs/Discovery_phase_paper.pdf.

fundamental ecosystem health in priority to allocating water for consumptive uses.[52] Further provisions required that a water sharing plan must establish environmental water rules.[53] In a recent challenge to a statutory water management plan, the NSW Court of Appeal found that the plan for the Gwydir River was inconsistent with the original provisions of the Act.[54] The court went on to consider whether the plan was thus invalidated. Applying a test developed by the High Court,[55] the NSW Court of Appeal held that the statutory provisions did not have the effect of invalidating a water management plan that was inconsistent with them. The High Court granted special leave to the Nature Conservation Council to appeal the decision. Quickly, the state government moved to amend the water legislation redefining environmental water, inserting a validating provision for management plans thus retroactively resolving the matter. Its actions made the appeal redundant.[56]

5. Water Planning in Australia

The NWI requires more transparent and comprehensive water planning that deals with key emerging issues. A cardinal principle that underpins planning is that water users, interest groups and the general community are to be involved as partners in catchment planning processes. Despite the attempt to consult and involve the public in water reform, the scale and pace of change has meant that implementation of COAG reform and the NWI has been contentious.

The latest water resource information figures are provided in Australian Water Resources 2005. In October 2006, a number of key issues were identified.[57] They are:

---

[52] NSW Water Management Act 2000, Section 5(3) (a) and (c).
[53] Ibid., Section 50.
[54] *Nature Conservation Council of NSW Inc v The Minister Administering the Water Management Act 2000*, [2005] NSWCA 9.
[55] *Project Blue Sky Inc v Australian Broadcasting Authority*, (1998) 194 CLR 355.
[56] For an analysis of the NSW legislative provision see Alex Gardner, 'Environmental Water Allocations in Australia', 23 *Environmental and Planning Law Journal* 208, 214 (2006), 215–219.
[57] *See* http://www.water.gov.au/publications/AWR2005_Level_1_Findings_061006.pdf.

- Water planning and management approaches are highly varied across Australia's 340 surface water management and 367 ground water management units
- At present, only 18 percent of surface water management units and 33 per cent of ground water management units have a draft or final management plan in place
- Despite physical connections between surface and ground water, most areas are not managed from an integrated water resources perspective
- Of the surface water management units with draft or final management plans only 22 percent considered ground water, while of the ground water management units with final or draft management plans only 65 percent considered surface water.

More recently, in its first Assessment of the National Water Initiative (NWI) the National Water Commission emphasised that effective water planning is fundamental to the NWI. How and when plans are developed and implemented is the critical driver for implementing the NWI entitlement and planning framework.[58] Besides the provision of public and private benefits through security, certainty and tradeability of entitlements and statutory recognition of water for the environment with at least equal security to private entitlements, the Assessment reiterated that procedural and substantive issues for water plans are:

- The need for a transparent process.
- The use of best available science including social-economic analysis.
- The need for community support of plans.
- The need to account for indigenous interests in water.
- The need to address over-allocation of water in plans by 2007, and for substantial progress towards adjusting these systems by 2010.

Over-allocation is an issue of foremost importance and the NWC takes the view that the process of planning itself needs to be robust

---

[58] Australian National Water Commission, National Water Initiative: First Biennial Assessment of Progress in Implementation, 2007 [hereafter NWC]. *Source:* http://www.nwc.gov.au/PUBLICATIONS/index.cfm.

otherwise the issue will not be adequately addressed. The Assessment of Progress repeatedly comments that:

> There appears to be a wide gap between putting in place NWI-compliant water plans that address over allocation and perceptions that, despite these plans, systems remain significantly stressed or over allocated. ... The NWI also recognises that determining the needs of consumptive and environmental uses in the water plans (clause 36) involves settling trade-offs in competing outcomes for water systems that will involve judgements informed by best available science, socio-economic analysis and community input. It is clear that in developing their plans all states are striving to achieve this balance.[59]

The NWC notes that while all states have mechanisms that involve indigenous representation in water planning, the explicit inclusion of indigenous interests in water plans is rare. States have taken a range of approaches. For example, NSW has provision for aboriginal cultural and community development access licences but whether any licences have been issued is not known. Queensland on the other hand provides for indigenous interests through environmental flows. In western Australia, it was reported that indigenous interests are identified through cultural value surveys.[60]

5.1 The Process of Water Planning

There is a large variation in water planning approaches adopted in the various States. For example NSW has a State wide strategic plan which set broad targets and policies, which are underpinned by more specific water sharing plans for specific reaches of rivers, as well as broader 'macro' plans for major catchments. This may be contrasted against Queensland which has a strategic Water Resource Plan for each catchment, implemented by Resource Operational Plans.[61]

---

[59] Ibid., 30, 35.
[60] Ibid., 40.
[61] Even within states, variations apply. For example, in Queensland, the Gulf Water Resource Plan 2007 covered eight catchments that drained to the Gulf of Carpentaria.

However Hamstead *et al.* in their recent assessment of current water planning processes practices in Australia consider that the water plan preparation process can be broken into the following steps:

(i) *Planning initiation*. This involves making the decision to undertake planning, establishing the planning processes, and organising the human resources required to drive the process.

(ii) *Situational analysis*. This step looks at the current status of resources, environmental and other public benefits, uses, and socio-economic factors as well as future threats, risks and opportunities.

(iii) *Setting directions*. Given the situational analysis, this step is where broad decisions are made on which way to go, including the objectives and outcomes that are being sought. It encompasses such things as vision statements, which are typically very broad, and outcomes or objectives which can be more specific.

(iv) *Identifying and assessing strategies*. This is usually achieved through a process of identifying and assessing options (benefits, impacts, mitigation measures).

(v) *Strategy selection*. This involves weighing up trade-offs (including socio-economic and equity factors) and deciding on a preferred approach. Arising from this are strategies/activities and measureable targets and actions.

(vi) *Building in adaptability*. This step identifies how implementation and outcomes will be monitored and what should happen if things do not work as expected (e.g., implementation failure, assumptions prove to be wrong, strategies ineffective, improved data, situational change). Arising from this is a monitoring strategy and triggers for adaptation or change.

(vii) *Plan approval*. For water planning, this is the final Ministerial endorsement that incorporates the outcomes of the process into a statutory framework.

The authors emphasise that the process is reiterative in nature, with later steps frequently requiring the revisiting of earlier steps. For example, plans can be sent back by the Minister at the final approval stage for further assessment and development of further options.[62]

5.2 The Challenges of Water Planning

It is recognised that water planning involves making decisions based on tradeoffs between competing outcomes. Planning systems need to be transparent in the sense that there is adequate opportunity for productive, environmental and other public benefit considerations be identified and considered in an open way.[63] The NWI recognises that settling these trade-offs require judgements informed by best available science, socio-economic analysis and community input.[64] As the acceptance of water plans in the community requires their confidence, a number of mechanisms were adopted for example independent community reference panels, use of independent scientific reports, public notification of draft plans, and ability to make public submissions. In its 2005 Audit, the National Water Commission has found that transparency is an issue in trade-offs between environmental and consumptive use, with concerns in NSW, SA, Tasmania and NT.[65] The recent study by Hamstead confirms this saying 'further exploration of alternative options for increasing the transparency and objectivity of trade-offs and final decision making' is required.[66]

The importance of having a clear policy framework in place prior to the commencement of water planning was yet another critical issue.[67] Communities become highly dissatisfied in the absence of clear frameworks for collaboration and agreement on the role of community sectors, and other partners in planning.[68] In

---

[62] M. Hamstead, C. Baldwin and V. O'Keefe, Water Allocation Planning in Australia – Current Practices and Lessons Learned (Report Commissioned by the Australian National Water Commission, April 2008) [hereafter Hamstead]. *Source:* http://www.nwc.gov.au/PUBLICATIONS/index.cfm.
[63] *See* NWI, note 16 above at para 5.
[64] *See* NWI, note 16 above at para 36.
[65] *See* NWC, note 58 above.
[66] Hamstead *et al.*, note 62 above, xii.
[67] Ibid., 151.
[68] Poh-Ling Tan, 'Legislating for Adequate Public Participation in Allocating Water in Australia' 31 *Water International* 12, 22 (2006).

some instances committee expectations of their role and influence have not matched those of water planners, or executive levels of government. Conflicting incidents of this nature have been experienced in Queensland and NSW.[69]

Science-based conflicts are characterised by a number of challenges. This includes obtaining all the necessary scientific information, providing scientific information in formats which are accessible to the public, disagreements as to whether the information is the best scientific information available. There is considerable difficulty in dealing with little, conflicting or inconclusive technical information. The latter is particularly the case with environmental allocations where the extent of outcomes from a proposed environmental flow is uncertain.

General concerns over the adequacy of public participation processes are high. There is a recognised need for all participants to be informed and effective at representing their interests, expressing their values and responding to complex problems. Often a lack of knowledge and expertise by participants, who have little understanding of the technical jargon used by water managers, may diminish their capacity to understand fully the various options presented for debate and decision.[70]

Indigenous people tend to be involved to varying degrees in water planning as part of a general public participation process. However, questions have been raised about the extent and adequacy of the mechanisms enabling their involvement. The NWI provides for direct representation of indigenous people in water planning process that should, where possible, allocate and account for water for native title purposes. Distinct problems relating to indigenous forms of representation, decision-making and communication need to be well understood within water policy and planning sectors and forums. It has been said that there is a

---

[69] D. Cleary, Community Involvement in the Water Planning Process (Paper presented at the Conference on Public Participation in the Australian Water Industry, Sydney, 19 August 2003) and G. Kuehne and B. Bjornlund, 'Frustration, Confusion and Uncertainty – Qualitative Responses from Namoi Valley Irrigators', 33 *Water* 78 (2006).

[70] Poh-Ling Tan, 'A Historical Introduction to Water Law Reform in NSW – 1975 to 1994', 19 *Environmental and Planning Law Journal* 445 (2002).

'chasm between the perceptions of the available opportunities for involvement and the reality experienced by indigenous people'.[71] Additionally, the lack of understanding of NWI processes by remote indigenous communities adversely affects their ability to participate effectively in water planning processes.[72]

Public participation in water planning is typified by imbalances in access to the debate – when access is allowed, the extent of access, financial support to enable access and the sorts of knowledge that are valued. Too often, specialist knowledge is the only knowledge accepted – with local knowledge particularly excluded from the debate. These power imbalances are not easily addressed by clever strategies late in the planning process. International experience strongly suggests that a clear acknowledgment and re-negotiation of these power imbalances by all parties early in collaborative planning processes lessens the chance of less empowered participant groups boycotting the process. This improves the credibility of planning outcomes and the chances of their acceptance and implementation by all parties.[73]

A project addresses these issues through the formulation and trial of participatory tools in two large case-studies in Northern Australia. Northern Australia has the world's most significant concentration of river catchments that still retain their ecological integrity.[74] Northern rivers provide valuable ecological services, internationally recognised biodiversity values and productive coastal fisheries. The Water Security Plan 2007 acknowledges the immense pressure to develop the water resources of this region. Water planning processes need to be legitimate, robust and reliable if rivers in Northern Australia are to avoid the worst of the mistakes affecting over-allocated rivers in southern catchments.

---

[71] Brian McFarlane, The National Water Initiative and Acknowledging Indigenous Interests in Planning (Paper presented at the National Water Conference, Sydney, 29 November 2004).

[72] Sue Jackson, 'Compartmentalising Culture: The Articulation and Consideration of Indigenous Values in Water Resources Management', 37/1 *Australian Geographer* 19 (2006).

[73] Archon Fung and Erik O. Wright, *Deepening Democracy: Institutional Innovations in Empowered Participatory Governance* (London: Verso, 2003).

[74] Collaborative Water Planning Project. *Source:* www.griffith.edu.au/centre/slrc/water.

Conclusion

In the two centuries since colonial occupation, river systems have suffered degradation to a point that it cannot be ignored. In this time wetlands were drained, natural habitats destroyed and native species have dwindled. Ecosystems lose resilience and become accidents waiting to happen. The current drought exacerbated by growing levels of water use continues the crisis.

Cooperative federalism brought about Australia's ambitious agenda for reform. The key decision was to accept that water needed to be sustainably managed. However as we have seen in the NSW's example, even when that principle was introduced in legislation, its implementation can be in jeopardy. States tend to resile from high statutory standards when difficulties are encountered. We see that the law is a weak tool in the face of legislative and political intervention.

High level policy making brought about the NWI in 2004, providing specific actions, standards and target dates. The role of the National Water Commission to oversee reform is important. Its reports have brought about refreshing openness into water management by states who are made to account for delay or lack of success in reaching goals. The political commitment in funding reform measures is a reflection of the importance of the issue.

A number of challenges lie ahead, for example, the ongoing improvement in the specification of entitlements. With drought and climate change, the most critical issue is the unpredictable nature of long term availability of water. In recognition of this, the NWI has made water planning the cornerstone of reform. Formulating transparent, comprehensive planning processes which addresses power imbalances, builds capacity of all participants including disadvantaged groups and allows for resolution of conflicts over scientific information and its interpretation, is arguably the biggest challenge for water reform.

17

Law and 'Development' Discourses About Water: Understanding Agency in Regime Changes

RADHA D'SOUZA

1. Introduction

Two events of significance for freshwater resources in the 'Third World' occurred in 1997. The World Bank (WB) set up the World Commission on Dams (WCD) in March 1997. In May 1997 the United Nations (UN) General Assembly adopted the United Nations Convention on Non-navigational Uses of International Watercourses in May 1997 (UN Water Convention).[1] The first development was the culmination of a sustained critique of large dams by environmental and social justice movements in the 'Third' and 'First' worlds alike.[2] The critique of large dams occurred in the context of the rise of neoliberal transformations within International Organisations (IO). The second was the culmination

---

[1] Convention on the Law of the Non-navigational Uses of International Watercourses, New York, 21 May 1997, *reprinted in* P. Cullet and A. Gowlland-Gualtieri eds., *Key Materials in International Environmental Law* (Aldershot: Ashgate, 2004), 481.

[2] For e.g., *see* Anthony H.J. Dorcey ed., Large Dams: Learning from the Past, Looking to the Future (Washington: World Bank, 1997); World Commission on Dams, Dams and Development: A New Framework for Decision-Making, Report of the World Commission on Dams (Earthscan: World Commission on Dams, 2000).

of sustained efforts to create a legal framework to resolve transboundary conflicts over freshwater and pave the way for transboundary institutions for water projects and dispute resolution. Development of the UN Water Convention spanned nearly all of the post-World War II period of economic 'development' and concluded against the context of rising concerns about 'water wars' and security.[3]

Both events were, *ex facie*, about dams and development but nevertheless ramifying legal, institutional, local and global changes for water regimes in the 'Third World'. Yet the discourses around the two events ran parallel without convergence or contestation, intra-discourse, seen at best, as a coincidence. There is nothing in the events per se that suggest the possibility that there might be anything more to the absence of connections in the discourses on the two events. This chapter argues that the insular yet related discourses on dams, development, water conflicts, and international water law render opaque a political programme for restructuring the international regime for regulating freshwater resources along neo-liberal principles. Thus the absence of discourse on the interconnections between the two events is problematic in its own right.

Understandings of structure–agency relations in social theory point to the ways in which social structures and social agency constrain and enable social change. What is less understood is the role of *concepts and ideas* that mediate the actions of social agents in structural change. For example critical responses to neoliberal transformations from scholars and activists juxtapose states and markets as antithetical institutions. In doing so, did they mirror the conceptual frameworks of neoliberal transformations albeit from different ends of the binaries? Could this conceptual framework be the reason why the interconnections went unnoticed thereby facilitating the very neoliberal regime changes that the critical voices opposed? This wider question is examined in the sections below by interrogating the two events, the contexts within which they occurred, the conceptual frameworks that informed the two discourses and the social outcomes in the water sector that followed.

---

[3] *See* Joyce R. Starr, 'Water Wars', 82 *Foreign Policy* 17 (1991); Juha I. Uitto and Alfred M. Duda, 'Management of Transboundary Water Resources: Lessons from International Cooperation for Conflict Prevention', 168 (4) *The Geographical Journal* 363 (2002).

2. Regime Changes, Human Agency and Neo-liberalism

The rise of neo-liberalism since the end of the Cold War has triggered pervasive transformations in regulatory regimes in a wide range of social sectors.[4] Water is no exception. Regime changes have occurred historically during certain periods either as a result of revolutionary social transformations, or far-reaching changes in the institutional mechanisms within the same constitutional order. Whatever the means, changes in regulatory regimes entail wide-ranging institutional transformations and relationships between institutions in a social system.

One conceptual challenge posed by the emergence of neo-liberalism globally is the problem of human agency in regime changes. Regime theories have been criticised, and rightly, for their tendency to subsume human agency and to construct regimes premised on empirical conjunction of events and facts within narrow positivist frameworks. Regimes need not be understood as a conjunction of facts and events, however, and human agency does not have to be excluded in accounts of regimes.[5] Regimes involve relatively enduring interrelationships between institutions. The stability is achieved through 'manufacturing consent' achieved through reconciling conflicting interests where necessary and establishing decisive hegemony by one or more interests in society where required.

The other conceptual challenge relates to transitions from one regime to another as with the transition from the post World War II world order to the post Cold War world order. Regime transitions, the period when one regime has broken down and another is in construction, are periods when the 'social whole' appears blurred and ideological debates by major social actors emphasise some strands in the structural changes underway over others. The processes of change are rationalised or resisted by different social actors using different types of arguments, usually economic arguments, political arguments or moral/ethical

---

[4] *See* John Braithwaite and Peter Drahos, *Global Business Regulation* (Cambridge: Cambridge University Press, 2000).

[5] Christopher Lloyd, 'Regime Change in Australian Capitalism: Towards a Historical Political Economy of Regulation', 42 (3) *Australian Economic History Review* 238 (2002).

arguments.[6] These arguments are grounded in the position of different social actors within the previous social order and the ways in which the changes impact upon them.

The dominance of positivism in law and empiricism in social sciences means, the arguments appear disaggregated and disconnected. The 'social whole' is rendered opaque as a result.[7] The fluidity during periods of transition means the nature of the 'social whole' can be grasped only after the regime has achieved some degree of stability. The systemic coherence of regimes thus becomes visible only retrospectively. Regime theories therefore tend to lapse into retrospective analysis of the institutional relationships within a social order that appear to discount the social agents that brought about the transformation. The challenge therefore is to be able to envision the structural and systemic ramifications of the arguments, economic, political and moral/ethical that social agents put forward in support of, or opposition to social and legal changes during periods of transition. The simultaneous insularity and complementarities in the WCD and UN Water Convention processes provide a useful vantage point to investigate the ways in which political arguments, economic arguments and moral/ethical arguments by different social actors on questions affecting water resources development, especially the controversies on large dams, made from their positions within social structures, contribute to our understanding of the way concepts and ideas of social actors mediate regime changes.

Two most significant concerns for law under capitalism remains managing competition between economic actors and managing social conflicts following from economic developments. In relation to water resources development the concerns have been about managing the apportionment of water to different riparian users and regions; and providing mechanisms for dispute resolution arising from water appropriation and use. Neo-liberal regime changes entail transferring both functions from the institutions of

---

[6] Phillip Darby, *Three Faces of Imperialism, British and American Approaches to Asia and Africa 1870–1970* (New Haven, London: Yale University Press, 1987).

[7] On envisioning the social whole under capitalism, *see* Susan Buck-Morss, 'Envisioning Capital: Political Economy on Display', 21 *Critical Inquiry* 434 (1995).

the state to market institutions. In classical liberal theory, the rule of markets was ensured by 'rule of law' wherein the role of the state was, in Adam Smith's words, akin to that of a 'night watchman'. Sir Henry Maine the legal theorist who extended classical liberalism to the colonies rationalised colonial law by arguing all societies evolved from status based social relations to contractual social relations.[8] In developing law for the colonies, Maine blended social Darwinism and liberal theory to create the basis of 'progress' as the rationale for colonial law and governance.

In the post World War II world order the philosophical and conceptual foundations of policies developed by international development organisations, notably the WB, have further developed classical approaches to colonial law and governance. In the water sector, the WB and other IOs actively promoted state regulation through state economic planning, state bureaucracies and bilateral and multilateral development assistance in the post World War II period to facilitate regimes of appropriation of labour and environment through industrial development, mechanised agriculture and infrastructure development.[9] In recent times, the transition from state to market regulation has seen the WB advance market regulation in the water sector through water users' associations based on private property regimes, market instruments using user pay principles [10] to facilitate appropriation of labour and environment through development of industry, agriculture and infrastructure. The policies aim to take developing countries further up the ladder of 'progress' seen as movement from 'status' to 'contract' based social relations through law reforms in line with what Sir Henry Maine envisioned for the colonies.

Under early capitalism, before the World Wars, more and more relations and transactions in society assumed the form of a contract between individual(s) and/or group(s) within the umbrella of the

---

[8] Sir Henry Sumner Maine, *Ancient Law: Its Connections with the Early History of Society and its Relation to Modern Ideas* (London: Routledge and Kegan Paul, 1909).

[9] For a more specific contextual study in India, *see* Radha D'Souza, *Interstate conflicts over Krishna Waters: Law, Science and Imperialism* (New Delhi: Orient Longman, 2006).

[10] *See* Salman M.A. Salman, *The Legal Framework for Water Users' Associations: A Comparative Study* (Washington DC: World Bank, 1997).

nation-state.[11] The legal form of contractual relations provides (a) the conceptual framework for social transactions, (b) the value framework for social transactions and (c) the sanctions framework (i.e., mechanisms for dispute resolution and penalties for non-compliance). In the post-World War II world order, contractual social relations were extended to the international arena. The extension occurred by transforming economic relations between states and between states and IOs to (semi)/contractual legal forms. During this period the institution of the state developed a 'split personality'. The functions of the state as an institutional player in the economy, through public enterprises, manufacturing and trade, was akin to 'private' institutions with monopoly status, and the political functions were cast in the mould of traditional 'public' law.[12]

The constitutional status of the International Economic Organisations (IEO) within the UN system notably the WB and the International Monetary Fund (IMF) was formalised through specialised agency agreements with the UN.[13] The IEOs with independent legal personality could develop contractual relations between the IEOs and states and between states inter se using instruments such as bilateral and multilateral aid agreements, contracts and memorandums using private law principles and dispute resolution mechanisms. Neoliberalism takes the contract form of social relations to new heights by restructuring the relations between corporations, states and social groups qua collective/corporate entities as contracting parties. In other words law under neo-liberalism creates new institutions with their own sets of rules and goals; and regulates the relationship between the institutions. Legal innovations under neo-liberalism involves developing new forms of enacting and enforcing law, new discourses for legitimating law and new institutions that will regulate relations between different types of collective entities and institutions, in the new language of neoliberal legalism – the 'stakeholders'.

---

[11] Michael E. Tigar and Madeleine R Levy, *Law and the Rise of Capitalism* (New York: Monthly Review Press, 1977).

[12] For theoretical viewpoints on the 'public/private' divide in law *see* 130 *University of Pennsylvania Law Review* 'Special Issue on Public Private Divide with discussion and debate' (1982).

[13] Radha D'Souza, *Interstate conflicts over Krishna Waters: Law, Science and Imperialism* (New Delhi: Orient Longman, 2006), at 294.

Viewed in this way the WCD and the UN Water Convention reconstitute different strands in the regime-changes for water along neoliberal lines – the first restructures relations between social agents within nation states internally; and the other between states and transnational organisations and corporate entities externally.

3. State versus Market Regulation in Law

To assert any connection between the two events, it is necessary to acknowledge that both events are designed to transform the legal regimes for water in different spheres. The WCD develops rules, principles, guidelines and policies to regulate appropriation and use of water within national jurisdictions. The UN Water Convention develops rules, principles, guidelines and policies to guide appropriation and use of transboundary water between states internationally. Acknowledging that law is involved in both the events makes it possible to begin by interrogating the law as a point of departure to understand the hiatus in the discourses about the two events and the political programme that underpins both.

Markets are a complex of laws. Legal frameworks for market regulation are premised on and support: (a) multiple institutions (b) communities formed around economic interests (c) incorporation of communities of economic interests into legal entities (d) laws to regulate the relations between different legal entities and (e) application of private law principles (from Roman law traditions) to regulate relations between the state and the legal entities.

States are also a complex of laws. Legal frameworks for state regulation are premised on and support: (a) monolithic institutions e.g., civil service (b) communities formed around rights and obligations towards the state and its institutions (c) a constitution as the founding document that governs relations between the state and citizens (d) application of public law principles (from Roman law traditions) to regulate relations between the state and citizens, corporate and natural.

It is widely accepted that neoliberal transformations involve rolling back the state, and is associated with liberalisation and privatisation in economics. Differences in the characteristics of 'the law' under state regulation and market regulation may be less apparent. In essence the difference lies in the institutional framework for 'the law' seen as a set of rules and principles. Markets undertake 'enactment' and 'enforcement' of law in very different ways from states. An extended period of state regulation of economic regimes has familiarised us with certain legal forms that are now seen as essential features of the law by many, especially social scientists. These features include: (a) conflating law with statute law (b) an instrumentalist view of law that sees state agencies achieving certain outcomes mandated through statutes, rules, regulation and policies (c) law as a set of imperatives for different social actors to abide by; (d) law as comprising two distinct domains, the 'public' the 'private' domains; (e) regulation through the institution of the civil service, the executive and in the final analysis the legislature, all operating under public law principles.

Market regulation, the characteristic feature of law under neo-liberalism, involves regulation through market institutions. Market institutions involve setting up authorities/agencies/organisations that operate under a distinct set of institutional rules autonomous from the state.[14] Rolling back the state thus entails autonomy from conventional rules that govern state institutions comprising the civil service, the executive and rules of parliamentary procedures. Legal instruments under market regulation routinely take the form of setting up regulatory authorities to regulate a specified field in market relationships – e.g., competition, inflation, currencies and so forth. The regulatory authorities set up norms for the actors within that field and take steps to ensure actors conform to the norms for that field. The type of instruments used to regulate the market may include voluntary codes, industry standards, dispute resolution mechanisms amongst others, all operating on private law principles. Social policies too are brought under market instruments.[15] Hence the emphasis in more recent times on 'corporate social responsibility', labour market regulation through

---

[14] For regulation of different sectors of economy and society using market principles, *see* John Braithwaite and Peter Drahos, *Global Business Regulation* (Cambridge: Cambridge University Press, 2000).

[15] For examples of regulatory regimes in specific sectors, *see* John Braithwaite and Peter Drahos, *Global Business Regulation* (Cambridge: Cambridge University Press, 2000).

inflation policies and new institutional models for tertiary education funding.[16]

State regulation rationalises economic regulation on the basis of 'public' good in the name of society. Thus state regulation retains the distinction between the economic sphere and the social sphere, the public and the private domains in law. Market regulation rationalises economic regulation on the basis of 'public' good but assumes economic policy *is* social policy and therefore benefits all of society. Market regulation therefore conflates the economic and social spheres, the public and private domains in law. Thus it is the institutional context of the law, and the type of legal instruments used in law, that marks the point of departure for law under state and market regulation.

However, both state and market regulation share common attributes of law under capitalism. The common attributes include: (a) privileging economic relationships over all other social relationships (b) sanctifying private property rights (c) creating and refining legal regimes, principles and instruments for appropriation of labour and environment (d) legal polices and instruments for alienation of people from land, water, minerals and other natural resources by turning them into commodities for exchange in the market-place (e) positive law underpinned by empiricism and positivism in social and physical sciences.

The differences in the institutional frameworks for the law encompass different modes of enactment, enforcement and legitimisation of the law; and different philosophies, theories and rationalisations of principles and rules. It is important to emphasise the convergences in the characteristics of 'the law' under state and

---

[16] *See* Michael Cavadino, 'Commissions and Codes: A Case Study in Law and Public Administration', *Public Law* 333 (1993); Carl Emery, 'Public Law or Private Law? – the Limits of Procedural Reform', *Public Law* 450 (1995); Mark Freedland, 'Government by Contract and Public Law', *Public Law* 86 (1994); M. Sornarajah, 'Good Corporate Citizenship and the Conduct of Multinational Corporations', *in* Jianfu Chen and Gordon Walker eds, *Balancing Act: Law, Policy and Politics in Globalisation and Global Trade* 224–250 (Annandale NSW: The Federation Press, 2004); Diana Woodhouse, 'Delivering Public Confidence: Codes of Conduct, a Step in the Right Direction', *Public Law* 511 (2003).

market regulation. All too often the differences understood without the convergences create gaps in knowledge that allow insular developments in different dimensions of the same social phenomenon. The absence of apparent connections between the WCD processes on one hand and the UN Water Convention proceedings on the other in the discourses on dams and development exemplify the insular processes and conceptual gaps in the transition from state to market regulation of water resources development and law.

4. The World Commission on Dams

Throughout the post World Wars' era large dams have been the foci of bilateral and multilateral development assistance under the aegis of UN organisations and 'Third World' developmental states.[17] This is because in the post World War II international political economy dams became inextricably tied to industrialisation and a new international division of labour based on cheap agricultural production, cheap labour, consumerism and transferring environmental costs to the 'Third World'. By mid 1990 there developed a widespread critique of large dams within the academe and outside. There were a number of strands to the critique. Popular movements of displaced people in 'developing' countries challenged developmental models promoted by international organisations, most prominently, the WB.[18] The environmental critique was the other.[19] Calls for accountability of

---

[17] The United Nations publication series from 1949 to the present, first as 'Flood Control Series', later continued as 'Water Resources Series' is useful to trace the changes in the priorities and approaches of bilateral and multilateral organisations to water resources and river basin development.

[18] *See* Amita Baviskar, *In the Belly of the River* (Delhi: Oxford University Press, 1995); William Fisher ed., *Toward Sustainable Development: Struggling Over India's Narmada River* (Armonk, New York: M.E. Sharpe, 1995); Aviva Imhof, *The Asian Development Bank's Role in Dam-Building in The Mekong* (Berkeley: International Rivers Network, 1997); Leonard Sklar and Patrick McCully, Damming The Rivers: The World Bank's Lending for Large Dams (Berkeley: International Rivers Network, 1994); Enakshi Ganguly-Thukral ed., *Big Dams, Displaced People: Rivers of Sorrow, Rivers of Change* (New Delhi: Sage Publications, 1992).

[19] *See* Sklar and McCully, note 18 above; Donald Worster, 'Water and the Flow of Power', 13(5) *The Ecologist* 168 (1983).

international development agencies;[20] and internal reviews of lending policies[21] followed the critique of development models.[22] In this context the WCD was a significant event in that it rallied different 'stakeholders' in water and attempted to arrive at a lowest common denominator on standards and processes that were acceptable to all the 'stakeholders'.[23] The WCD was necessitated by the widespread critique of large dams. The critique of large dams was not the only factor that necessitated the WCD however. Without minimising the importance of the critique of large dams based on development models in the post World War II period of state-centred development, it is necessary to interrogate the structural transformations that were underway which provided the context for the WCD.

Briefly recapping the institutional arrangements for regulating the global economy at the end of World War II, the Bretton Woods agreement envisioned the creation of three institutions – the International Bank for Development and Reconstruction, (IBRD) later WB, to regulate banking, lending and finance; the IMF to regulate fiscal matters, exchange rate mechanisms and balance of payment matters between states; and the International Trade Organisation (ITO) to regulate global trade. Of the three functions, international trade did not acquire an independent institutional

---

[20] *See* Dana Clark, Jonathan Fox and Kay Treakle eds., *Demanding Accountability: Civil-Society Claims and the World Bank Inspection Panel* (New York: Rowman & Littlefield, 2003).

[21] *See* Bradford Morse and Thomas Berger, *Sardar Sarovar: Report of the Independent Review* (Ottawa: The Independent Review, 1992); *See* Anthony H.J. Dorcey ed., Large Dams: Learning from the Past, Looking to the Future (Washington: World Bank, 1997).

[22] For critiques of development from a range of perspectives, *see* Arturo Escobar, *Encountering Development: The Making and Unmaking of the Third World* (New Jersey: Princeton University Press, 1995); Colin Leys, *The Rise and Fall of Development Theory* (Bloomington: Indiana University Press, 1996); David B. Moore and Gerald J. Schmitz eds, *Debating Development Discourse: Institutional and Popular Perspectives* (New York: St. Martin's Press Inc., 1995); Wolfgang Sachs ed., *The Development Dictionary: A Guide to Knowledge as Power* (London: Zed Books, 1992).

[23] *See* Navroz K. Dubash *et al.*, A Watershed in Global Governance? An Independent Assessment of the World Commission on Dams (Washington: World Resources Institute, 2001).

framework and legal persona in international law. The ITO was never formed for reasons that may not be necessary to discuss here; and during the interim trade matters were regulated through the Interim Committee of the International Trade Organisation (ICITO). The ICITO was an *ad hoc* body recognised by the UN as a *de facto* specialised agency.[24] Without the constraints imposed by legal terms of incorporation, the ICITO through the General Agreement on Tariffs and Trade (GATT) could become the site from where neoliberal reforms came to be carried out internationally.

The transformations bore the imprint of trade from the inception and with it the deepening and expansion of contractual relations in every sphere of national and international relations. Spearheaded by the domain of trade, the theoretical justifications and rationale borrowed concepts and ideas from classical liberalism: philosophical, political and legal. Classical liberalism was, however, modified and adapted to the context of large, multifaceted organisational structures in the economy, military and politics, supported by complex institutional relationships that had developed in the post World War II era. The 'invisible hand of the market' returned centre-stage but within very different institutional contexts than that prevailing in the eighteenth and nineteenth centuries. Critical responses to these developments were reactive in that theory and practice took the form of 'anti-market' thinking and the critique, with exceptions, came to be framed within the binary of 'market versus state' with different positions in between the two ends of the spectrum.

The Marrakesh Agreement in 1994 ended the *ad hoc* status of ICITO and the World Trade Organization (WTO) emerged in its place as a new IO with a constitution and an independent legal personality. In other words, the WTO became an independent institutional player in its own right. Unlike other IOs set up in the context of the World Wars, the WTO became a global regulator unconstrained by the post World War role for states in the economy, domestic and international. That the functions of the WTO was to

---

[24] United Nations, Ad Hoc Committee on the Restructuring of the Economic and Social Sectors of the United Nations System: Relations of the General Agreement on Tariffs and Trade with the United Nations, General Assembly, UN Doc. A/AC.179/5 (1976).

restructure institutional relationships between states, between IOs and between states and IOs is borne out by a ministerial declaration signed in December 1993 towards the end of the Uruguay Round, the last round of GATT negotiations under the ICITO. The Declaration spells out the brief for the WTO which was to be set up the following year. It may be useful to quote the Declaration at some length.

> 2. [...] Ministers note the role of the World Bank and the IMF in supporting adjustment to trade liberalization, including support to net food-importing developing countries facing short-term costs arising from agricultural trade reforms. [...]
>
> 3. Ministers recognize, however, that difficulties the origins of which lie outside the trade field *cannot be redressed through measures taken in the trade field alone*. This underscores the importance of efforts *to improve other elements of global economic policymaking to complement the effective implementation of the results achieved in the Uruguay Round.*
>
> 4. The interlinkages between the different aspects of economic policy require that the international institutions with responsibilities in each of these areas follow consistent and mutually supportive policies. The World Trade Organization *should therefore pursue and develop* cooperation with the international organizations responsible for monetary and financial matters, while respecting the mandate, the confidentiality requirements and the necessary autonomy in decision-making procedures of each institution, and avoiding the imposition on governments of cross-conditionality or additional conditions. *Ministers further invite the Director-General of the WTO to review with the Managing Director of the International Monetary Fund and the President of the World Bank, the implications of the WTO's responsibilities for its cooperation with the Bretton Woods institutions, as well as the forms such cooperation might take, with a view to achieving greater coherence in global economic policymaking.*[25]

---

[25] Ministerial Declaration: Trade Negotiations Committee, 'Declaration on the Contribution of the World Trade Organization to Achieving Greater Coherence in Global Economic Policymaking', Declarations adopted by the Trade Negotiations Committee on 15 December 1993, General Agreements of Tariffs and Trade, Uruguay Round (emphasis added).

What is important is this – once global restructuring of institutional relationships from state to market regulation entailed in neo-liberal transformations had begun, there was no way a sector as important as water could remain outside the transformative processes underway.[26] Comprehending the role of agency in the 'social whole' that is in the making requires understanding how different social actors responded to the initiatives to restructure the regulatory regime for water and why.

A meeting of different 'stakeholders' including representatives from dam industry, governments, academia, NGOs and civil society groups involved in anti-dam movements, convened by the WB and the World Conservation Union (IUCN) on March 1997, resolved to set up the WCD, a body representative of the 'stakeholders', with two objectives: (a) to review the effectiveness of large dams and assess alternatives for water resources and energy development; and (b) to develop internationally acceptable criteria, guidelines and standards for planning, design, appraisal, construction, operation, monitoring and decommissioning of dams.[27]

Methodologically, the work programme of the WCD was comprehensive in that it was based on a WCD Knowledge Base drawn from eleven case studies, seventeen thematic reviews, surveys of one hundred and twenty-five dams in fifty-six countries, four regional consultations in Africa, Middle East, East and Southeast Asia, Latin America and South Asia, nine-hundred and fifty submissions from seventy-nine countries and input from WCD Forum at which seventy organizations were represented. The thematic reviews were grouped under five categories: (i) social and distributional issues (ii) environmental issues (iii) economic and financial issues (iv) options assessment and (v) governance and institutional processes, and supported by over a hundred commissioned papers. The WCD Knowledge Base, thus, encapsulates a spectrum of diverse, conflicting and contradictory

---

[26] Again this is not the place to engage the rise of neoliberal restructuring within the important centres of capital signified by Reganomics, Thatcherism and such, and the restructuring of the relations between the centres and international organisations in the UN system. It is sufficient to note that such an engagement is possible.

[27] *See* World Commission on Dams, note 2 above at 2.

views and policy debates on dams and water resources at that point in time. The synthesis of divergent views of the 'stakeholders', the thesis and antithesis entailed in their discourses, finds a point of convergence in the way all 'stakeholders' conceptualise the law. This convergence in the way law is conceptualised is significant for 'manufacturing consent' for the regimes changes in the regulation of water. We return to regime transformations for the water sector below, but before that it may be useful to examine the other important strand in the regime change for water, the UN Water Convention.

5. The UN Convention on Non-navigational Uses of International Watercourses

The UN Water Convention was the culmination of a number of parallel strands of developments relating to regulation of water resources in the post World Wars era. The development of international law on transboundary waters parallels the emergence of large dams and spans the length of the post World Wars era.[28] The 1923 Geneva Convention on the development of hydraulic power affecting more than one nation developed by the League of Nations was limited and its further development thwarted by the events of the Depression and World War II. After the end of World War II, the constitutive strands that led to the UN Water Convention include: (a) the need for a legal framework for transboundary waters felt by private international lawyers who were required to provide legal services for the expanding dam industry (b) Article 13(1)(a) of the UN Charter that gave the mandate to codify international law and to ensure peaceful settlement of disputes and promote cooperation under Articles 1 and 2 (c) the involvement of UN IOs, economic, developmental, and scientific, in water resources development which created harmonisation of principles and practices and laid the basis for a UN convention (d) the emergence of environmental law and the duties of states to prevent

---

[28] *See* Ludwik A. Teclaff, 'Fiat or Custom: The Checkered Development of Inter-national Water Law', 31 *Natural Resources Journal* 45 (1991); Ludwik A. Teclaff, *The River Basin in History and Law* (The Hague: Martinus Nijhoff, 1967).

transboundary pollution and to promote environmental practices developed by IOs; and (e) concerns about environmental security and water as a possible source of security threats especially since the 1990s, that provided the rationale for international law on trans boundary waters.

From 1945 a growing number of river water disputes and an expanding dam industry provided the impetus for legal initiatives from private organisations of law professionals and experts most notably in the United States (US). The US chapter of the International Law Association initiated the formation of the Rivers Committee of the International Law Association (ILA) and in 1954 proposed that 'a committee to study the rights and obligations between states as to inland waters' be appointed. [29]. The ILA developed the Helsinki Rules on the Uses of Waters of International Rivers 1966 (Helsinki Rules) that provided a conceptual framework for regulation of rivers and utilisation of freshwaters and conflict resolution arising from water projects. It became, de facto, the international law on trans boundary water for nearly three decades. Not surprisingly the orientation of the Helsinki Rules was to facilitate global water industry and transboundary projects.

Although the UN General Assembly adopted a resolution in 1959 to study the problems relating to the utilisation of international rivers in order to determine if codification of the law by the International Law Commission (ILC) was required, the resolution appointing the ILC to codify the law was adopted only in 1970. 'Developing' countries had had limited influence or role in the development of Helsinki Rules and the legitimacy of the ILA rules remained open at best. When Finland (a country with little interest in dams or 'development' or international rivers) moved a resolution to adopt the Helsinki Rules as UN law, i.e., as public international law, the objections from 'developing' countries forced

---

[29] *See* Slavlo Bagdanovic, 'International Law of Water Resources: Contributions of the International Law Association (1954–2000)', *in* Patricia Wouters and D. S. Vinogradov ed., *International and National Water Law and Policy Series* (London: Kluwer Law International, 2001); Charles B. Bourne, 'The International Law Association's Contribution to International Water Resource Law', 36 *Natural Resources Journal* 155 (1996).

the UN to adopt the resolution for codification of the law on watercourses in 1970.[30] The context of the 1970s was important.

The 1970s saw the emergence of 'North'/'South' tensions, the rise of 'dependency theories' within the Economic and Social Council of the UN, calls for a New International Economic Order, and the UN Conference on Trade and Development (UNCTAD) as institutional vehicle to address unequal economic relations in the post-World Wars' era. During the three UN Development Decades[31] states and international organisations were the principal actors on transboundary water resource development. 'Private' interests, including industry, agriculture, electricity producers and other consumers and users depended heavily on states and IOs to safeguard their interests. Governance over water during this period was largely through administrative mechanisms and state bureaucracies on the one hand and IOs and UN bureaucracies on the other. In other words, both IOs and States followed 'rule by men'. The codification mandate complemented the 'development' mandate in the UN Charter and prepared the ground for 'rule by law' on a global scale.

The rise of the environmental movements, especially after the 1972 UN Conference on Environment and Development's Stockholm Declaration, the 1987 World Commission on Environment and Development's Brundtland Report and the rise to prominence of environmental policies in the IOs eroded the state sovereignty principle in law and developed new ways of conceptualising international law wherein the sanctity of state sovereignty was watered down by the sanctity of the 'whole earth'. The end of the Cold War also saw the rise of new security concerns and new ways framing military and defence issues. Environmental security concerns rose to prominence as a result and 'water wars' became a topic for public debate. In turn both these strands of development contributed to developing the UN Water Convention.

---

[30] For an account of the development of the UN Water Convention, *see* Attila Tanzi and Maurizio Arcari, 'The United Nations Convention on the Law of International Watercourses: A Framework for Sharing', (London: Kluwer Law International, 2001).

[31] 1960–1970: First Development Decade; 1971–1980: Second Development Decade; 1981–1990: Third Development Decade.

The contentious nature of the proceedings of the ILC in codifying international law on transboundary waters which prolonged the finalisation of the UN Water Convention, and later its ratification by states, suggests *real* contradictions in relations over water internationally between states.[32] Most of the reservations came from 'Third World' states. After nearly 30 years of deliberations the UN General Assembly adopted the UN Water Convention in 1997. The UN Water Convention does not yet have the required number of signatories to bring it into effect. Like the WCD report, the UN Water Convention too fructified against the backdrop of the global rise of neoliberalism.

The Helsinki Rules had profound influence on the development of the UN Water Convention and on interstate and intrastate water regimes.[33]. In turn, although technically a framework convention, the normative ramifications of the convention are significant.[34] The influence of the UN Water Convention is profoundly ideological and conceptual in that it conceptualises the legal and institutional framework for dam projects, promotes regional and economic integration, defines 'equitable' and 'reasonable' utilisation, and most importantly, provides the legal basis for transnational institutions, mechanisms for dispute resolution, management of water conflicts and water security. In other words it defines legal relations over water between different global actors.

The conceptualisation of relations over water in the UN Water Convention informs the work of IOs such as the WB, the UN Environment Programme (UNEP) and other agencies on sustainable development policies and lending for dams. The convention creates a space for third party interventions in the work of IOs such as the WB and the GEF.[35] The principles provide the legal basis for

---

[32] For voting patterns on the UN Water Convention, *see* Patricia Wouters, The Legal Response to International Water Scarcity and Water Conflicts: The UN Watercourses Convention and Beyond (2003). *Source:* http://www.africanwater.org/pat_wouters1.htm.

[33] Stephen McCaffrey, 'International Organizations and the Holistic Approach to Water Problems', 31 *Natural Resources Journal* 139 (1991).

[34] *See* Tanzi and Arcari, note 30 above at 24–32.

[35] Alfred M. Duda and David La Roche, 'Joint Institutional Arrangements for Addressing Transboundary Water Resources Issues – Lessons for the GEF', 21 (2) *Natural Resources Forum* 127 (1997).

resolution of intrastate water conflicts within domestic jurisdictions in a federal state. It is therefore significant that in the WCD proceedings, the UN Water Convention, a framework convention, went largely unchallenged and accepted by all 'stakeholders' as a matter of course.[36] The 'equitable utilisation' principle, the cornerstone of the UN Water Convention, is controversial as it raises questions about social values, values in selection of technologies, conceptualising corporations-state-citizen relations and what constitutes 'human development' and 'sustainable development'[37]; in other words the very issues at the heart of the WCD proceedings. The critique of large dams in social sciences and by social movements stops within national boundaries. It does not extend to international law and the global legal regime that underpins large dams and sustains commodified relations over water between users, appropriators and 'stakeholders'.

Instrumentalist conceptualisation of 'development' grounded in empirical approaches of the WCD and the positivist approaches of the ILC do not suggest anything suspect in the absence of any apparent connections between the two events that are so closely tied to dams and 'development'. Both approaches de-contextualise the legal and institutional developments from the overarching backdrop of the global rise of neoliberalism. The problem of two parallel yet apparently unconnected developments in relation to water resources arises only if the problematic is re-framed as: is it possible that two major developments relating to dams and development, both of major significance to regulation of rivers, both having their genesis in post-war developments, both emerging against the backdrop of neoliberal reforms globally, are unconnected? Reframing the question in that way opens up conceptual spaces to draw out the common grounds between the two proceedings and to bridge the gaps in the discourses over large dams in social sciences and international law on development of water resources.

---

[36] *See* P. Millington, River Basin Management: Its Role in Major Water Infrastructure Projects, Thematic Review v. 3 (Cape Town: World Commission on Dams, 2000).

[37] *See* D'Souza, note 9 above at 464–467.

6. Creating New Regimes: What the WCD and the UN Water Convention Do

1. *The WCD Process and the New Water Regime*

The main rationale for the WCD was, as the title of the report suggests, developing a 'new framework for decision making'. It proposes three broad criteria to promote five core values – equity, sustainability, efficiency, participatory decision-making and accountability – all core components of 'democratic development'. The criteria are:

(a) A rights-and-risks approach as a practical and principled basis for identifying all legitimate stakeholders in negotiating development choices and agreements
(b) Seven strategic priorities and corresponding policy principles for water and energy resources development – gaining public acceptance, comprehensive options assessment, addressing existing dams, sustaining rivers and livelihoods, recognising entitlements and sharing benefits, ensuring compliance, and sharing rivers for peace, development and security; and
(c) Criteria and guidelines for good practices related to the strategic priorities, ranging from life-cycle and environmental flow assessments to impoverishment risk analysis and integrity pacts.[38]

The WCD reaffirms the view that dams have made important contributions to human development; that the social and environmental costs of dams have been considerable; that technological alternatives to sustainable development of water resources need more attention, that efficiency of projects need improving and 'inefficient' projects need to be dealt with, that financial viability of projects need closer monitoring and lastly and most significantly for the law, the WCD Report finds that:

> By bringing to the table all those whose rights are involved and who bear the risks associated with different options for water

---

[38] *See* World Commission on Dams, Dams and Development: A New Framework for Decision-Making, Report of the World Commission on Dams (Earthscan: World Commission on Dams, 2000), 5.

and energy resources development, the conditions for a positive resolution of competing interests and conflicts are created.[39]

Summarising the work of the WCD it can be said that there were two different but related 'stakes' involved in the WCD process. One was the 'stakes' that different 'stakeholders' had in the appropriation and use of water. It included the interests of the urban and rural poor in the 'Third World' evicted from land and deprived of means of subsistence, as well as environmental concerns in the 'First' and 'Third Worlds'. The other was the 'stakes' that IOs and 'First World' states had in ensuring a smooth transition from a state to market regime for regulation of relations over water. This involved removing water from the 'citizen-state' framework of regulation and inserting it into 'stakeholders-markets' framework of regulation.

Not surprisingly the WCD framed the debate as 'pro versus anti large dams' and invited all 'stakeholders' to participate in the proceedings. By participating in the proceedings the 'stakeholders' ceased to claim water as citizens with ties to a place, a location, a nation; and instead claimed water as 'non-state actors' with 'stakes' in the water markets. For the purposes of the regime transformation it did not matter what positions the 'stakeholders' took on the 'pro versus anti large dam' controversy. Indeed many 'stakeholders' including states and non-state actors criticised the WCD report from different standpoints.[40] Regulatory regimes create a field for

---

[39] Ibid., 7.

[40] *See,* Jayanta Bandyopadhyay, 'A Critical Look at the Report of the World Commission on Dams in the Context of the Debate on Large Dams on the Himalayan Rivers', 18 (1) *Water Resources Development* 127 (2002); Jeremy Bird, 'Nine Months after the Launch of the World Commission on Dams Report', 18 (1) *Water Resources Development* 111 (2002); Ryo Fujikura and Mikiyasu Nakayama, 'Study on Feasibility of the WCD Guidelines as an Operational Instrument', 18 (2) *Water Resources Development* 301 (2002); Ramaswamy Iyer, *Water: Perspectives, Issues, Concerns* (New Delhi: Sage Publications, 2003); B. N Navalawala, 'World Commission on Dams: Biased?', 36 (12) *Economic and Political Weekly* 1008 (2001); Thayer Scudder, 'The World Commission on Dams and the Need for a New Development Paradigm', 17 (3) *Water Resources Development* 329 (2001); C.D. Thatte, 'Aftermath, Overview and an Appraisal of Past Events Leading to Some of the Imbalances in the Report of the World Commission on Dams', 17 (3) *Water Resources Development* 343 (2001).

non-state actors to 'stake' their claims. Within that field, how effectively 'stakeholders' defend their 'stakes' depends on their ability for institutional innovation, alliances with other 'stakeholders' and above all common interests in the appropriation and use of water.

The WCD process was subjected to a 'social audit' soon after it was completed. The 'non-state actors', the World Research Institute, Lokayan and Lawyers' Environmental Action Team, all non-governmental 'epistemic communities', carried out the audit. Their work was supported by the Ford Foundation, the Royal Dutch Ministry of Foreign Affairs, the Swedish International Development Co-operation Agency, the United States Agency for International Development (USAID) and MacArthur Foundation, who were states, quasi government organisations, industry foundations and trusts with 'stakes' in regulatory mechanisms for water markets.[41] The 'social auditors' reported:

> In this report, we look at the efforts of the WCD and its initiators to create political space for diverse access to the process through:
>
> - full representation of relevant stakeholder groups on the Commission
> - independence from external influence
> - transparency to ensure the Commission's accountability to stakeholders' concerns, and
> - inclusiveness of a range of views in compiling the knowledge base.
>
> We assess how the WCD put these principles into practice and the effect of this experience on stakeholder perceptions of the WCD's legitimacy as the process unfolded. This approach was made possible by the time frame of our assessment, which was concurrent with the WCD.
>
> We pay close attention to the political and practical trade-offs that the WCD faced in its efforts to create *a representative, independent, transparent, and inclusive process.*[42]

---

[41] *See* Navroz K. Dubash *et al.*, A Watershed in Global Governance? An Independent Assessment of the World Commission on Dams (Washington: World Resources Institute, 2001).

[42] Ibid., 3 (emphasis added).

The 'social auditors' were not inquiring into whether the recommendations of the WCD were consistent with the interests of the poor in the 'Third World' and the global environment in whose name the 'anti-large dams' campaigners spoke. Instead they were concerned primarily with 'stakeholder perceptions of WCD legitimacy' and in 'a representative, independent, transparent and inclusive process'. What was really at stake here was the legitimacy of new types of law-making entailed in market regulation in a sector of economy that had become especially disillusioned with the inequitable appropriation and use of water.

Likewise, for the WB too the substance of the issues in the 'pro versus anti large dam' controversy was less important than the processes for decision making. What was important was the willingness of the 'stakeholders' to recognise and participate in the new water regime. Assessing the work of the WCD, the WB states:

> The focus of much controversy regarding the WCD Report has centred on the twenty-six 'guidelines', which have been interpreted by some proponents and critics of the Report as a proposed new set of binding standards. The World Bank's conclusion on the guidelines is best summarized by the Chair of the WCD, who has explained that 'our guidelines offer guidance – not a regulatory framework. They are not laws to be obeyed rigidly. They are guidelines with a small 'g'.' Individual governments and/or private sector developers may wish to test the application of some of the WCD guidelines in the context of specific projects. In such cases, the World Bank will work with the government and developer on applying the relevant guidelines in a practical, efficient and timely manner.[43]

The WB's statement clarifies that the WCD guidelines were 'not a regulatory framework. They are not laws to be obeyed rigidly. They are guidelines with a small 'g'.[44] This indicates that the WCD guidelines should not be seen as state regulation; they are not to be seen as 'state law' enforced through public law instruments of rights and sanctions within a citizen-state framework. Rather the

---

[43] World Bank, The World Bank Position on the Report of the World Commission on Dams, 2002. *Source:* http://www.talsperrenkomitee.de/info/Official_World_Bank_Response_to_the_WCD_Report.pdf.

[44] Ibid.

guidelines are principles that will inform institutional players in the water markets; and the flexibility of the principles will allow institutional players to 'stake' their claims in the marketplace. In other words the state will be 'rolled back' to allow the market to regulate; and the neoliberal legal form of 'flexible principles' will guide transactions over water. The WB developed an Action Plan comprising six complementary areas based on the WCD report, amongst them:

[...]
- Continuing to emphasize institutional reform for more efficient use of water and energy;

[...]
[...]
- Practicing a proactive and development-oriented approach to international waters; and

[...] [45]

What is important is that the WCD processes would be replicated by the WB for all projects hereafter. The WB states:

> The World Bank remains committed to implementation of its operational policies to ensure that: key stakeholders are systematically identified and involved in project planning and implementation; upstream meaningful consultations are held with affected groups to guide project decision making, and their views and preferences are reflected in the plans developed as an integral part of the project.[46]

Not surprisingly since the WCD process was completed water privatisations, river privatisations and corporate players in the water markets regime have increased greatly.[47] The 'stakeholders' who spoke for the 'Third World' poor and the global environment now voice concerns about water privatisation and the expansion

---

[45] Ibid.
[46] Ibid.
[47] *See* Anton Earle, International Water Companies, 2001. *Source:* http://www.african water.org/int_companies1.htm#where; Public Citizen, Will the World Bank Back Down? Water Privatization in a Climate of Global Protest (Washington D.C: Public Citizen, 2001).

of corporate interests in the water sector.[48] The WB's earlier shift of emphasis to legal and institutional issues to develop markets instruments in the water sector[49] is reaffirmed and given a green signal by the WCD. There is a proliferation of different industry, scientific and other water organisations all seeking to play in the market field of 'stakeholders'. All of these developments are consistent with principles of market regulation and neoliberalism.[50] The developments suggest the convergence achieved through the WCD process was about law-making and 'manufacturing consent' for market regulation. It was never about resolving the conflicts of interests between 'stakeholders'. Under market regulation it is the markets that do 'justice' between 'stakeholders' acting through their institutions. In the final analysis law and regulation are about processes, procedures and practices that regulate conduct/ transactions between different individuals/groups and institutions in society.

Undoubtedly the 'stakeholders' who spoke for the poor and the environment, did so because of their frustrations with the 'citizen-state' model of state regulation where the state did not do justice to the poor and the environments. They took their chances in the 'stakeholder-market' model of regulation in the hope that they might be able to play a better role in the water markets to

---

[48] Maude Barlow and Tony Clarke, *Blue Gold: The Fight to Stop the Corporate Theft of the World's Water* (New York: The New Press, 2002); Vandana Shiva, *Water Wars: Privatization, Pollution, and Profit* (Cambridge: South End Press, 2002).

[49] *See* Sayed Kirmani and G. Le Moigne, *Fostering Riparian Cooperation in International Basins* (Washington DC: World Bank, 1997); Harvey Olem and Alfred M. Duda, 'International Watercourses: The World Bank Looks Towards a More Comprehensive Approach to Management', 31 (8) *Water Sciences and Technology* 345 (1995); Carol V. Rose, 'The 'New' Law and Development Movement in the Post Cold War Era: A Vietnam Case Study', 32 (1) *Law & Society Review* 93 (1998); Salman M.A Salman and Kishor Uprety, *Conflict and Cooperation on South Asia's International Rivers: A Legal Perspective* (Washington D.C.: The World Bank, 2002).

[50] Joachim Blatter and Helen Ingram, 'States, Markets and Beyond: Governance of Transboundary Water Resources', 40 *Natural Resources Journal* 339 (2000); Radha D'Souza, 'The 'Third World' and Socio-legal Studies: Neo-liberalism and Lessons from India's Legal Innovations', 14 (4) *Social & Legal Studies* 487 (2005).

bring justice to those on whose behalf they spoke. In so far as both models of regulation are designed to facilitate appropriation of water for industry, for profit-maximisation, for increased rate of return on investments, the 'stakes' of the poor and the environment invite attention to the substance of water regimes: for whom and for what and how appropriation occurs. The substance of water appropriation transcends questions about the legal forms and processes for appropriation and use.

2. *The UN Water Convention and the New Water Regime*

The UN Water Convention, a framework convention, undertakes to codify the law on international watercourses. The mandate to codify international law derives from Articles 1(4) on 'harmonizing the actions of nations' and 13(1) (a) on 'encouraging the progressive development of international law and its codification' in the UN Charter. The UN Water Convention acknowledges the special needs of 'developing' countries. It reaffirms the need for sustainable utilisation of waters and rivers to ensure development, conservation, management and protection of international watercourses, the need for international co-operation, the Rio Declaration of 1992 and Agenda 21, and existing bilateral and multilateral agreements.[51]

Typical of statutes, the UN Water Convention defines terms and concepts. Article 2 (d) defines a regional economic integration organisation as an:

> organisation constituted by sovereign States of a given region, to which its member *States have transferred competence in respect of matters governed by this Convention* and which has been duly authorised in accordance with its internal procedures, to sign, ratify, accept, approve or accede to it.[52]

Thus, Article 2(d) provides for creation of supranational organisations for regulation and management of rivers. Once

---

[51] Convention on the Law of the Non-navigational Uses of International Watercourses, New York, 21 May 1997, *reprinted in* P. Cullet and A. Gowlland-Gualtieri eds., *Key Materials in International Environmental Law* 481 (Aldershot: Ashgate, 2004).

[52] Ibid., Article 2(d) (emphasis added).

formed, within the present neo-liberal institutional context, these supranational organisations will further roll back the states by requiring them to 'transfer ... competence' on certain aspects of management of water resources to the global institution. The transnational organisation envisioned under Article 2(d) operating within the wider context of relations between 'First' and 'Third Worlds' in the new International Relations environment of 'globalisation' and the new conceptions of market let 'development' spearheaded by the WTO and other IOs informed by neoliberalism, will facilitate removal of more aspects of water resources management outside the framework of citizen-state relations based on rights and sanctions. The new global institutions, with their own internal rules, objectives, procedures and practices with a legal personality will become institutional players in the water markets in their own right independent of the states that formed the transboundary regional organisation. It may be noted here that the Mekong Agreement in 1995 set up the Mekong River Commission as a transboundary regional organisation. It gave renewed impetus to transboundary dam projects on the Mekong River which had commenced in the 1950s and came under cloud during the Cold War and at the same time revived local tensions against perceived loss of state sovereignty to the regional organisation.[53, 54]

Part II of the UN Water Convention sets out the general principles governing use of river waters and covers the substantive rights and obligation of states. Articles 5, 6 and 10 are the most significant and controversial principles. Article 5(1) develops the principle of 'equitable and reasonable utilisation' and requires that:

---

[53] *See* Chris Sneddon and Coleen Fox, 'Rethinking Transboundary Waters: A Critical Hydropolitics of the Mekong Basin', 25 *Political Geography* 181 (2006); Karren Bakker, 'The Politics of Hydropower: Developing the Mekong', 18 *Political Geography* 209 (1999).

[54] To the contrary, on the Indus River, during the Cold War the peace was kept through the interventions of IOs and Western States, especially the US. The end of the Cold War has renewed tensions. *See* Radha D'Souza, 'Water Resources Development and Water Conflicts in Two Indian Ocean States', *in* Timothy Doyle and Melisa Risely eds, *Regional Security and the Environment in the Indian Ocean Region* (New Jersey: Rutgers University Press, at 157–170).

... international watercourses shall be used and developed by watercourse States with a view to attaining optimal and sustainable utilisation thereof and benefits there from.[55]

The legal concept of 'equitable utilisation' is problematic.[56] The concept involves assessing the role and competing interests of different 'stakeholders'. Under the UN Water Convention processes the 'stakeholders' are global players, states, intergovernmental organisations and IOs acting as economic actors at a time when the role of the states within national jurisdictions has been rolled back to varying degrees. The status of other global 'stakeholders': the dam industry, power generation industry, epistemic communities, and water trading industries are privileged because their place is secured by the way equity in water appropriation and use is conceptualised. To determine 'equitable utilisation' the preamble provides the guidelines. The meaning must be derived from the United Nations Conference on Environment and Development of 1992, the Rio Declaration and Agenda 21. It follows that the meaning and application of the principle of 'equitable utilisation' must be derived from developments in global economic and 'development' policies pursuant to the interagency cooperation initiatives after the WTO was formed as discussed above. In doing this the WCD principles and guidelines will undoubtedly provide 'objective' and authoritative basis for determining what is or is not 'equitable utilisation'.

Article 6 enumerates the factors relevant to equitable and reasonable utilisation. The factors to be considered include the social and economic needs of the states, the populations dependent on watercourses, the effects of developments, amongst others. Article 6 does not create a weighting mechanism for the relative importance of the factors, or a hierarchy of priorities. In fact Article 10 explicitly states that 'no use of international watercourse enjoys inherent priority over other uses'. The key point here is that the global water regime that the UN Water Convention formalises as international law predetermines the conditions for water appropriation and use within nation-states and within national law. The global water regime that predetermines the appropriation

---

[55] Article 5(1), Convention on the Law of the Non-navigational Uses of International Watercourses, New York, 21 May 1997.

[56] *See* Radha D'Souza, *Interstate conflicts over Krishna Waters: Law, Science and Imperialism* (New Delhi: Orient Longman, 2006), 464–471.

and use of water within natural boundaries went unchallenged because 'epistemic communities' speaking on behalf of the environment and the global poor were unable to make the connections between the UN Water Convention processes and the WCD processes. Those connections could only be made by anchoring both the developments to the wider context of developments in capitalism and imperialism and the ways in which the wider processes expropriate the poor and the environment.

At the global level, legal theory hangs on to the principle that states represent their populations. If their populations comprise diverse and competing interests the states must sort out those differences within domestic jurisdictions. This is a circular argument because states have been rolled back, global institutions have emerged as major players, neoliberalism has changed the rules of the game, and states have limited leeway to manage competing domestic interests. For the less economically powerful water users like subsistence farmers or the urban poor who must rely on their political power within a constitutional framework of national law, the willing participation of their spokespersons in rewriting the rules of the game and their willing repositioning as 'stakeholders' in the global market is not exactly empowering.

Part III of the UN Water Convention sets out the obligations on the part of States when planning water projects. Part IV provides for protection, preservation and management of rivers, Part V for emergency situations and Part VI for dispute resolution during armed conflict and project related disputes and provides for arbitration and/or submitting the dispute to the International Court of Justice. In other words Part III creates a normative framework for the role of the states in 'development' and management of water resources and dispute resolution between states. Article 33 of the UN Water Convention includes the conventional mechanisms for dispute resolution mechanisms based on consensual decisions by states. Article 33 extends the conventional principles for invoking dispute resolution mechanism in international law in significant ways.[57] Article 33(3) provides that if the state parties are unable to settle their disputes within six months, then one of the state parties

---

[57] Attila Tanzi and Maurizio Arcari, 'The United Nations Convention on the Law of International Watercourses: A Framework for Sharing', *in* Patricia Wouters and D.S. Vinogradov eds., *International and National Water Law and Policy Series* (London: Kluwer Law International, 2001).

may request a fact finding commission to be appointed unilaterally. Article 33 also provides for a range of non-judicial third-party settlement procedures including mediation, arbitration and negotiations. The WB is imminently placed in a position to play the role of mediator. A number of UN organisations like the Global Environmental Facility (GEF) a financial body supports the idea the WB's role as mediator in transboundary water disputes.[58] These developments dovetail the WB's thinking on a greater role of the WB in mediation and dispute resolution. A mediation and conciliation role for the WB will invest it with a quasi-regulatory role between 'stakeholders'.[59]

Thus the UN Water Convention process which began in 1959 languished until the Stockholm Convention in 1972 gave it a reluctant start; but picked up momentum and culminated in a UN convention against the context of rising neo-liberalism and wide ranging neo-liberal reforms with profound ramifications for the international relations context within which 'Third World' states operated. The UN Water Convention bears the stamp of the market-oriented conceptions of economic development and the place of water and states in it.

7. Conclusion

To sum up, the UN Water Convention creates a framework for decision making and conflict resolution between states on transboundary waters. It creates the legal framework for supranational organisations that facilitates dam construction[60], in

---

[58] *See* Duda and Roche, note 35 above.

[59] The WB set up the Inspection Panel in 1993 as an autonomous forum for investigating and resolving disputes about the impact of dams on local populations. This step takes the WB's dispute resolution role to the social arena, a further step from its arbitration role in investment disputes under the International Centre for Settlement of Investment Disputes (ICSID) set up, also as an autonomous body, in 1965.

[60] *See* Peter Beaumont, 'The 1997 UN Convention on the Non-navigational Uses of International Watercourses: Its Strengths and Weaknesses from a Water Management Perspective and the Need for New Workable Guidelines', 16 (4) *Water Resources Development* 475 (2000); Mikiyasu Nakayama, Mikiyasu, 'Successes and Failures of International Organisations in Dealing with International Waters', 13 (3) *Water Resources Development* 367 (1997).

other words creates new institutional players in the water markets with powerful interests in sustaining large dams. The WCD recommendations create a framework for decision making and conflict resolution between 'stakeholders' within the state by addressing questions of social equity and environmental sustainability within the framework of neo-liberal economic development. Both are informed by the same core values, concepts, ideas; both are committed to developing processes with legitimacy, for use and appropriation of water on the one hand and conflict resolution mechanisms on the other, between states and between 'stakeholders'. Both processes are directed at building institutions capable of engaging and facilitating market transactions in the appropriation and use of water. Taken together, the WCD and the UN Water Convention are complementary processes that seek to redefine new public and private spheres, create new roles for states and 'stakeholders' in relations to waters and rivers. Together the two frameworks seek to create a new regime by:

- providing for supranational organisations for utilisation and management of water based on core concept of the river basin as a 'natural' unit of regulation[61]
- creating a framework to take the regulation of waters and rivers to the next stage of legal and institutional development: from a bureaucratic administrative form of governance typical of the post World War II period to regulation by market institutions, mechanisms and principles; in other words from take water from 'rule of men' to 'rule of law', from state to market mechanisms of governance
- creating communities of 'stakeholders' in water based on market principles, institutions and instruments
- redefining the relations between states, IOs, corporations and supranational organisations within a rights-based framework in the public sphere
- providing for international interstate institutions by requiring the states to cede some of their powers in relation to rivers to IOs committed to facilitating water resources development for industrialisation, agriculture and power generation through private actors

---

[61] For a critique of what is entailed in this concept, *see* D'Souza, note 9 above.

- redefining the relations between citizens interspersed within a rights-based framework in the private sphere.

A legal regime is a much broader concept in that it includes a variety of statutes, policies, concepts, values, goals, instruments and mechanisms of governance that taken together define social relations over water (or any other social relations) in society and prescribes the ways in and the extent to which different segments of society will participate in the regime. Law is about relations.[62] Law casts different social actors into normative roles and thereby creates behavioural expectations that facilitate repeated transactions required for social relationships to work. Law under neoliberalism casts different institutional actors into normative frameworks that regulate institutional responses, behaviour and repeated transactions. In this law under neoliberalism enables a classical liberal world view to operate on enlarged scales, with enlarged ramifications for inequality, dispossession, and social and environmental conflicts.

Taken together, the WCD and the UN Water Convention appear complementary processes that seek to redefine relations over waters and rivers in the 'Third World' and between the 'First' and 'Third Worlds' along neo-liberal principles; and creates frameworks for institutional developments of market regulated regimes for water resources. The social actors engaged in the regime changes do not however make the connections between the two events. Disciplinary orientations, immediate sectoral interests, and minimising the importance of theory and philosophy in discourses on law and social policy, especially in the 'Third World' and in international law, prevent envisioning of the 'social whole' that is in the making.

The tragedy lies in the fact that the 'epistemic communities' speaking for the dispossessed, the environment, for distributive justice and human values, participated willingly and contributed to a regime change that could produce results that are the very opposite of the reasons that prompted their involvement and interventions. De-contextualised analysis unconstrained by history or geography disengages the analysis of water resources from the wider processes of transformations in capitalism, new forms of

---

[62] Alan Hunt, *Explorations in Law and Society: Toward a Constitutive Theory of Law* (New York: Routledge, 1993).

colonialism and ways in which structuring and restructuring of social orders occur.[63] Narrow empiricists' approach to social and natural phenomena, narrow positivist approaches to law, reductionist methodologies and disciplinary closures cast a veil over social relations over water. The veil conceals the politics of water as the WCD/UN Water Convention processes show. There is by now an extensive critique in social theory and philosophy on all of the approaches. Why the philosophical and theoretical critique eludes critical engagement on water issues by 'epistemic communities' speaking on behalf of the dispossessed and the environment must be addressed another time.

---

[63] Radha D'Souza, 'Re-territorialising and Re-centering Empires: The Connivance of Law and Geography', *in* Jay Gao, Richard Le Heron and June Logie eds., *Windows on a Changing World: Proceedings of the 22nd New Zealand Geographical Society* 324–327 (Auckland: New Zealand Geographical Society, 2003).

18

Marginal Remarks Concerning Water Policy Regimes; Governance, Rights, Justice, and Development: An Epilogue

Upendra Baxi

The Editors of this volume (Philippe Cullet, Alix Gowlland-Gualtieri, Roopa Madhav, and Usha Ramanathan) deserve warm appreciation for assembling in this work a range of distinguished contributions concerning the nature and scope of 'water law' in national and international perspectives. Because I believe that nearly everything worth saying has been here well said, I tried to resist the request of Philippe Cullet and Usha Ramanathan to contribute an additional chapter. I failed; hence, this Epilogue which merely constitutes a cluster of remarks concerning the question of relation between theory, policy, law, and movement, the will to human rights, forms of ideological critique, and the affairs of justice.

Towards Understanding Water Policies: A Preliminary Excursus

'Policy' emerges as a ubiquitous notion, and a key category, and yet also as a disquieting sphere. One may (here following Radha D'Souza) deploy the difficult notion of 'regimes' in speaking about 'water policy regimes' (WPR). This requires some analytical unpacking. One may proceed to do at several levels of analysis such as the distinctions typically furnished by the purpose, intent, or the overall *telos* of water policy regimes often unhelpfully described in terms of the trichotomy – water for life, industry, and development, or proceed to perform a level-of-analysis task. In the latter perspective, one distinguishes among: (a) macro water policies provide a general approach to water governance (often expressed as WRM – water resource management- policies; as 'water sector reform' policy (WSRP); (b) meso water policies are those specifically concerned with a particular sector/vector of water management or governance or (c) micro-policies, primarily though in terms not so much of policy enunciation but rather as strategies of implementation at the ground or the local level.

No matter what may be our privileged acronyms best describing WPR, several related questions arise; and I may without recourse to the many histories of the politics of naming only suggest a modest range of concerns. Obviously this work richly suggests *first* that WPR remain marked by hybridity, at least in terms of the *sources of origins*. *Second*, the relations between the three levels of analysis (macro/meso/micro) may either be predetermined or radically contingent; put another way, the relations may be either viewed in the paradigm of dominating hierarchy or in terms of dialectical flows.

Third, while the 'macro' enunciations of WPR present the question of their relatedness to the specificity of water-related *telos*, the WPR deployed as a handy catch-all tool may diminish a critical awareness of the linkages between water and related environmental policy regimes (such as land use, forests, biodiversity, and environmental policy). The question here, of course, relates to the sensible ways of assertion of the historic need to accentuate the relative autonomy of the water-specific policies, without losing sight of the inter-connectedness amidst different macro-regimes of environmental policy frameworks.

Fourth, the ontological presumption that WPR may only speak to the tasks of water governance raises several concerns. Whenever historically available or somehow at hand, apex justices also engage the tasks of *adjudicatory water policy enunciation*. Increasingly, WRP choices and frameworks remain much affected by processes of constitutional adjudication (in some leading postcolonial societies such as India, South Africa, and Brazil).¹ Further, policies may not remain an exclusive realm of state practice; social and human rights movements also re-engage in distinctive water policy making.² WPR is however and overall addressed to water governance and development policies.

Fifth, WPR may remain normative but not always ethically so. The goals and means of any specific WPR involve prescription of standards of conduct relating to the postulation of goals and prescribed means for their attainment. But neither the goals nor means may go beyond their enunciation in fairly techno-efficient ways. Put another way, governance WPR may well remain *amoral*, that is uninformed by any substantive conceptions of human rights, social justice, or regard for the interspecies dignity. Surely, many extant regimes of WPR entail a fuller attention to the wisdom of Michel Foucault's distinction between government and governmentality as concerned with 'men [human beings], their interrelationships, their links, their *imbrication with things that are wealth, resources, subsistence, the territory* with its specific

---

[1] In India, and perhaps not only there, water rights and executive policy performances also constitute a rather distinctive sphere of adjudicatory policy-making. Put another way, while the Indian Supreme Court remains, on the whole, deferential to acts of the supreme executive water policy making, it also exercises its co-equal powers to shape some human rights based executive water policies. Of the many examples, I may here and perfunctorily cite the following illustrations. It creatively interprets the Article 21 guarantees of the right to life and liberty to shaping a new regime of human water rights. This offers a complex and contradictory narrative, indeed. At least in relation to 'big' dams, judicial activism performs some noble scavenging functions directed towards the fashioning constitutional regimes of after-the-event compensation and rehabilitation of the vast masses of the hapless public project affected Indian citizenry. It also triggers thus some historically tardy and perhaps entirely belated formulations of national rehabilitation and resettlement policies. The Court also plays a major proactive role, as for example recently illustrated but its directive mandating a fashioning of an inter-river resources regime, which has triggered a national debate over the relative weights of water expertise available to a hegemonic executive in contrast to a 'juristocracy'. This remains not altogether a typically Indian question, but rather a question of wider comparative adjudicative provenance.

[2] *See* Belén Balanyá *et al.* eds., *Reclaiming Public Water: Achievements, Struggles and Visions from Around the World* (Amsterdam: Transnational Institute (TNI) and Corporate Europe Observatory (CEO), 2005).

characteristics, *climate, irrigation,* fertility, and so on' and humans in their 'relation to things that are *custom, habits, acts of acting and thinking...* in relation to those other things, that might be *accidents, and misfortunes such as famines, epidemics, and death...*' (emphases added).[3] Foucault may be illuminating read scrupulously with reference to each of the italicised phrase-regimes as articulating the complexity of *amoral* water governance.

Sixth, understandably then the steadily proliferating companion species best named as 'watercrits' – critics of WPR, an assemblage of theory/movement diversities – respond to the need for development of standards of critical morality of water governance. Watercrits proceed to do this as the work in your hands, as well a growing companion literature. I designate these rather summarily as follows – the source of origin, integrity in administration, policy-law complexes, and the question of justice. Our judgements will vary on these and related axes.

The Source of Origin Approach

In a movement from the policy texts towards its various contexts many eminent watercrits proceed with an exploration of the dominant political ideologies. These are never easy to read, and what may after all wish to signify by 'ideologies' never remains compellingly clear. However, the watercrits (and of course I count myself among these) identifying do not always agree in the 'source of origin'. There remain at work at least the following approaches. The identification of source of origins of WPR in terms of neoliberal ideology and auspices; the distinctive approaches to human rights and global social policy – enunciations; and the more specifically history-centred narratives. Here, perforce a narrative silhouette will have to suffice.

Many contributions here testify to the prowess of international and regional financial institutions, the Northern intergovernmental and international aid and development agencies, and their terminal beneficiaries primarily multinational corporations) who also remain dominant actors configuring global WPR. Activist analyses also provide narratives of the continuing misappropriation by industry

---

[3] Michel Foucault, 'Power', *in* James D. Fabion ed., *The Essential Works of Michel Foucault – Volume 3* at 208–209 (New York: The New Press, 2000).

of the scare water resources for hyperprofit.[4] In these narratives, the range of meanings ascribed to WPR via standard distinctions such as 'water for life', 'water for industry' and 'water for development' stand collapsed into a general matrix generating forms of late capitalist ideologies. 'Water for life' becomes rather indistinguishable from 'water for industry' when especially policy acts of privatisation of water and sanitation and water resources regard water as a commodity, or an economic resource rather than a series of 'public goods'.[5] And of course 'water for development' raises concerns about the ideology of developmentalism.[6] Perhaps, a crucial site here concerns the difficult relationship between WSRP and the discourse of 'sustainable' development'.

A second approach stresses not so much the ideologies inherent to human rights but rather the globally formulated consensus concerning the values, norms, and standards furnished by the contemporary human rights. Central to this remain claims and concerns about equitable and effective access to the means of existence (such as water, sanitation, food, health, and shelter rights) as well as to the means of livelihood but also for dignitary interests, specifically inclusive of regard for gender-based right to dignity.[7] Human rights approaches prioritise 'water for life' and rearticulate

---

[4] *See* in particular, Maude Barlow and Tony Clarke, *Blue Gold: The Fight to Stop the Corporate Theft of the World's Water* (New York: The New Press, 2002), Vandana Shiva, *Water Wars: Privatization, Pollution, and Profit* (Cambridge, MA: The South End Press, 2002) and Vandana Shiva, Resisting Water Privatisation, Building Water Democracy (Paper presented at the World Water Forum, Mexico City, March 2006). *See further* Patrick Bond, 'Water Commodification and Decommodification Narratives: Pricing and Policy Debates from Johannesburg to Kyoto to Cancun and Back', 15/1 *Capitalism, Nature, Socialism* 7 (2004).

[5] *See* Inge Kaul *et al.* eds., *Providing Public Goods: Managing Globalisation* (Oxford: Oxford University Press, 2003).

[6] These refer to discourses variously named as the ideology of developmentalism, dependant development, alternative *to* development and alternative, *another* development, and more recent to post-development. *See* Upendra Baxi, *Human Rights in a Posthuman World: Critical Essays* 76–123 (Delhi and Oxford: Oxford University Press, 2007).

[7] For gendered and first nations peoples perspectives, *see* Kuntala Lahri-Dutt ed., *Fluid Bonds: Views on Gender and Water* (Kolkata: Stree Publications, 2006); *see* The United Nations, *Water for Life Decade: 2005–2015* (New York: The United Nations Department of Public Information, 2005).

the right to development or human rights based development as collective human right to shape – that is to articulate and audit-WPR, as an additional aspect of water rights contributing further to the development of the right to development.[8]

Each of these approaches raises not just some critical questions concerning ideology, but also about the detail and design of water governance in particular and governance in general. This brings me to a register of history-related concerns – that of understanding the practices of culturally, even civilisationally embedded, violent structural denials to access to water as well as water resources. In this context, it needs to be fully recalled that in contrast with Mohandas Gandhi, who invented the ethical technology of civil disobedience for civil society reform via temple-entry movements for India's Untouchables, Ambedkar launched the Mahad Satayagraha directed against the denial of access to village wells by the millennially excluded communities.[9] He thus fashioned the jurisprudence of basic human needs to access to water as the grundnorm of all human rights. Other stories elsewhere (the histories of human chattel slavery, or denial of water access to women in some customary regimes, for example) need to be archived and foregrounded at least in terms that anticipate some contemporary struggles for water justice as a basic virtue of social movements, arrangements, and institutions.

No doubt, as an empirically verified global social fact violent social exclusion in access to water and water based resources, now remains best presented in terms of contestation over global ownership of water and water-based resources as a multinational corporation asset.[10] Yet, a wider framework of the 'political ecology

---

[8] *See* Baxi, note 6 above, at 124–155.

[9] *See* Upendra Baxi, 'Justice as Emancipation: The Legacy of Babasaheb Ambedkar', *in* Upendra Baxi and Bhikhu Parekh eds., *Crisis and Change in Contemporary India* 122–149 (New Delhi: Sage, 1995). *See also* Upendra Baxi ed., *Law and Poverty: Critical Essays*, Chapters VI and VIII (Bombay: N.M. Tripathi, 1988).

[10] The chapter in this volume by Francesco Costamagna and Francesco Sindico indicates that 'international investment law and WTO are posed to play a greater role in regulating water-related issues in the near future'. *See* Lyla Mehta and Birgit la Cour Madsen, 'Is the WTO After Your Water? The General Agreement on Trade in Services (GATS) and Poor People's Right to Water', 29/2 *Natural Resources Forum* 154 (2005).

of water'[11] also points towards the need to understand water scarcity as also constructed by war and warlike activities, including acts of 'civil strife', and cross-border invasions- which grievously affect access to water and water rights. The question here draws attention to the insufficiency of standards of international humanitarian law in the context of access to need/right to water and access to water-based resources for communities engulfed in situations of 'civil' strife and the struggles by the International Red Cross, for example, towards states of minimal amelioration. Likewise the current situations of the War *'on'* and *'of' terror*[12] present some new dimensions; if the latter evokes the spectre of 'bio-terrorism', which may in the future consist in practices of belligerent acts of water-toxic warfare, the former, in 'hot pursuit' of cross-border international 'terrorists' and actual practices of 'regime change' may pose grave threats to the water security regimes of the thus affected populaces.

The importance of a critique of neoliberal WPR, may indeed be enhanced by recourse to a far more inclusive range of factors. Well we may ask whether the time of the postcolonial is now historically over?[13] Moving beyond, how may watercrits reconcile their pragmatic deployments of the available enunciation of human rights

---

[11] Political ecology signifies 'an analysis of the mutually constitutive interrelationships between the discursive, social and material dimensions of environmental change and socioeconomic restructuring' at 53. *See also* Karen Bakker, 'A Political Ecology of Water Privatization', 70 *Studies in Political Economy* 35 (2003). *See* Adam Davidson-Harden, Anil Naidoo and Andi Harden, 'The Geopolitics of the Water Justice Movement', 11 *Peace Conflict and Development* (2007). *Source:* http://www.peacestudiesjournal.org.uk/docs/PCD%20Issue% 2011_Article_Water%20Justice%20Movement_Davidson%20Naidoo%20Harden.pdf.

[12] *See*, Baxi, note 6 above at 156–196.

[13] The range of questions remains formidable here. For example: Do the performances of colonial WPR remain an affair of choice rather than necessity? In what may the times of the postcolonial WPR then consist? How may watercrits proceed to distinguish between acts of globally imposed WPR and the diffusion of regulatory normative frameworks and cultures? Are 'free markets' in water culturally embedded or universal, and either way what difference may this make to a postcolonial perspective on WPR, distinct from the globalising perspective? How may the postcolonial remake of customary rights and relations in water differ from some master narrative frames of water marketisation and privatisation?

(contd.)

to water alongside some genuine concerns marking the discourse about universality/relativity of human rights?[14] Perhaps a way out is furnished by the claim that human rights far from constituting the gifts of the West to the Rest rather remain authored by peoples in struggle and communities in resistance everywhere? Further, how may watercrits avoid grave narrative risks of presenting the ideologies of 'neoliberalism' as it were a smooth, and also a monolithic, surface bereft of normative and material contradictions? To fully speak to these remains necessarily a task for another day, even as regards the struggles for enunciation of a more humane regime of and for water rights.

The task of critique is to construct some approaches toward a Real Utopia of fair and efficient water governance regimes. A major question here is not just about constraining governance towards human rights values, norms, and standards but annexing these as well to state-like actors (beyond the entirety effete languages of corporate social responsibility) via the 2005 Draft UN Norms concerning human rights responsibilities of Multinational Corporations and Business Entities.[15] Further, such a utopic frame needs to address more fully the questions concerning animal rights and the rights of natural entities to continue to exist (as illustrated by the discourses concerning endangered and extinct species, biodiversity preservation, and more complexly by the current discourses concerning global warming). These narrative shifts certainly imperil the more narrowly constructed watercrits discursive platforms, oriented to the specificity of human water rights and claims for just water governance policies at all pertinent sites.

Integrity in WPR Administration

This approach worries less about sources of WPR origins but rather remains concerned with the integrity of WPR administration. Were

---

What contributions may be said to have been made by postcolonial diplomatic feats, and the struggles of the Third World peoples, to any re-fashioning of the role and place of water human rights in the newer enunciations of international WPR?

[14] *See* Upendra Baxi, *The Future of Human Rights* 115–199 (Delhi and Oxford: Oxford University Press, 3rd ed., 2008).

[15] Ibid., at 234–275.

watercrits ever to agree on criteria of a 'good' WPR, the potential of its betrayal ought surely to remain a central concern? 'Integrity' for the most part in WPR discourse refers to corruption in the administration of water policy. The World Commission on Dams (WCD) has rather fully archived these unconscionable costs in relation to large irrigation projects; so now does a recent report of Transparency International in relation to related aspects of administration of WPR.[16] Both propose integrated policy frameworks, however contestable. Watercrits have indeed succeeded in shaping the discourse of integrity; yet WPR corruption remains integral to governance corruption in its entirety. Put another way, the concerns here are not merely about forms of developmental governance corruption as marking various orders of governance deficits (enacting the logics and languages of restoration of sovereignty) but rather about the birthing of a new human right to immunity from corruption in high public places enacting the logics and languages of popular sovereignty. It also remains important to note that these concerns may help redefine the trichotomy about water for life, industry, and development. The best narrative practices of watercrits further fully acknowledge that these incipient normative tendencies owe their origins to some heroic struggles of long suffering peoples.

Granting all this, and within the space-constraints, I may here only pose a few simple questions. *First*, how may we understand the very notion of governance corruption, beyond the conventional understandings of bribery, nepotistic ethnoclientism, and manipulation of hermeneutic markets (interpretive discretion) in ways favouring the highest global bidder in contrast with a structural understanding of corruption as 'state capture'[17] – that often mandates the politics of governance immunity and impunity as perhaps a necessary evil for the practices of developmental governance? *Second*, how may watercrits best proceed to de-link 'national' from 'global' markets for WPR corruptibility as also to

---

[16] *See* Transparency International, *Global Corruption Report 2008: Corruption in the Water Sector* (Cambridge: Cambridge University Press, 2008).

[17] *See* for an elaboration of this notion, Joel S. Hellman, Geraint Jones and Daniel Kaufmann, Seize the State, Seize the Day: State Capture, Corruption and Influence in Transition (Washington DC: The World Bank, Policy Research Working Paper No. 2444, 2000).

re-link these into some unitary narratives? *Third,* and related, *how* may we narrate forms of governance corruption as special or distinctive features of water regime governmentality? *Fourth,* in what ways (how) and to what effect (why) may watercrits encraft/re-employ the narratives of the insurgent moment of peoples' movements pursuing some forms of ethical cleansing of water governance conduct? To say that all neoliberal programmes of policy action remain ridden with the crisis constituted by a corrupt sovereign is of course crucial but only to the extent that one may also propose a *programschrift* for the moves ahead.

Water Governance Policy/Law Complexes

The question of law/policy mix remains a vexed one. While one may prefer to read all legislations as statements of public policy, there also exist and even proliferate macro-policy enunciations untranslated into operative domain of the state law conceived as LAIE (legislation, administration, interpretation, and enforcement). Such policies are often said to furnish poor guides to action as any sincere student of the Part IV of the Indian Constitution (enunciating the Directive Principles of State Policy) surely knows well! Much the same may be said concerning the more recent enunciations of the Indian WPR.[18] In what ways does this passage from 'policy' to 'law' matter? It is clear that water laws differ significantly from 'mere' statements of policy, if only because these aside from articulating codes of legal (as distinct form human rights) obligations also construct certain deliberative public sphere involving state actors other than the supreme executive, engaging lawyerly contention, acts of adjudicatory power, and variegated modes of civil society articulatory interventions.

For example, most Indian national water and environmental related policy regimes remain even today solely within the province

---

[18] *See* for example, Ramaswamy R. Iyer, *Towards Water Wisdom: Limits, Justice, Harmony* (New Delhi: Sage, 2007). A new 'policy' on rehabilitation and resettlement of project affected peoples is now broadcast over the state–monopoly the All India Radio in the melodious yet also desperate slogan: 'Nunytam Visthhpan, Adhiktam Punarnivas' (minimum displacement, maximal rehabilitation!), I need not, as practising watercrit, here comment any further on this terminal obscenity of state advertising power.

of executive power, constituting its sovereign prerogative. Major irrigation projects, even when they entail at times adjudicative decisions in sharing of inter-state waters, remain unalloyed acts of executive sovereign discretion, unaccompanied by any substantial regard for either the constitutionally mandated or human right normative necessities. No overt legislative regime is still thought necessary, though the reason of the constitution and of its rights and justice provisions obviously suggest that all major public projects ought to fully proceed under detailed legislative frameworks and provisions.[19]

Watercrits need further to attend to the dynamic histories of public/private regimes of water laws/regulation. The register of difference stands admirably summated by Karen Bakker in terms of 'a genealogy of efficiency' contrasting the eras of nationalised with the contemporary performances of privatisation of WPR. The difference remains crucial. In contrast with the 'principle of social equity' underscoring the benefit principle contrasted with the earlier 'ability to pay principle', what now rules the roost, as it were, is the 'economic principle' (the principle that 'users of a utility service should pay, as nearly as possible, the costs that they individually impose on the system ...)'.[20] This registers of contrasts between 'economic equity' and 'equalisation' remains important for many reasons. For one thing, water privatisation (like related measures of utilities privatisation) regards citizens almost entirely as consumers (as 'buying a commodity' rather than political beings with a right to demand service.)[21] For another, water privatisation signifies a shift from access to water as a 'precondition of participation in collective social activity' towards 'access to material goods as a means of need-fulfilment.'[22]

The concerns go beyond water-pricing polices. Indeed, a veritable transformation of the imagery of state and law also occurs

---

[19] *See* Upendra Baxi, 'Notes on Constitutional and Legal Aspects of Displacement and Rehabilitation', *in* Walter Fernandez and Enakshi Ganguly Thakral eds, *Development, Displacement and Rehabilitation* (New Delhi: Indian Social Institute, 1989).

[20] *See* Karen J. Bakker, 'Paying for Water: Water Pricing Policy and Equity in England and Wales', 26/2 *Transactions Institute of British Geography* 143, 147 (2001).

[21] Ibid., 156.

[22] Ibid., 157.

with increased privatisation of access to water and to water-based resources. 'Law' itself, and now more upfront, becomes a medium of allocation of common community resources to the owners of means of production; in the process it increasingly becomes technical regulation rather than remaining an affair of redistributive political process (to here recall the germinal distinction inaugurally framed by E.B. Pashukanis in terms of 'technical' – regulation – and political – law.) The 'state', in turn, redirects its human rights obligations to promote the human rights of all human begins now to promote a trade-related, market-friendly human rights paradigm protecting the collective 'human' rights of the owners of globalised means of production.[23]

Indeed, when these water policies are converted into legislation, they further reinforce the transformation of citizens into consumers and accentuate the jurisprudence of basic needs over that of basic human rights. Law reform becomes a vessel containing a series of de-/ re-/ regulatory processes and correspondingly the potential for judicial water activist interpretation itself gets structurally adjusted in several ways. As cultures placing a high value on governance techno-efficiency develop, the discourse of human rights pursuit via legislation and adjudication becomes increasingly fragmented. Understandably, the logics of consumer rights as human rights develops apace; so do to varied extents in many law-regions the practises of judicial activism designed to protect and promote 'environmental' human rights. In the process, human rights and social movement activism itself begins to assume many new *avatars*, most if not of all which must now find ways to restore old paradigms of law, or fashion new approaches to Paradise Regained.

To say all this is really to summate a situation in which all watercrits find themselves in this moment of hyperglobalisation. To say this also entails a more qualified reception of my sustained defence of the best practices of Indian judicial activism and my acts of advocacy insisting that all major public projects ought to be taken out of the province of sovereign executive discretion and placed under legislative auspices. To acknowledge this tension does not necessarily entail rejection of either this reception or advocacy but rather names some theoretical/reflexive burdens for the communities of radical or counterhegemonic watercrits.

---

[23] Baxi, note 14 above at 234–302.

Moving a little further, we perhaps need to purse a bit further watercrits' reflexive awareness that WPR are not things of nature but rather constitute complex artefacts of water governance, almost always liable to deconstruction and destabilisation by counterhegemonic water theory and movement. Ardent watercrits remain aware that WPR constitute complexes of intra-governmental communicative signals oriented towards coordinating governmental action, or forms of state conduct, always a prerequisite for effective governance, whether 'federal' or 'unitary'. Watercrits seek to disturb these already parlous states of unity/unification; I say 'parlous' because intra-governmental communication remains always beset by some inherent tensions amidst the amalgams of otherwise fixed strategic interests. Not merely water law/policy complexes remain riven by within governance inter-agency turf wars but these also stand confronted by the practices of competitive 'liberal' politics of water and environmental governance practices and networks. If so, watercrits also play severally the games of disrupting this regime of communicative signals! In the main this occurs, and may continue to occur, at least in two distinct though related ways.

Watercrits may thus produce a lot of channel 'noise' ambushing and even often disarticulating the circuit of a corrupt sovereign's communicative signals, contesting some orders of global sovereign's policy enunciations, by proceeding to an exposé of what I have elsewhere narrated as the rule-of-law and human rights costs, almost always excluded by the economist calculi of cost-benefit analysis.[24] Second watercrits by acts of privileging the *organic* (experience-based and tradition tested community) knowledges concerning water as a resource and water-based resources call into deep question the state-certified ways of privileging *erudite* (specialist, science/technology) knowledges. And this merely constitutes a series of first steps!

Towards Water Justice

'Justice' remains a rather silent category in these and related imageries of policy-making, analysis, and evaluation, outside the

---

[24] Upendra Baxi, 'Rehabilitation and Resettlement: Some Human Rights Perspectives', *in* Council for Social Development, *The Social Development Report 2008* (Delhi: Oxford University Press, 2008).

domain of procedural justice – the legitimacy of the claim that water sector reform proposals should be based on fullest access to information, and regimes of disclosure, the full provision of equal opportunity for hearing and participation by all those actually or likely to be affected by the 'old' or 'new' WPR, and further the continual recognition of the claims of social audit of polices thus enunciated. The struggle for procedural justice remains crucial because it serves the ends of reflexive and participatory process in the making, remaking, and unmaking WPR.

Some watercrits may (and indeed actually do) go so far as to insist that procedural justice is all that we may aspire towards because it promotes variegated articulation of new political identities, and the responsibility for self-determination, in a situation where massive diessenus may remain the rule rather than exception for any fashioning of agreement on the ends of distributive justice. On such approaches to water policy, what remains a critical resource is a just regard for the dignity of public deliberative discourse.

The question thus concerns the ways in which water 'publics' ought to be constituted (the normative question) or how they remain actually so (the empirical question.) The latter question remains rather felicitously addressed in terms, for example, by the travails of the quest for an 'integrated public policy framework proposed by the WCD (World Commission on Dams); while I agree with the acute critique of the WCD offered in this volume by Radha D'Souza, the question concerns the ways in which the platforms of identity politics thus constructed may ever fully address the tasks of the politics of redistribution.[25] It will constitute an egregious error even to think that forms, and practices, of descriptive realism come to us in some 'neat' narrative packages![26]

Granting fully that historically constituted experience of water injustice has been a more effective resource for combating structural political injustices than the acts of theorising about justice the normative question opens up an encyclopaedic genre! If so, it is

---

[25] Baxi, note 14 above at 115–159. *See also* Upendra Baxi, 'What Happens Next is Up to You: Human Rights at Risk in Dams and Development', 16/6 *American University International Law Journal* 1507 (2001).

[26] Jane Kelsey, *Serving Whose Interests? The Political Economy of Trade in Services Agreements* (London: Routledge-Cavendish, 2008).

simply not enough to point to the poverty of justice theorising by acts of naïve reading that question, for example, why a 'modern Aristotle', John Rawls fails to mention access to water as a primary moral good; surely this remains somewhat besides the point. Watercrits critique commoditisation of water in any event without a significant recourse to metaphysical notions of justice; for them the languages of human rights remain adequate. Further, John Rawls possessed the moral courage to imagine the tasks of distributive justice in terms of justice-across-generations. However, watercrits find much implicit use for John Rawls' notion of inter-generational justice as virtue of basic structure of society and of state conduct. The time-dimension (justice across generations) may not always be best addressed by procedural notions of water justice.

Surely, any talk about water justice needs fully to take into account the feature, especially when 'policy' also becomes a kind of conceptual theory in itself as entailing relationship between 'goals' and 'policies, programs, projects, decisions, options, means, or other alternatives that are available for achieving the goals'; it also remains a relationship that is established by a mix of *'intuition*, authority, statistics, observation, deduction, *guesses*, or by *other means.'*[27] The italicised words do (and perhaps going thus beyond the authorial intention of Stuart Nagel) perform a good deal of conceptual and historical work. Policies may be rational only in so far as these prescribe a relationship between 'goals' and 'means'. But, as Rawls memorably reminds us not all that may be fully described as *rational* may yet be *'reasonable'*.[28] Put my way, the realm of polices is never always a realm of practical reason; these goals and means, constructed rather than given, remain also affairs of political *desire* and *passion*, whether of state managers or the activists/movement folks. To perforce here summarily conclude, most needed are 'unsustainable' WPR critiques that seek to contradict the *unreason* of globalisation with a full regard for the growing ethical sentimental *'reason'* of human rights.

---

[27] Stuart Nagel, 'Conceptual Theory and Policy Evaluation', 6/3 *Public Administration & Management: An Interactive Journal* 71 (2001) and Louise White, 'Values, Ethics and Standards in Policy Analysis', *in* Stuart Nagel ed., *Encyclopaedia of Policy Studies* (New York: Marcel Dekker, 1994).

[28] John Rawls, *Political Liberalism* (New York: Columbia University Press, 1993).

How may we read water sector policies' – whether as enactments of governance desires or as flowing from the shared collective perceptions casting duties upon the State – is indeed a difficult question. There exist no doubt historic triggers for the act of policy-making as acts of governance as we partially witness on the Indian scenario of water policy enunciations. For a long while, the colonial executive in India, as is well-known, remained preoccupied with water resources as directed to means of enhancement of land revenues; further its policies (as for example enshrined in the Easements Act, and elsewhere) remained concerned with considerations either of the Crown sovereignty over India's natural wealth as colonial possessions or with the sacrosanct protection of the rights of property and contract. Water sector policy here interestingly assumed the shape of law or direct administrative regulation. So, it may even be said, it remains now given the continuation of colonial laws and regulations and the signatures of the postcolonial sovereign power.

By the latter phrase, I designate the fact not merely of continuation of colonial forms of legality but more importantly the notion that 'development' remains overall the function of the supreme executive, in some distinctive claims of the ethical superiority of human rights neutral modes of governance. Thus, most Indian national water and environmental related policy regimes remain even today solely within the province of executive power, constituting its sovereign prerogative. Major irrigation projects, even when they entail at times adjudicative decisions in sharing of inter-state waters, remain unalloyed acts of executive sovereign discretion, unaccompanied by any constitutionally mandated or human right normative necessities. No overt legislative regime is still thought necessary, though the reason of the constitution and of its rights and justice provisions obviously suggest that all major public projects ought to fully proceed under detailed legislative frameworks and provisions.[29]

The general point here is just this: the sovereign executive discretion may proceed just as arbitrarily as it may desire as concerns what I have elsewhere narrated as the rule-of-law and human rights costs, almost always excluded by the economist

---

[29] *See* Baxi, note 19 above.

calculi of cost-benefit analysis.³⁰ The state of affairs/things would be alright in this mode of proceeding, but for the fact that the Indian Supreme Court, and various High Courts, still remain susceptible to factor in human rights and rule of law costs in the variegated pursuit of the 'water sector policy' reform.

Further, the time, manner, and circumstance of major water policy enunciations remains the expedient time of the everyday practices of liberal competitive politics, un-governed by any order of fidelity either to the considerations of distributive justice or towards the dignity of human rights. Thus, it also comes to pass those not all executive acts of policy-making concerning water, and related natural resources stand always legislatively enacted. When otherwise, the legislative translation of nationally enunciated policy regimes seem to impoverish the political imagination as repeatedly manifest concerning governance over water, environment, forests, and biodiversity. Further, the accelerated transformation of constitutional governance objectives towards the promotion and protection of the powers and even human rights of the communities of multinational corporations and direct foreign investors – or what I have elsewhere named as the movement from universal human rights of all human beings into trade-related, market-friendly human rights paradigm³¹ – seems to clone certain patterns of disciplinary and regulatory globalisation.³² Essaying a critique of this cloning remains an arduous task, indeed! Increasingly, justice now means exacting standards of fairness

---

³⁰ *See* Baxi, note 24 above.

³¹ *See* generally, Baxi, note 14 above.

³² *See* Stephen Gill, The Constitution of Global Capitalism (Paper presented to a Panel: The Capitalist World, Past and Present at the International Studies Association Annual Convention, Los Angeles, 2000). *Source:* http://www.theglobalsite.ac.uk/press/010gill.pdf; Stephen Gill, *Power and Resistance in the New World Order* (New York: Palgrave-McMillan, 2003); Frank J. Garcia, 'The Global Market and Human Rights: Trading Away the Human Rights Principle', 25 *Brooklyn Journal of International Law* 51 (1999); David Schneiderman, 'Investment Rules and the New Constitutionalism', 25/3 *Law & Social Enquiry* 757 (2000) and David Schneiderman, *Constitutionalizing Economic Globalisation: Investment Rules and Democracy's Promise* (Cambridge: Cambridge University Press, 2008).

towards agents and mangers of global techno-capital flows because they seem to promise a better future for 'efficient' management of the future of water resources. Without framing it this way many contributions to this volume challenge this revisionist notion of water justice in which standards of global efficiency replace the norms of substantive/distributive justice. (See specially Radha D'Souza in the companionship of Francesco Costamagna and Francesco Sindico, Andrés Olleta, Karen Coelho, and Philippe Cullet.) The question of course concerns how may we read these, from any vantage point!

All this raises the question of forms of desire at stake in the practices of the 'subaltern' or popular water politics questing towards human rights based regime of access to water and water-based resources. Put differently, in the realms of water law, policy, reform, and rights (though perhaps more comprehensively as well on the planes/pains of the human right to food, shelter, livelihood, and health, for example) what may turn out to be decisive are not so much the ranges of policy/instrumental reason, but rather the future histories of affects of politics of dominant and subaltern desires. In turn all this, also invites attention to acts of 'doing' water rights theory, critique, and movement.

Further, India presents a rather unusual salience of adjudicatory policy-making concerning the right to water, water as a resource, and water-based resources as integral aspects/dimensions of the human rights to 'life' and 'liberty'. This howsoever understood meandering ways of judicial action and some by now settled dispositions of human rights judicial activism have contributed substantially to the tasks of both monitoring executive/legislative natural resources policies but also to the judicially mandated human rights based practices of Indian governance and developmentalism. A cameo study (here provided by Sujith Koonan) speaks volumes about the one-step-forward-several-steps backward itineraries of Indian adjudicatory policy making.

More generally put, in relation to water policy, law, and rights regime the conventional realms of policy studies will never suffice at least in terms of understanding of the borderlands (or wastelands, if you so insist) between the sovereign executive acts of policy-making discretion and no less sovereign acts of activist adjudicatory policy-making. At stake, further remains not the usually unproductive realm of dialoguing about the antimajoritarian character of adjudicatory policy-making (so well-beloved of Anglo-American juristic performances) but rather the making of judicial process and institutions as at the very least, adjunct representative institutions. Neither the Indian nor the South African experience, for example, offered here make much narrative sense outside civil society/people's movement as reconstituting activist adjudicatory ways of conceiving public power as so crystallising myriad forms of social trust.

But 'rationality' of policies is never always a realm of practical reason; these goals and means, constructed rather than given, remain also affairs of political desire or passion, whether of state managers or the activists/movement folks. To take a further and momentous step, the so-called practical reason moves under the auspices of political sentiment/passion as well. Most 'water sector reform' policies analysed as well as critiqued in this volume suggest the *disorders of neoliberal governance desires.*

Theory/Movement

Theory remains a suspiciously dreadful word for water rights human rights and social movement type (though not only for these) movements. Clearly, there remains operative certain orders of 'implicit' theory when activist/movement critique confronts the World Dam Commission Report, the WTO and related regimes as these concern access to water as an integral aspect of the human rights to life and livelihood, some perfidious (because in the short and long term benefiting, via acts of water privatisation, the might of the MNCs). It is for this and related reasons that the ideologies of 'sustainable development' regimes remains always contested at micro-levels of human rights and social action (see the contributions by Priya Sangameswaran, Roopa Madhav, Mattia Celio, Videh Upadhyay, A. Gurunathan and C.R. Shanmugham, Karen Coelho, and Sujith Koonan).

Yet, these may not proceed at all without some implicit 'theory' and if so the indictment of theory lies in the performance of the forms of *theory-aversion* and acts of *resistance to theory*.[33] The theory/

---

[33] *See* Baxi, note 6 above at 1–29.

movement diction remains hard to satisfactorily describe. For example, and not only in relation to 'water', we need to understand both in any event, one may well ask: What may any 'water' theory – loosely designating some shared protocols of conversation based on some general propositions about 'water' – look like? Of course, there are at hand entire ranges of biodiversity concerning 'theory'! We have many kinds of 'theory' – *epistemological* (questions about how may one arrive at valid knowledge) *ontological* (concerning the questions of Being), *political* (raising concerns about traditional notions of sovereign power and more recently recasting of these in terms of biopower/biopolitics), *cultural* (that is meanings crafted by material and non-material social practices of creating/crafting shared meanings, beliefs, sentiments, and symbols), and *ethical* (concerning at least forms of virtuous conduct of the rulers and ruled).

At least as I read this gifted work, it remains centred upon issues of 'policy' as constituting a kind of ordering of many different orders of implicit theory-constructions. Ramaswamy Iyer's contribution to this volume remains notable at least for grasping an ontological approach to rivers, articulating species-being claims of natural rights of river to immunity from being constituted as a 'drain', and as ordaining the river rights to 'space' and flows. For the most part, however, this volume remains focussed on policy as a serial ordering of 'theory'. Perhaps, all policy discourses are based on some such performances, as Radha D'Souza so insightfully suggests.

Surely, even with regard to these stimulating contributions, we may ask two useful questions: one, how may we distinguish between 'implicit' and 'explicit' acts of theorising and two, how may we still somehow maintain some bright lines between 'theory' and 'policy' critique, and advocacy. I do not offer these questions in any adversarial mode because I believe that critique of existing theoretical approaches to 'water' (rights, policies, management, governance, etc.) and at all levels of water governance and management (sub-national, national, supranational, and globalising policy and institutional regimes) may stand further to benefit by some reflexivity on theory/practice distinction, even when we may want to mean by 'theory' some orders of shared protocols of conversation somehow based on some generally agreed propositions.

By 'movement' I wish to signify standpoints/auspices for critique and resistance to the dominant regimes. A principal merit which I endorse, of the many splendid contributions to this volume (see Section 4) is simply this: describing dominance in water governance need not remain at all an excruciating task, given the narratives of spectacular, even if at the same moment somewhat scattered hegemonies of the international, and regional financial institutions as well of global social policy enactments (such as the tall talk of the UN Millennial Development Goals).

Further, neither theory nor movement may altogether escape some situated authorial constraints; understandably, almost all contributors to this work occupy distinct subject-positions in the global production of knowledges and also some shared platforms of ethical concerns concerning ways of water governance and management. In sum, various ontological and epistemological issues/concerns crowd the landscapes of 'water' theory and movement.

Bibliography

1. Books

Bagdanovic, Slavlo. *International Law of Water Resources: Contributions of the International Law Association (1954–2000)* (London, The Hague, Boston: Kluwer Law International, 2001).

Bakshi, P. M. *The Constitution of India* (Delhi: Universal Law Publishing Co., 2006).

Balanyá, Belén et al. eds. *Reclaiming Public Water – Achievements, Struggles and Visions from Around the World* (Amsterdam: Transnational Institute and Corporate Europe Observatory, 2nd ed., 2005).

Barlow, Maude and Tony Clarke. *Blue Gold: The Fight to Stop the Corporate Theft of the World's Water* (New York: The New Press, 2002).

Baviskar, Amita. *In the Belly of the River* (Delhi: Oxford University Press, 1995).

Beach, Heather L. et al. *Transboundary Freshwater Dispute Resolution: Theory, Practice and Annotated References* (Tokyo: United Nations University Press, 2000).

Birnie, Patricia W. and Alan E. Boyle. *International Law and the Environment* (Oxford: Oxford University Press, 2002).

Biswas, Asit K., Olli Varis and Cecilia Tortajada eds. *Integrated Water Resources Management in South and South-East Asia* (New Delhi: Oxford University Press, 2005).

Braithwaite, John and Peter Drahos. *Global Business Regulation* (Cambridge: Cambridge University Press, 2000).

Briscoe, John and R.P.S. Malik. *India's Water Economy: Bracing for a Turbulent Future* (New Delhi: The World Bank and Oxford University Press, 2006).

Claridge, Gordon and Bernard O'Callaghan. *Community Involvement in Wetland Management: Lessons from the Field* (Kuala Lumpur: Wetlands International, 1997).

Clark, Dana, Jonathan Fox and Kay Treakle eds. *Demanding Accountability: Civil-Society Claims and the World Bank Inspection Panel* (New York: Rowman & Littlefield, 2003).
Clark, Robert Emmet ed. *Water and Water Rights*, Vol. I (Indiana: The Allen Smith Company Publishers, 1967).
Conca, Ken. *Governing Water: Contentious Transnational Politics and Global Institution Building* (Cambridge: MIT Press, 2006).
Connell, D. *Water Politics in the Murray Darling* (Annadale: Federation Press, 2007).
Craven, Matthew. *The International Covenant on Economic, Social and Cultural Rights: A Perspective on its Development* (Oxford: Oxford University Press, 1995).
Cullet, P. and A. Gowlland-Gualtieri eds. *Key Materials in International Environmental Law* (Aldershot: Ashgate, 2004).
Cullet, P., A. Gowlland-Gualtieri, R. Madhav, and U. Ramanathan eds. *Water Law for the 21st Century: National and International Aspects of Water Law Reforms in India* (Abingdon: Routledge, forthcoming 2009).
Curiel, Pedro Brufao. *La revisión ambiental de las concesiones y autorizaciones de aguas* (Zaragoza: Colección Nueva Cultura del Agua No. 18, 2008).
Darby, Phillip. *Three Faces of Imperialism, British and American Approaches to Asia and Africa* (New Haven, London: Yale University Press, 1987).
Divan, Shyam and Armin Rosencranz. *Environmental Law and Policy in India: Cases, Materials and Statutes* (Oxford: Oxford University Press, 2001).
D'Souza, Radha. *Interstate Conflicts over Krishna Waters: Law, Science and Imperialism* (New Delhi: Orient Longman, 2006).
Dwivedi, Gaurav, Rehmat and Shripad Dharmadhikari. *Water: Private, Limited: Issues in Privatisation, Corporatisation and Commercialisation of Water Sector in India* (Badwani: Manthan Adhyayan Kendra, 2006).
Dyson, Megan, Ger Bergkamp and John Scanlon eds. *Flow: The Essentials of Environmental Flows* (Gland: IUCN, 2003).
Escobar, Arturo. *Encountering Development: The Making and Unmaking of the Third World* (New Jersey: Princeton University Press, 1995).
Espeland, Wendy. *The Struggle for Water: Politics, Rationality, and Identity in the American Southwest* (Chicago: University of Chicago Press, 1998).
Finger, M. and J. Allouche. *Water Privatisation: Trans-National Corporations and the Re-Regulation of the Water Industry* (London and New York: Spon Press, 2002).
Fisher, William ed. *Toward Sustainable Development: Struggling Over India's Narmada River* (Armonk, New York: M.E. Sharpe, 1995).
Fitzmaurice, Malgosia and Olufemi Elias. *Watercourse Co-operation in Northern Europe – A Model for the Future* (The Hague: T.M.C. Asser, 2004).

Fung, Archon and Erik O. Wright. *Deepening Democracy: Institutional Innovations in Empowered Participatory Governance* (London: Verso, 2003).
Ganguly-Thukral, Enakshi ed. *Big Dams, Displaced People: Rivers of Sorrow, Rivers of Change* (New Delhi: Sage Publications, 1992).
Getches, David H. *Water Law in a Nutshell* (Minnesota: West Group, 1997).
Gill, Stephen. *Power and Resistance in the New World Order* (New York: Palgrave-McMillan, 2003).
Glazewski, Jan. *Environmental Law in South Africa* (Durban: LexisNexis Butterworths, 2005).
Gulati, Ashok, Ruth Meinzen-Dick and K.V. Raju eds. *Institutional Reforms in Indian Irrigation* (New Delhi: Sage Publications, 2005).
Hanumantharao, V., N.K. Acharya and M.C. Swaminathan eds. *Andhra Pradesh at 50: A Data-Based Analysis* (Hyderabad: Data News Features, 1996).
Hartman, Loyal M. and Don Seastone. *Water Transfers: Economic Efficiency and Alternative Institutions* (Baltimore: Johns Hopkins Press, 1970).
Hildering, Antoinette. *International Law, Sustainable Development and Water Management* (Delft: Eburon Academic Publishers, 2004).
Hooja, Rakesh, Ganesh Pangare and K.V. Raju eds. *Users in Water Management* (Jaipur and New Delhi: Rawat Publications, 2002).
Hu, Desheng. *Water Rights: An International and Comparative Study* (London: IWA Publishing, 2006).
Hunt, Alan. *Explorations in Law and Society: Toward a Constitutive Theory of Law* (New York: Routledge, 1993).
International Water Management Institute. Molden, David ed. *Water for Food, Water for Life – A Comprehensive Assessment of Water Management in Agriculture* (London: Earthscan, 2007).
Iyer, Ramaswamy. *Water: Perspectives, Issues, Concerns* (New Delhi: Sage Publications, 2003).
Laski, Harold J. *A Grammar of Politics* (London: George Allen & Unwin, 5th ed., 1967).
Leys, Colin. *The Rise and Fall of Development Theory* (Bloomington: Indiana University Press, 1996).
Lincoln, D. et al. *Important Bird Areas in Africa and Associated Islands. Priority Sites for Conservation* (Newbury: Pisces Publications/BirdLife International 2001).
Joshi, L.K. and Rakesh Hooja ed. *Participatory Irrigation Management: Paradigm for the 21st Century* (New Delhi: Rawat Publications, 2000).
Kaul, Inge et al. eds. *Providing Public Goods: Managing Globalisation* (Oxford: Oxford University Press, 2003).
Kelsey, Jane. *Serving Whose Interests? The Political Economy of Trade in Services Agreements* (London: Routledge-Cavendish, 2008).

Maine, Sir Henry Sumner. *Ancient Law: Its Connections with the Early History of Society and its Relation to Modern Ideas* (London: Routledge and Kegan Paul, 1909).

Mason, E. and R. Asher. *The World Bank since Bretton Woods* (Washington: The Brookings Institute, 1973).

Moore, David B. and Gerald J. Schmitz eds. *Debating Development Discourse: Institutional and Popular Perspectives* (New York: St. Martin's Press, 1995).

Morse, Bradford and Thomas Berger. *Sardar Sarovar: Report of the Independent Review* (Ottawa: The Independent Review, 1992).

Mosse, David. *The Rule of Water: Statecraft, Ecology and Collective Action in South India* (New Delhi: Oxford University Press, 2003).

Narayanamoorthy, A. and R.S. Deshpande. *Where Water Seeps: Towards a New Phase in India's Irrigation Reforms* (New Delhi: Academic Foundation, 2005).

Nayeem, M.A. *The Splendour of Hyderabad: The Last Phase of an Oriental Culture (1591-1948 A.D.)* (Hyderabad: Hyderabad Publishers, 2002).

Petrella, Riccardo. *The Water Manifesto: Arguments for a World Water Contract* (London: Zed, 2001).

Rao, V. Sitararama *Law Relating to Water Rights* (Hyderabad: Asian Law House, 1996).

Sachs, Wolfgang ed. *The Development Dictionary: A Guide to Knowledge as Power* (London: Zed Books, 1992).

Sadeleer, Nicolas de. *Environmental Principles: From Political Slogans to Legal Rules* (Oxford: Oxford University Press, 2002).

Sainath, P. *Everybody Loves a Drought: Stories from India's Poorest Districts* (New Delhi: Penguin, 1996).

Saleth, Maria and Ariel Dinar. *The Institutional Economics of Water: A Cross-Country Analysis of Institutions and Performance* (Cheltenham: Edward Elgar, 2004).

Saleth, Maria R. *Water Institutions in India: Economics, Law, and Policy* (New Delhi: Commonwealth Publishers, 1996).

Salman, Salman M.A. and Kishor Uprety. *Conflict and Cooperation on South Asia's International Rivers: A Legal Perspective* (Washington D.C.: The World Bank, 2002).

Sands, Philippe. *Principles of International Environmental Law* (Cambridge: Cambridge University Press, 2003).

Schneiderman, David. *Constitutionalizing Economic Globalisation: Investment Rules and Democracy's Promise* (Cambridge: Cambridge University Press, 2008).

Shiva, Vandana. *Water Wars: Privatisation, Pollution and Profit* (Cambridge, Mass: South End Press, 2002).

Shue, Henry. *Basic Rights: Subsistence, Affluence and U.S. Foreign Policy* (Princeton: Princeton University Press, 1980).
Singh, Chhatrapati. *Water Rights and Principles of Water Resource Management* (Bombay: Tripathi, 1991).
Sivaramakrishnan, K.C., A. Kundu, and B.N. Singh. *Handbook of Urbanization in India* (New Delhi: Oxford University Press, 2005).
Stone, Ian. *Canal Irrigation in British India: Perspectives on Technological Change in a Peasant Economy* (Cambridge: Cambridge University Press, 1984).
Tanzi, Attila and Maurizio Arcari. *The United Nations Convention on the Law of International Watercourses: A Framework for Sharing* (London, The Hague, Boston: Kluwer Law International, 2001).
Tarlock, Dan A. ed. *Water Transfers in the West: Efficiency, Equity, and the Environment* (Washington: National Academy Press, 1992).
Teclaff, Ludwik A. *The River Basin in History and Law* (The Hague: Martinus Nijhoff, 1967).
Tigar, Michael E. and Madeleine R. Levy. *Law and the Rise of Capitalism* (New York: Monthly Review Press, 1977).
United Nations. *Water for People – Water for Life* (Paris: UNESCO, 2003).
United Nations. *Water – A Shared Responsibility* (Paris: UNESCO, 2006).
United Nations Development Programme. *Human Development Report 2006, Beyond Scarcity: Power, Poverty and the Global Water Crisis* (New York: Palgrave Macmillan, 2006).
Winpenny, James. *Managing Water as an Economic Resource* (London: Routledge, 1994).
World Bank. *World Development Report 1994: Investing in Infrastructure* (New York: Oxford University Press, 1994).
World Bank. *World Development Report 1997: The State in a Changing World* (Washington: World Bank, 1997).

2. Book Chapters

Ahmed, Sara. 'Why is Gender Equity a Concern for Water Management?', in Sara Ahmed ed. *Flowing Upstream: Empowering Women through Water Management Initiatives in India* 1 (Ahmedabad: Centre for Environment Education, 2005).
Amenga-Etego, R.N. and S. Grusky. 'The New Face of Conditionalities: The World Bank and Water Privatization in Ghana', in D. McDonald and G. Ruiters eds. *The Age of Commodity: Water Privatization in Southern Africa*, 275 (London: Earthscan, 2005).
Clark, S.D. and I.A. Renard. 'Constitutional, Legal and Administrative Problems', in H.J. Firth and G. Sawer. *The Murray Waters: Man, Nature and a River System* (Sydney: Angus Robertson, 1974).

Coe, Joe J. and Noah Rubins. 'Regulatory Expropriation and the *Tecmed* Case: Context and Contributions', *in* Todd G. Weiler ed. *International Investment Law and Arbitration: Leading Cases from the ICSID, NAFTA, Bilateral Treaties and Customary International Law*, 599 (London: Cameron May, 2005).

Coelho, Karen. 'Unstating 'the Public': An Ethnography of Reform in an Urban Water Utility in South India', *in* David Mosse and David Lewis eds. *The Aid Effect: Giving and Governing in International Development*, 171 (London: Pluto Press, 2005).

Deedat, Hameda and Eddie Cottle. 'Cost Recovery and Prepaid Meters and the Cholera Outbreak in KwaZulu-Natal: A Case Study in Madlebe', *in* David A. McDonald and John Pape eds. *Cost Recovery and the Crisis of Service Delivery*, 81 (Cape Town: Human Sciences Research Council Publishers, 2002).

D'Souza, Radha. 'Re-territorialising and Re-centering Empires: The Connivance of Law and Geography', *in* Jay Gao, Richard Le Heron and June Logie eds. *Windows on a Changing World: Proceedings of the 22nd New Zealand Geographical Society*, 324 (Auckland: New Zealand Geographical Society, 2003).

D'Souza, Radha. 'Water Resources Development and Water Conflicts in Two Indian Ocean States', *in* Timothy Doyle and Melisa Risely eds. *Regional Security and the Environment in the Indian Ocean Region*, 157 (New Brunswick, New Jersey and London: Rutgers University Press, 2008).

Eide, Asbjørn. 'Economic, Social and Cultural Rights as Human Rights', *in* Asbjørn Eide, Catarina Krause and Allan Rosas eds. *Economic, Social and Cultural Rights – A Textbook* 9 (Dordrecht, Boston, London: Martinus Nijhoff Publishers, 2nd ed., 2001).

Flynn, Sean and Danwood Mzikenge Chirwa. 'The Constitutional Implications of Commercializing Water in South Africa', *in* David A. McDonald and Greg Ruiters eds. *The Age of Commodity, Water Privatization in Southern Africa*, 59 (London, Sterling: Earthscan, 2005).

Garrido, Alberto. 'Analysis of Spanish Water Law Reform', *in* B.R. Bruns, C. Ringler and R. Meinzen-Dick eds. *Water Rights Reform: Lessons for Institutional Design*, 219 (Washington, DC: International Food Policy Research Institute, 2005).

Gilhuis, Piet *et al.* 'Negotiated Decision-Making in the Shadow of the Law', *in* Boudewijn de Waard ed. *Negotiated Decision-Making*, 219 (The Hague: BJu, 2000).

Glazer, Nathan. 'Individual Rights Against Group Rights', *in* G. Mahajan ed. *Democracy, Difference and Social Justice*, 416 (Oxford: Oxford University Press, 1998).

Glazewski, J. 'Environmental Rights and the New South African Constitution', *in* A. Boyle and M. Anderson, *Human Rights Approaches to Environmental Protection*, 177 (Oxford: Clarendon Press, 1996).

Govinadrajalu, K. 'Industrial Effluent and Health Status – A Case Study of Noyyal River Basin', *in* Martin J. Bunch *et al.* eds. *Proceedings of the Third International Conference on Environment and Health*, 15 (Chennai: Department of Geography, University of Madras and Faculty of Environmental Studies, York University, 2003).

Hollings, C.S. 'What Barrier? What Bridges?', *in* L.H. Gunderson *et al. Barriers and Bridges to the Renewal of Ecosystems and Institutions*, 9 (New York: Columbia University Press, 1995).

Just, Richard E. and Sinaia Netanyahu. 'International Water Resource Conflicts: Experience and Potential', *in* Richard E. Just and Sinaia Netanyahu eds. *Conflict and Cooperation on Trans-boundary Water Resources*, 1 (Boston: Kluwer Academic Publishers, 1998).

Khalfan, Ashfaq and Anna Russell. 'The Recognition of the Right to Water in South Africa's Legal Order', *in* Henri Smets ed. *Le Droit à l'Eau dans les Législations Nationales*, 121 (Nanterre: Académie de l'Eau, 2005).

Klein, Eckart. 'General Comments: Zu Einem Eher Unbekannten Instrument des Menschenrechtsschutzes', in Jörn Ipsen and Edzard Schmidt-Jortzig eds. *Recht - Staat - Gemeinwohl, Festschrift für Dietrich Rauschning*, 301 (Cologne: Heymann, 2001).

Kok, Anton. 'Privatisation and the Right to Access to Water', *in* Koen De Feyter and Felipe Gómez Isa eds. *Privatisation and Human Rights in the Age of Globalisation*, 259 (Antwerp: Intersentia, 2005).

Kok, Anton and Malcolm Langford. 'The Right to Water' *in* Stuart Woolman *et al.* eds. *Constitutional Law of South Africa*, 56B (Lansdowne: Juta, 2nd ed., 2005).

Lang, John Temple. 'Legal Certainty and Legitimate Expectations as General Principles of Law', *in* Ulf Bernitz and Joakim Nergelius eds. *General Principles in European Community Law*, 163 (The Hague: Kluwer Law International, 2000).

Liberti, Lahra. 'Investissements et droits de l'homme', *in* Philippe Kahn and Thomas Wälde eds. *Les aspects nouveaux du droit des investissements internationaux/New Aspects of International Investment Law*, 791 (Leiden-Dordrecht: Martinus Nijhoff Publishers, 2007).

Liebenberg, Sandra. 'The Interpretation of Socio-Economic Rights', *in* Stuart Woolman *et al.* eds. *Constitutional Law of South Africa*, 33 (Lansdowne: Juta, 2nd ed., 2005).

Loras, Antonio Fanlo. 'Water Resources Management in Spain', *in* S. Marchisio, G. Tamburelli and L. Pecoraro eds. *Sustainable Management and Rational Use of Water Resources: A Legal Framework for the Mediterranean*, 149 (Italy: Instituto di Studi Giuridici Internazionali, 1999).

MacPherson, W.J. 'Economic Development in India under the British Crown, 1858–1947', in A.J. Youngson ed. *Economic Development in the Long Run*, 144 (London: G. Allen & Unwin, 1972).

Mann, Howard. 'The Rights of States to Regulate and International Investment Law', in UNCTAD, *The Development Dimensions of FDI: Policy and Rule-Making Perspectives*, 211 (Geneva-New York: UNCTAD, 2003).

McDonald, David A. 'The Theory and Practice of Cost Recovery in South Africa', in David A. McDonald and John Pape eds. *Cost Recovery and the Crisis of Service Delivery*, 17 (Cape Town: Human Sciences Research Council Publishers, 2002).

McDonald, David A. 'The Bell Tolls for Thee: Cost Recovery, Cutoffs, and the Affordability of Municipal Services in South Africa', in David A. McDonald and John Pape eds. *Cost Recovery and the Crisis of Service Delivery*, 161 (Cape Town: Human Sciences Research Council Publishers, 2002).

McDonald, David A. and Greg Ruiters, 'Theorizing Water Privatization in Southern Africa', in David A. McDonald and Greg Ruiters eds. *The Age of Commodity, Water Privatization in Southern Africa*, 13 (London, Sterling: Earthscan, 2005).

McInnes, P. 'Entrenching Inequalities: The Impact of Corporatization on Water Injustices in Pretoria', in D.A. McDonald and G. Ruiters eds. *The Age of Commodity: Water Privatization in Southern Africa*, 99 (London: Earthscan, 2005).

Mehta, Lyla. 'Problems of Publicness and Access Rights: Perspectives from the Water Domain', in Lyla Mehta ed. *Providing Global Public Goods: Managing Globalization*, 556 (Oxford: Oxford University Press, 2003).

Mistelis, Loukas A. 'Confidentiality and Third Party Participation: *UPS v Canada* and *Methanex Corp. v United States*', in Todd G. Weiler ed. *Investment Law and Arbitration: Past Issues, Current Practice, Future Prospects*, 169 (Ardsley: Transnational Publishers, 2004).

Molden, David, R. Sakthivadivel and Samad Madar. 'Accounting for Changes in Water Use and the Need for Institutional Adaptation', in Charles L. Abernethy ed. *Intersectoral Management of River Basins*, 73 (Colombo and Feldafing: International Water Management Institute and German Foundation for International Development, 2001).

Mollinga, Peter. 'Power in Motion: A Critical Assessment of Canal Irrigation Reform in India', in Rakesh Hooja, Ganesh Pangare and K.V. Raju eds. *Users in Water Management*, 265 (Jaipur and New Delhi: Rawat Publications, 2002).

Muralidhar, S. 'The Right to Water: An Overview of the Indian Legal Regime', in Eibe Riedel and Peter Rothen eds. *The Human Right to Water*, 65 (Berlin: Berliner Wissenschafts-Verlag, 2006).

Nelliyat, Prakash. 'Public-Private Partnership in Urban Water Management: The Case of Tirupur', *in* Barun Mitra, Kendra Okonski and Mohit Satyanand eds. *Keeping the Water Flowing*, 149 (New Delhi: Academic Foundation, 2007).

Peter, J. Raymond. 'Irrigation Reforms in Andhra Pradesh', *in* Rakesh Hooja, Ganesh Pangare and K.V. Raju eds. *Users in Water Management*, 59 (Jaipur and New Delhi: Rawat Publications, 2002).

Reddy, Ratna V. 'Irrigation: Development and Reforms', *in* C.H. Hanumantha Rao and S. Mahendra Dev eds. *Andhra Pradesh Development: Economic Reforms and Challenges Ahead*, 170 (Hyderabad: Centre for Economic and Social Studies, 2003).

Riedel, Eibe. 'The Human Right to Water', *in* Klaus Dicke ed. *Weltinnenrecht – Liber Amicorum Jost Delbrück*, 585 (Berlin: Duncker & Humblot, 2005).

Russell, Sage. 'Minimum State Obligations: International Dimension', *in* Danie Brand and Sage Russell eds. *Exploring the Core Content of Socio-Economic Rights: South African and International Perspectives*, 11 (Pretoria: Protea, 2002).

Singh, Chhatrapati. 'Water Rights in India', *in* Chhatrapati Singh ed. *Water Law in India* (New Delhi: Indian Law Institute, 1992).

Sornarajah, M. 'Good Corporate Citizenship and the Conduct of Multinational Corporations', *in* Jianfu Chen and Gordon Walker eds. *Balancing Act: Law, Policy and Politics in Globalisation and Global Trade*, 224 (Annandale NSW: The Federation Press, 2004).

Waart, Paul J.I.M. de. 'Securing Access to Safe Drinking Water through Trade and International Migration', *in* Edward H.P. Brans *et al.* eds. *The Scarcity of Water: Emerging Legal and Policy Responses*, 101 (London: Kluwer Law International, 1997).

Wälde, Thomas. 'The Specific Nature of Investment Arbitration', *in* Philippe Kahn and Thomas Wälde eds. *Les aspects nouveaux du droit des investissements internationaux/New Aspects of International Investment Law*, 43 (Leiden: Martinus Nijhoff Publishers, 2007).

Weiss, Edith Brown. 'Water Transfers and International Trade', *in* Edith Brown Weiss, Laurence Boisson de Chazournes and Nathalie Bernasconi-Osterwalder eds. *Fresh Water and International Economic Law*, 61 (Oxford: Oxford University Press, 2005).

Whitcombe, E. 'Irrigation', *in* D. Kumar and M. Desai eds. *The Cambridge Economic History of India*, 737 (New Delhi: Orient Longman, Volume II, 2004).

White, Louise. 'Values, Ethics and Standards in Policy Analysis', *in* Stuart Nagel ed., *Encyclopaedia of Policy Studies* (New York: Marcel Dekker, 1994).

3. Articles and Journal Articles

Abrahams, Abe. 'Orange River Mouth Transboundary Ramsar Site: Green Scene', 55, *African Wildlife*, 46 (2001).

Adger, W.N. *et al.* 'Advancing a Political Ecology of Global Environmental Discourses', 32, *Development and Change*, 687 (2001).

Alston, Philip. 'Out of the Abyss: The Challenges Confronting the New U.N. Committee on Economic, Social and Cultural Rights', 9, *Human Rights Quarterly*, 332 (1987).

Anderson, Mark D. *et al.* 'Waterbird Populations at the Orange River Mouth from 1980–2001: A Re-assessment of its Ramsar Status', 74, *Ostrich*, 159 (2003).

Bakker, Karren. 'The Politics of Hydropower: Developing the Mekong', 18, *Political Geography*, 209 (1999).

Bandyopadhyay, Jayanta. 'A Critical Look at the Report of the World Commission on Dams in the Context of the Debate on Large Dams on the Himalayan Rivers', 18 (1), *Water Resources Development*, 127 (2002).

Beaumont, Peter. 'The 1997 UN Convention on the Non-navigational Uses of International Watercourses: Its Strengths and Weaknesses from a Water Management Perspective and the Need for New Workable Guidelines', 16 (4), *Water Resources Development*, 475 (2000).

Bernasconi-Osterwalder, Nathalie. 'Who Wins and Who Loses in Investment Arbitration? Are Investors and Host States on a Level Playing Field? The Lauder/Czech Republic Legacy', 6 *Journ. World Investm. & Trade*, 69 (2005).

Bhaduri, Amit and Arvind Kejriwal. 'Urban Water Supply: Reforming the Reformers', 40/53 *Economic and Political Weekly*, 5543 (2005).

Bijoy, C.R. 'Kerala's Plachimada Struggle: A Narrative on Water and Governance Right', 41/41, *Economic and Political Weekly*, 4332 (2006).

Bilchitz, David. 'Giving Socio-Economic Rights Teeth: The Minimum Core and its Importance', 119, *South African Law Journal*, 484 (2002).

Bilchitz, David. 'Towards a Reasonable Approach to the Minimum Core: Laying the Foundations for Future Socio-Economic Rights Jurisprudence', 19 *South African Journal on Human Rights*, 1 (2003).

Bird, Jeremy. 'Nine Months after the Launch of the World Commission on Dams Report', 18 (1), *Water Resources Development*, 111 (2002).

Bjornlund, Henning and J. McKay. 'Do Permanent Water Markets Facilitate Farm Adjustment and Structural Change within Irrigation Communities?', 9, *Rural Society*, 3 (1999).

Blatter, Joachim and Helen Ingram. 'States, Markets and Beyond: Governance of Transboundary Water Resources', 40, *Natural Resources Journal*, 339 (2000).

Bourne, Charles B. 'The International Law Association's Contribution to International Water Resource Law', 36, *Natural Resources Journal*, 155 (1996).
Brower II, Charles H. 'Investor-State Disputes under NAFTA: The Empire Strikes Back', 40 *Columbia J. Transn. L.*, 43 (2001).
Brunnée, Jutta and Stephen J. Toope. 'Environmental Security and Freshwater Resources: A Case for International Ecosystem Law', 5 *Yb. Int'l Envtl. L.*, 41 (1994).
Buck-Morss, Susan. 'Envisioning Capital: Political Economy on Display', 21, *Critical Inquiry*, 434 (1995).
Carruthers, Fiona. 'Politics Muddies $13bn Water Reform Plan', *The Australian Financial Review*, 19 May 2006.
Cavadino, Michael. 'Commissions and Codes: A Case Study in Law and Public Administration', *Public Law*, 333 (1993).
Celio, Mattia and Mark Giordano. 'Agriculture-Urban Water Transfers: A Case Study of Hyderabad, South-India', 5 *Paddy and Water Environment*, 229 (2007).
Choudhury, Barnali. 'Evolution or Devolution? Defining Fair and Equitable Treatment in International Investment Law', 6, *Journ. World Investm. & Trade*, 297 (April 2005).
Conley, Alan H. and Peter H. van Niekerk. 'Sustainable Management of International Waters: The Orange River Case', 2 *Water Policy*, 131 (2000).
Connell, Daniel and Stephan Dovers. 'Tail Wags Dog – Water Markets and the National Water Initiative', July–September *Public Administration Today*, 18 (2006).
Costamagna, Francesco. 'L'impatto del GATS sull'autonomia regolamentare degli stati membri nei servizi idrici ed energetici', 19 *Diritto del commercio internazionale*, 501 (2005).
Cullet, Philippe. 'Water Law Reforms. Analysis of Recent Developments', 48 *Journal of the Indian Law Institute*, 206 (2006).
Derman, Bill and Anne Ferguson. 'Value of Water: Political Ecology and Water Reform in Southern Africa', 62/3 *Human Organisation*, 277 (2003).
Deshpande, R.S. and A. Narayanamoorthy. 'Issues before Second Irrigation Commission of Maharashtra', 36(12) *Economic and Political Weekly*, 1034 (2000).
Dharmadhikary, Shripad. 'Maharashtra: Water Regulatory – A Flawed Model for Water Regulation', *India Together* (23 May 2007).
Dolzer, Rudolph. 'Indirect Expropriations: New Developments?', 11 *NY Univ. Env. L. J.* 64 (2003).
Dolzer, Rudolph. 'The Impact of International Investment Treaties on Domestic Administrative Law', 27 *NY Univ. Journal of Int'l Law & Pol.*, 953 (2005).

Dolzer, Rudolph. 'Fair and Equitable Treatment: A Key Standard in Investment Treaties', 39 *The International Lawyer*, 87 (2005).

D'Souza, Radha. 'The 'Third World' and Socio-legal Studies: Neo-liberalism and Lessons from India's Legal Innovations', 14 (4) *Social & Legal Studies*, 487 (2005).

Duda, Alfred M. and David La Roche. 'Joint Institutional Arrangements for Addressing Transboundary Water Resources Issues – Lessons for the GEF', 21 (2) *Natural Resources Forum*, 127 (1997).

Emery, Carl. 'Public Law or Private Law? the Limits of Procedural Reform', *Public Law*, 450 (1995).

Erasmus, Gerhard and Debbie Hamman. 'Where Is the Orange River Mouth? The Demarcation of the South African/Namibian Maritime Boundary', 13 *South Afr. Yb. Int'l L.* 49 (1987–1988).

Fantini, Enrico. 'The Human Rights to Water: Recent Positive Steps and the Way Ahead', 2, *Pace Diritti Umani*, 123 (2005).

Francis, R. 'Water Justice in South Africa: Natural Resources Policy at the Intersection of Human Rights, Economics, and Political Power', 18 *Georgetown Intl Envtl L.Rev.* 149 (2005).

Franck, Susan D. 'The Legitimacy Crisis in Investment Treaty Arbitration: Privatising Public International Law Through Inconsistent Decisions', 73, *Fordham J. Int'l L.* 1521 (2005).

Franck, Thomas. 'The Emerging Right of Democratic Governance', 86, *Am. J. Int'l L.* 46 (1992).

Freedland, Mark. 'Government by Contract and Public Law', *Public Law*, 86 (1994).

Fujikura, Ryo and Mikiyasu Nakayama. 'Study on Feasibility of the WCD Guidelines as an Operational Instrument', 18 (2), *Water Resources Development* 301 (2002).

Gabru, N. 'Some Comments on Water Rights in South Africa', 1, *Potchefstroom Electronic L.J.* 1 (2005).

Gantz, David A. 'The Evolution on FTA Investment Provisions: From NAFTA to the United States – Chile Free Trade Agreement', 19 *Am. Univ. Int'l L. Rev.* 679 (2004).

Gardner, Alex. 'Environmental Water Allocations in Australia', 23 *Environmental and Planning Law Journal*, 208 (2006).

Gleick, Peter. 'Basic Water Requirements for Human Activities: Meeting Basic Needs', 21 *Water International* 83, (1996).

Gleick, P.H. 'The Human Right to Water', 1 *Water Policy*, 487 (1998).

Griff, Angus and David Crowe, 'NATS Threaten PM's $10bn Water Plan', *The Australian Financial Review*, 9 March 2007.

Hardberger, Amy. 'Whose Job Is It Anyway?: Governmental Obligations Created by the Human Right to Water', 41, *Texas International Law Journal*, 541 (2006).

Harten, Guus Van and Martin Loughlin. 'Investment Treaty Arbitration as a Species of Global Administrative Law', 17 *Europ. J. Int'l L.* 121 (2006).

Hervic, Joelle. 'Water, Water Everywhere?', 77/1 *Florida Bar Journal*, 49 (2003).

Howe, Charles W., Dennis R. Schurmeier and Douglas W. Shaw Jr. 'Innovative Approaches to Water Allocation: The Potential for Water Markets', 22, *Water Resources Research*, 439 (1986).

Jackson, Sue. 'Compartmentalising Culture: The Articulation and Consideration of Indigenous Values in Water Resources Management', 37/1 *Australian Geographer*, 19 (2006).

Kameri-Mbote, Patricia. 'The Use of the Public Trust Doctrine in Environmental Law', 3/2 *Law, Environment and Development Journal*, 197 (2007).

Keller, Jack, Andrew A. Keller and Grant Davids. 'River Basin Development Phases and Implications for Closure', 33, *Journal of Applied Irrigation Science* 145 (1998).

Kessides, Ioannis N. 'The Challenges of Infrastructure Privatisation', 1 2005 *DICE Report*, 19 (2005).

Kidd, M. 'Not a Drop to Drink: Disconnection of Water Services for Non-Payment and the Right of Access to Water', 20 *South African J. Human Rts*, 119 (2004).

Koehler, Cynthia L. 'Water Rights and the Public Trust Doctrine: Resolution of the Mono Lake Controversy', 22, *Ecology Law Quarterly*, 541 (1995).

Kuehne, G. and B. Bjornlund. 'Frustration, Confusion and Uncertainty – Qualitative Responses from Namoi Valley Irrigators', 33, *Water*, 78 (2006).

Kwaterski, Scanlan Melissa. 'The Evolution of the Public Trust Doctrine and the Degradation of Trust Resources: Courts, Trustees and Political Power in Wisconsin', 27, *Ecology Law Quarterly*, 135 (2000).

Liebenberg, S. 'The Value of Human Dignity in Interpreting Socio-Economic Rights', 21 *South African J. Human Rts* 1 (2005).

Liebenberg, Sandy. 'The National Water Bill – Breathing Life into the Right to Water', 1, *Economic and Social Rights Review*, 1 (1998).

Lloyd, Christopher. 'Regime Change in Australian Capitalism: Towards a Historical Political Economy of Regulation', 42 (3), *Australian Economic History Review*, 238 (2002).

Loftus, A. and D.A. McDonald. 'Of Liquid Dreams: A Political Ecology of Water Privatisation in Buenos Aires', 13/2 *Environment & Urbanization*, 179 (2001).

Lowe, Vaughan. 'Regulation or Expropriation?', 55, *Curr. Leg. Prob.* 447 (2002).

MacDonnell, Lawrence J. 'Rules Guiding Groundwater Use in the United States', 1 *Indian Juridical Review*, 43 (2005).

Mayeda, Graham. 'Playing Fair: The Meaning of Fair and Equitable Treatment in Bilateral Investment Treaties', 41 *Journ. World Trade*, 273 (2007).

McCaffrey, Stephen. 'International Organizations and the Holistic Approach to Water Problems', 31, *Natural Resources Journal*, 139 (1991).

McIntyre, Owen. 'The Emergence of an 'Ecosystem Approach' to the Protection of International Watercourses under International Law', 13, *Rev. European Community Int'l Envtl L* , 1 (2004).

Meinzen-Dick, Ruth S. and M.S. Mendoza. 'Alternative Water Allocation Mechanisms: Indian and International Experiences', 31 *Economic and Political Weekly*, 25 (1996).

Molden, David and M.G. Bos. 'Improving Basin Water Use in Linked Agricultural, Ecological, and Urban Systems', 51 *Water Science and Technology*, 147 (2005).

Molinari, Alejo. 'Lessons from Latin America: The Potential of Local Private Operators', 1/1 *Water Utility Management International*, 20 (2006).

Mostert, Hanri and Peter Fitzpatrick. 'Law Against Law: Indigenous Rights and the Richtersveld Cases', 2, *Law, Social Justice & Global Development J.*, 1 (2005).

Muchlinski, Peter. 'Caveat Investor'? The Relevance of the Conduct of the Investor Under the Fair and Equitable Standard', 5 *Int'l & Comp. L. Quarterly*, 527 (July 2006).

Nakayama, Mikiyasu Mikiyasu, 'Successes and Failures of International Organisations in Dealing with International Waters', 13 (3), *Water Resources Development*, 367 (1997).

Navalawala, B.N. 'World Commission on Dams: Biased?', 36 (12) *Economic and Political Weekly*, 1008 (2001).

Newcombe, Andrew. 'The Boundaries of Regulatory Expropriation in International Law', 20, *ICSID Review – FILJ*, 1 (2005).

Olem, Harvey and Alfred M. Duda. 'International Watercourses: The World Bank Looks Towards a More Comprehensive Approach to Management', 31 (8), *Water Sciences and Technology*, 345 (1995).

Partzsch, Lena. 'Wasser in der Krise – Das Beispiel Südafrika', 196, *Solidarische Welt*, 4 (2007).

Paulsson, Jan. 'Indirect Expropriation: Is the Right to Regulate at Risk?', 3, *Transn. Dispute Management*, 1 (April 2006).

Peel, Jacqueline and Lee Godden. 'Australian Environmental Management: A 'Dams' Story', 28, *UNSW Law Journal*, 668 (2005).

Petrova, Violeta. 'At the Frontiers of the Rush for the Blue Gold: Water Privatisation and the Human Right to Water', 31, *Brook. J. Int'l L.* 577 (2006).

Phadke, Anant. "Thiyya Andolan' in Krishna Valley', 39(8), *Economic and Political Weekly*, 775 (2004).

Phansalkar, Sanjiv and Vivek Kher, 'A Decade of the Maharashtra Groundwater Legislation: Analysis of the Implementation Process', 2/1, *Law, Environment and Development Journal*, 67 (2006).

Prasad, Naren. 'Privatisation Results: Private Sector Participation in Water Services after 15 Years', 24, *Development Policy Review*, 669 (2006).

Prasad, Naren. 'Privatisation of Water: A Historical Perspective', 3/2 *Law, Environment and Development Journal*, 217 (2007).

Purkey, Andrew and Clay Landry. 'A New Tool for New Partnerships: Water Acquisitions and the Oregon Trust Fund,' 12, *Water Law*, 5 (2001).

Raman, Ravi K. 'Corporate Violence, Legal Nuances and Political Ecology: Cola War in Plachimada', 40 (25) *Economic and Political Weekly* 2481 (2005).

Rawal, Vikas. 'Non-Markets Interventions in Water-Sharing: Case Studies from West Bengal, India', 2, *Journal of Agrarian Change*, 545 (2002).

Rose-Ackermann, Susan and Jim Rossi. 'Disentangling Regulatory Takings', 86 *Virginia L. Rev.* 1435 (2000).

Rose, Carol V. 'The 'New' Law and Development Movement in the Post-Cold War Era: A Vietnam Case Study', 32 (1), *Law & Society Review*, 93 (1998).

Rooijen, Daan J. van, Hugh Turral and Trent Wade Biggs. 'Sponge City: Water Balance of Mega-City Water Use and Wastewater Use in Hyderabad, India', 54, *Irrigation and Drainage*, 1 (2005).

Ruet, Joël, Marie Gambiez and Emilie Lacour. 'Private Appropriation of Resource: Impact of Peri-Urban Farmers Selling Water to Chennai Metropolitan Water Board', 24 *Cities*, 110 (2007).

Sacerdoti, Giorgio. 'Bilateral Treaties and Multilateral Instruments on Investment Protection', 269, *Recueil des Cours (RCADI)*, 251 (1997).

Salzman, James. 'Creating Markets for Ecosystem Services: Notes from the Field', 80, *New York University Law Review*, 870 (2005).

Sánchez-Moreno, Maria McFarland and Tracy Higgins. 'No Recourse: Transnational Corporations and the Protection of Economic, Social, and Cultural Rights in Bolivia', 27, *Fordham Int'l L.J.* 1663 (2004).

Savenije, Hubert H.G. and Pieter van der Zaag. 'Conceptual Framework for the Management of Shared River Basins, With Special Reference to the SADC and EU', 2, *Water Policy*, 9 (2000).

Sax, Joseph L. 'The Public Trust Doctrine in Natural Resource Law: Effective Judicial Intervention', 68, *Michigan Law Review*, 471 (1970).

Schreuer, Christoph. 'Fair and Equitable Treatment in Arbitral Practice', 6 *Journ. World Investm. & Trade* 357, (June 2005).

Schreuer, Christoph H. 'The Concept of Expropriation under the ECT and other Investment Protection Treaties', 2 *Trans. Disp. Manag.* 1 (November 2005).

Scott, Craig and Philip Alston. 'Adjudicating Constitutional Priorities in a Transnational Context: A Comment on Soobramoney's Legacy and Grootboom's Promise', 16 *South African Journal on Human Rights* 206 (2000).

Scudder, Thayer. 'The World Commission on Dams and the Need for a New Development Paradigm', 17 (3), *Water Resources Development*, 329 (2001).

Shah, Tushaar *et al*. 'Sustaining Asia's Groundwater Boom: An Overview of Issues and Evidence', 27, *Natural Resources Forum*, 130 (2003).

Smith, George and Michael Sweeny. 'Public Trust Doctrine and Natural Law: Emanations within a Penumbra', 33, *Boston College Environmental Affairs Law Review*, 307 (2006).

Smith, Julie A. and J. Maryann Green. 'Free Basic Water in Msunduzi, KwaZulu-Natal: Is It Making a Difference to the Lives of Low-Income Households?', 7, *Water Policy*, 443 (2005).

Sneddon, Chris and Coleen Fox. 'Rethinking Transboundary Waters: A Critical Hydropolitics of the Mekong Basin', 25, *Political Geography*, 181 (2006).

Snodgrass, Elizabeth. 'Protecting Investors' Legitimate Expectations: Recognising and Delimiting a General Principle', 21, *ICSID Rev. – FILJ* 1 (2006).

Starr, Joyce R. 'Water Wars', 82, *Foreign Policy*, 17 (1991).

Steinberg, C. 'Can Reasonabless Protect the Poor? A Review of South Africa's Socio-economic Rights Jurisprudence', 23, *South African L.J.* 264 (2006).

Stein, Robyn. 'Water Law in a Democratic South Africa: A Country Case Study Examining the Introduction of a Public Rights System', 83, *Texas L. Rev.* 2167 (2005).

Swart, M. 'Left Out In The Cold? Crafting Constitutional Remedies for the Poorest of the Poor', 21, *South African J. Human Rts*, 215 (2005).

Tan, Poh-Ling. 'Irrigators Come First: A Study of the Conversion of Existing Allocations to Bulk Entitlements in the Goulburn and Murray Catchments, Victoria', 18, *Environmental and Planning Law Journal*, 154 (2001).

Tan, Poh-Ling. 'A Historical Introduction to Water Law Reform in NSW – 1975 to 1994', 19, *Environmental and Planning Law Journal*, 445 (2002).

Tan, Poh-Ling. 'Legislating for Adequate Public Participation in Allocating Water in Australia', 31, *Water International*, 12 (2006).
Teclaff, Ludwik A. 'Fiat or Custom: The Checkered Development of International Water Law', 31, *Natural Resources Journal*, 45 (1991).
Tewari, D.D. 'A Brief Historical Analysis of Water Rights in South Africa', 30, *Water International*, 184 (2005).
Thatte, C.D. 'Aftermath, Overview and an Appraisal of Past Events Leading to Some of the Imbalances in the Report of the World Commission on Dams', 17 (3), *Water Resources Development*, 343 (2001).
Turral, Hugh, T. Etchells and Hector Malano et al. 'Water Trading at the Margin: The Evolution of Water Markets in the Murray-Darling Basin', 41, *Water Resources Research* (2005).
Uitto, Juha I. and Alfred M. Duda. 'Management of Transboundary Water Resources: Lessons from International Cooperation for Conflict Prevention', 168 (4), *The Geographical Journal*, 363 (2002).
Upadhyay, Videh. 'Water Management and Village Groups: Role of Law', 37(49), *Economic and Political Weekly*, 4907 (2002).
Upadhyay, Videh 'Command Area Development: Restructured Guidelines', 40(29), *Economic and Political Weekly*, 3119 (2005).
Verschuuren, Jonathan. 'The Case of Transboundary Wetlands Under the Ramsar Convention: Keep the Lawyers Out!', 19 *Colo. J. Int'l Envtl. L. & Pol'y*, 49 (2008).
Vicuña, Francisco Orrego. 'Regulatory Authority and Legitimate Expectations: Balancing the Rights of the State and the Individual under International Law in a Global Society', 5/3 *International Law FORUM du droit international*, 188 (2003).
Visser, Jaap de, Edward Cottle and Johann Mettler. 'Realising the Right of Access to Water: Pipe Dream or Watershed', 7, *Law, Democracy and Development*, 27 (2003).
Voigt, Christina. 'WTO Law and International Emissions Trading: Is there Potential for Conflict', 2, *Carbon and Climate Law Review*, 54 (2008).
Wade, Robert. 'The Information Problem of South Indian Irrigation Canals', 5, *Water Supply and Management*, 31 (1981).
Wade, Robert. 'The System of Administrative and Political Corruption: Canal Irrigation in South India', 18, *Journal of Development Studies*, 287 (1982).
Wälde, Thomas. 'Law, Contract and Reputation in International Business: What Works?', 3, *CEPMLP Internet Journal* (1998).
Wälde, Thomas and Stephen Dow. 'Treaties and Regulatory Risk in Infrastructure Investment: The Effectiveness of International Law Disciplines versus Sanctions by Global Markets in Reducing the Political and Regulatory Risk for Private Infrastructure Investment', 34, *J. World Trade* 1 (2000).

Wälde, Thomas and Abba Kolo. 'Environmental Regulation, Investment Protection and 'Regulatory Taking' in International Law', 50, *Int'l & Comp. Law Quarterly*, 811 (2001).

Wesson, M. *'Grootboom* and Beyond: Reassessing the Socioeconomic Jurisprudence of the South Africa Constitutional Court', 20, *South African J. Human Rts*, 284 (2004).

Wilkinson, John C. 'Muslim Land and Water Law', 1, *Journal of Islamic Studies*, 54 (1990).

Woodhouse, Diana. 'Delivering Public Confidence: Codes of Conduct, a Step in the Right Direction', *Public Law*, 511 (2003).

Woodhouse, Eric J. 'The 'Guerra del Agua' and the Cochabamba Concession: Social Risk and Foreign Direct Investment in Public Infrastructure', 39 *Stanford J. Int'l L.* 295 (2003).

Worster, Donald. 'Water and the Flow of Power', 13(5). *The Ecologist*, 168 (1983).

4. Judicial and Quasi-judicial Decisions

ADF Group Inc v United States, ICSID Case No. ARB/(AF)/00/1, Award of January 2003, 18 *ICSID Review – FILJ* 195 (2003).

Aguas Argentinas, S.A., Suez, Sociedad General de Agua de Barcelona, S.A. and Vivendi Universal, S.A. v The Argentine Republic, ICSID Case No. ARB/03/19, Order in response to a petition for transparency and participation as *amicus curiae* of 19 May 2005.

Alagar Iyengar and 12 Others v State of Tamil Nadu, 2002-4-L. W 498.

Annaswami Naicken and Others v C. Manicka Mudaliar and Others, AIR 1937 Madras 957.

Azurix Corp. v The Argentine Republic, ICSID Case No. ARB/01/12, Award of 14 July 2006.

Biwater Gauff (Tanzania) Ltd. v. United Republic of Tanzania, ICSID Case No. ARB/05/22, Procedural order No. 3 of 19 September 2006.

Canada — Measures Affecting Exports of Unprocessed Herring and Salmon, GATT Panel Report, Doc. L/6268–35S/98, 22 March 1988.

CMS Gas Transmission Company v. The Argentine Republic, ICSID Case No. ARB/01/8, Final Award of 12 May 2005, 44, *Int'l Leg. Mat.* 205 (2005).

C.N. Marudhanayagam Pillai v Secretary of State for India, 1939, MLJ, 176.

Compañia de Aguas del Aconquija S.A. and Vivendi Universal S.A. v. Argentine Republic, ICSID Case No. ARB/97/3, Award of 20 August 2007.

European Communities – Measures Affecting Asbestos and Asbestos-Containing Products, Report of the Panel, Doc. WT/DS135/R, 2000.

F.K. Hussain v Union of India, AIR 1990 Ker, 321.

Government of the Republic of South Africa and Others v Grootboom and Others, Constitutional Court of South Africa, Judgment of 4 October 2000, 2000 (11), BCLR 1169 (CC).

Hanuman Prasad v Mendwa, AIR 1935 All 876.
Het Singh v Anar Singh, AIR 1982 All 468.
Highveldridge Residents Concerned Party v Highveldridge Transitional Local Council [2002] 6 SA 66 (T).
Hindustan Coca Cola Beverages Private Limited v The Perumatty Grama Panchayat and Ors., W.A. No. 2125 of 2003.
Hindustan Coca-Cola Beverages (P) Ltd. v. Perumatty Grama Panchayat, M. Ramachandran and K.P. Balachandran (JJ), 7 April 2005.
Indian Council for Enviro-Legal Action and Ors. v Union of India, Supreme Court of India, (1996)3 SCC 212.
International Thunderbird Gaming v The United Mexican States, UNCITRAL (NAFTA), Separate Opinion (Professor T. Wälde) of 26 January 2006.
In the matter between: Lindiwe Mazibuko, Grace Munyai, Jennifer Makoatsane, Sophia Malekutu, *Vusimuzi Paki (Applicants) and The City of Johannesburg, Johannesburg Water (Pty) Ltd*, the Minister of Water Affairs And Forestry (Respondents) (July 2006).
Ipur Grama Panchayat v Government of Andhra Pradesh, 2000 (4) ALT 678.
Karnataka Industrial Area Development Board v Kenchappan (2006) 6 SCC 371.
Kesava Bhatta v Krishna Bhatta, AIR 1946 Mad 334.
Korea – Measures Affecting Imports of Fresh, Chilled and Frozen Beef, WTO Appellate Body Report, WT/DS161/AB/R, 2000.
Lindiwe Mazibuko et al. v The City of Johannesburg et al., High Court (Witwatersrand Local Division) of South Africa, Judgment of 30 April 2008.
Manqele v Durban Transitional Metropolitan Council, High Court (Durban and Coast Local Division) of South Africa, Judgment of 7 February 2001, (2002) 2 All SA 39 (D).
Manquele v Durban Transitional Metropolitan Council [2001] JOL 8956 (D).
M.C. Mehta v Kamal Nath (1997) 1 SCC 388.
M.C. Mehta v Union of India, AIR 1987 SC 1086.
Metalclad Corporation v United Mexican States, ICSID Case no. ARB(AF)/97/1, Award of 30 August 2000, 40 *Int'l Leg. Mat.* 36 (2000).
M. I. Builders Pvt. Ltd. v Radhey Shyam Sahu (1999) 6 SCC 464.
Intellectual Forum v State of Andhra Pradesh (2006) 3 SCC 549.
Minister of Health and Others v Treatment Action Campaign and Others, Constitutional Court of South Africa, Judgment of 5 July 2002, 2002 (10) BCLR 1033 (CC).
Narmada Bachao Andolan v. Union of India, Writ Petition (Civil) No. 319 of 1994, Supreme Court of India, Judgment of 18 October 2000, AIR 2000 SC 3751.
Natarajan v Ramapuram Panchayat, Thanjavur District, 1999 (1) MLJ 598.
Nature Conservation Council of NSW Inc v The Minister Administering the Water Management Act 2000, [2005] NSWCA 9.

Noyyal River Ayacutdars Protection Association and Others v Government of Tamil Nadu, Writ Petition Nos. 29791 and 39368, decided on 22 December 2006 by the Madras High Court.

Pope & Talbot Inc. v The Government of Canada, Interim Award of 26 June 2000, UNCITRAL, 40 Int'l Leg. Mat. 258 (2001).

Project Blue Sky Inc v Australian Broadcasting Authority, (1998) 194 CLR 355.

Ramachandra Iyer v Narainasami, 1892(2) MLJ 279; 1893 ILR Mad 333.

Research Foundation Science Technology Natural Resource Policy v Union of India, Supreme Court of India, Writ Petition No. 657 of 1995, Order dated 14 October 2004.

Residents of Bon Vista Mansions v Southern Metropolitan Local Council, High Court (Witwatersrand Local Division) of South Africa, Judgment of 5 September 2001, 2002 (6) BCLR 625 (W).

Saluka Investment BV (The Netherlands) v The Czech Republic, UNCITRAL, Partial Award on Jurisdiction and Liability of 17 March 2006, § 301.

Secretary of State v P.S. Nageswara Iyer & others, AIR 1936 Madras 1923.

South Africa v Grootboom [2000] 11 BCLR 1169 (CC).

State of Tamil Nadu v Sudalli Pothinadar, 1999 -1-L. W. 129.

Subhash Kumar v State of Bihar, AIR 1991 SC 420.

Tecnicas Medioambientales Tecmed S.A. v The United Mexican States, ICSID Case No. ARB(AF)/00/2, Award of 29 May 2003, 43 Int'l Leg. Mat. 133 (2004).

The Perumatty Grama Panchayat v State of Kerala and Ors., The Kerala High Court, Original Petition (Civil) No. 34292 of 2003, Judgement of 16 December 2003.

The Perumatty Grama Panchayat v State of Kerala and Ors., W.A. 1962 of 2003.

The Perumatty Grama Panchayat v State of Kerala and Ors, W.A. No. 215 of 2004.

The Perumatty Grama Panchayat v Secretary to Government, Local Self Governance, Government of Kerala, W.A. No. 12600 of 2004.

Tippetts, Abbett, McCarthy, Stratton v TAMS-AFFA Consulting Engineers of Iran, 6 Iran-US CTR 219.

United States – Import Prohibition of Certain Shrimp and Shrimp Products, WTO Appellate Body Report, Doc. WT/DS58/AB/R, 1998.

United States – Import Prohibition of Certain Shrimp and Shrimp Products - Recourse to Article 21.5 of the DSU by Malaysia, WTO Appellate Body Report, WT/DS58/AB/RW, 2001.

Valluri Adinarayana v Ramudu, SA No 931 of 1911, XXIV MLJ 17.

Vellore Citizen's Welfare Forum v Union of India, Supreme Court of India, (1996) 5 SCC 647.

Venkatagiriyappa v Karnataka Electricity Board & Others, 1999 (4) Kar LJ 482.

5. Reports and other Documents

Agarwal, A. and S. Narain eds. Dying Wisdom: Rise, Fall and Potential of India's Traditional Water Harvesting System (New Delhi: Centre for Science and Environment, Fourth Citizens Report, 1997).

Alcazar, L., M.A. Abdala and M.M. Shirley. The Buenos Aires Water Concession (Washington: World Bank Policy Research Working Paper No. 2311, 2000).

Artana, D., F. Navajas and S. Urbiztondo. Regulation and Contractual Adaptation in Public Utilities: The Case of Argentina (Washington: IADB, Paper IFM–115, 1998).

Asian Development Bank. Rehabilitation and Management of Tanks in India – A Study of Select States (Manilla: ADB, Publication Stock No. 122605, 2006).

Austria, Martinez P. and P. van Hofwegen eds. Synthesis of the 4th World Water Forum (Copilco El Bajo: Comisión Nacional de Agua, 2006).

Beekman, Hans, I. Saayman and S. Hughes. Vulnerability of Water Resources to Environmental Change in Southern Africa (Pretoria: CSIR, 2003).

Bhaktavatsalam, T. Politics of Opposition in Andhra Pradesh: 1983-88 (Guntur: Nagarjuna University, M.Phil. Dissertation, 1991).

Bond, P. The Battle over Water in South Africa (AfricaFiles, 2003).

Camdessus, Michel and James Winpenny. Financing Water For All (Report of the World Panel on Financing Infrastructure, 2003).

Chapagain, Ashok K., Arjen Y. Hoekstra and Huub H.G. Savenije. Saving Water through Global Trade, UNESCO Value of Water Research Report Series No. 17 (Delft: UNESCO–IHE, 2005).

Cowan, G.I. and G.C. Marneweck. South African National Report to the Ramsar Convention 1996 (Pretoria: Department of Environmental Affairs and Tourism, 1996).

Datar, Chhaya and Ajith Kumar. Rural Drinking Water in Maharashtra (Tuljapur: Tata Institute of Social Sciences, 2001).

Department of Water Affairs and Forestry. Internal Strategic Perspective for the Orange River System: Overarching (Pretoria: DWAF, 2004).

DHAN Foundation. Study on Customary Rights and their Relation to Modern Tank Management in Tamil Nadu, India (Madurai: DHAN Foundation, 2004).

Dinar, Ariel, Mark Rosegrant and Ruth Meinzen-Dick. Water Allocation Mechanisms. Principles and Examples (Washington, DC: World Bank Policy Research Working Paper 1779, 1997).

Dorcey, Anthony H.J. ed. Large Dams: Learning from the Past, Looking to the Future (Washington: World Bank, 1997).

Douglas, Vermillion ed. The Privatisation and Self Management of Irrigation (Final Report submitted to the GTZ Germany by the International Irrigation Management Institute, Colombo, Sri Lanka, 1996).

Dubash, Navroz K. et al. A Watershed in Global Governance? An Independent Assessment of the World Commission on Dams (Washington: World Resources Institute, 2001).

Dupuy, Pierre-Marie. Le droit à l'eau, un droit international? (Florence: EUI Working Paper No. 2006/06, 2006).

Government of Andhra Pradesh. Report of the Committee on Drawing Additional Water to Twin Cities from Srisailam or Nagarjunasagar or Other Projects (Hyderabad: Government of Andhra Pradesh, 1973).

Gowlland-Gualtieri, Alix. Legal Implications of Trade in 'Real' and 'Virtual' Water Resources (Geneva: IELRC, Working Paper 2008-01, 2008).

Grusky, Sara. The IMF, the World Bank and the Global Water Companies: A Shared Agenda (International Water Working Group, 2001).

Hamstead, M., C. Baldwin and V. O'Keefe. Water Allocation Planning in Australia – Current Practices and Lessons Learned (Report Commissioned by the Australian National Water Commission, April 2008).

Hearne, Robert R. and William Easter. Water Allocation and Water Markets. An Analysis of Gains-from-Trade in Chile (Washington: World Bank Technical Paper 315, 1995).

Hemson, D. et al. Still Paying the Price: Revisiting the Cholera Epidemic of 2000-01 in South Africa (Human Sciences Research Council, Occasional Papers Series Number 8, February 2006).

Hodgson, S. Legislation on Water Users, Organizations – A Comparative Analysis (Rome: FAO, FAO Legislative Study 79, 2003).

Holden, Paul and Mateen Thobani. Tradable Water Rights: A Property Rights Approach to Resolving Water Shortages and Promoting Investment (Washington: World Bank Policy Research Working Paper 1627, 1997).

Howard, Guy and Jamie Bartram. Domestic Water Quantity, Service Level and Health (Geneva: World Health Organization, WHO Doc. WHO/SDE/WSH/03.02, 2003).

Hyderabad Urban Development Authority. Hyderabad 2020: Draft Master Plan for Hyderabad Metropolitan Area (Hyderabad: 2003).

IBON Databank and Research Center. Water Privatization: Corporate Control versus People's Control (Manila: IBON, 2005).

Imhof, Aviva. The Asian Development Bank's Role in Dam-Building in The Mekong (Berkeley: International Rivers Network, 1997)

Jones, Gary et al. Ecological Assessment of Environmental Flow Reference Points for the River Murray System: Interim Report Prepared by the

Scientific Reference Panel for the Murray-Darling Basin Commission, Living Murray Initiative (Canberra: Cooperative Research Centre for Freshwater Ecology, 2003).

Kerala Ground Water Department. Report on the Monitoring of Wells in and Around the Coca Cola Factory in Plachimada, Kannimari, Palakkad district (Kerala Ground Water Department, September, 2003).

Kerala Ground Water Department. Report on the Monitoring of Water Levels and Water Quality in Wells in and Around the Hindustan Coca Cola Factory at Plachimada, Palakkad District (Thiruvananthapuram: Kerala Ground Water Department, 2006).

Kirmani, Sayed and G. Le Moigne. Fostering Riparian Cooperation in International Basins (Washington, DC: World Bank, 1997).

Law Commission of India. Report on Proposal to Constitute Environmental Courts, 186th Report (New Delhi: Law Commission of India, 2003).

Lele, Sharachchandra, A.K. Kiran Kumar and Pravin Shivashankar. Joint Forest Planning and Management in the Eastern Plains Region of Karnataka: A Rapid Assessment (Bangalore: Centre for Interdisciplinary Studies in Environment and Development, Technical Report, 2005).

Madhav, Roopa. Irrigation Reforms in Andhra Pradesh: Whither the Trajectory of Legal Changes? (Geneva: IELRC Working Paper 2007-04, 2007).

Malloch, Steven. Liquid Assets: Protecting and Restoring the West's Rivers and Wetlands through Environmental Water Transactions (Arlington, Canada: Trout Unlimited, 2005).

Mann, Howard and Konrad Von Moltke. NAFTA's Chapter 11 and the Environment (Winnipeg: International Institute for Sustainable Development, IISD Paper, 1999).

Mann, Howard. Private Rights, Public Problems: A Guide to NAFTA's Controversial Chapter on Investor Rights (Winnipeg: International Institute for Sustainable Development, 2001).

Mehta, Lyla and Birgit La Cour Madsen. Is the WTO after Your Water? The General Agreement on Trade in Services (GATS) and the Basic Right to Water (Brighton: Institute of Development Studies, 2003).

Millington, P. River Basin Management: Its Role in Major Water Infrastructure Projects, (Cape Town: World Commission on Dams, 2000).

Moench, Marcus and S. Janakarajan. Water Markets, Commodity Chains and the Value of Water (Chennai: Madras Institute of Development Studies, MIDS Working Paper No.172, June 2002).

Mohr, Alison. Governance through 'Public Private Partnerships': Gaining Efficiency at the Cost of Public Accountability? (International Summer

Academy on Technology Studies – Urban Infrastructure in Transition, 2004).
Molle, François. Development Trajectories of River Basins – A Conceptual Framework (Colombo: International Water Management Institute, Research Report No. 72, 2003).
Moltke, Konrad Von. An International Investment Regime? Issues of Sustainability (Winnipeg: International Institute for Sustainable Development, IISD Paper 16, 2000).
Morely, Dannielle ed. NGOs and Water – Perspectives on Freshwater, Issues and Recommendations of NGOs (London: United Nations Environmental and Development Forum, 2000).
Muller, M. Sustaining the Right to Water in South Africa (Geneva: UNDP Human Development Report Office, Occasional Paper 2006/29, 2006).
National Institute of Urban Affairs (NIUA). Status of Water Supply, Sanitation and Solid Waste Management in Urban Areas (New Delhi: NIUA, June 2005).
Niekerk, Lara van and Huizinga Piet. Guidelines for the Mouth Management of the Orange River Estuary (Stellenbosch: CSIR, 2005).
Oblitas, Keith, J. Raymond Peter, Gautam Pingle *et al.* Transferring Irrigation Management to Farmers in Andhra Pradesh, India (Washington: World Bank Technical Paper 449, 1999).
OECD. Indirect Expropriation and the Right to Regulate in International Investment Law (Working paper on international investment no. 2004/4, 2004).
P.A. Consulting Group. Water Reuse Preliminary Concept and Feasibility Study (Hyderabad: P.A. Consulting Group, 2004).
Panickar, Meena. State Responsibility in the Drinking Water Sector. An Overview of the Indian Scenario (Geneva: International Environmental Law Research Centre, Working Paper 2007-06, 2007).
Rosegrant, Mark and Renato Gazmuri Schleyer. Tradable Water Rights: Experiences in Reforming Water Allocation Policy (Washington: Applied Study Prepared for the Bureau for Asia and the Near East of the U.S. Agency for International Development , 1994).
Saleth, Maria. Strategic Analysis of Water Institutions in India: Application of a New Research Paradigm (Colombo: International Water Management Institute, Research Report No. 79, 2004).
Salman, Salman M.A. The Legal Framework for Water Users' Associations: A Comparative Study (Washington DC: World Bank, 1997).
Salman, Salman M.A. The Human Right to Water: Legal and Policy Dimensions (Washington: World Bank, 2004).
Sangameswaran, Priya. Review of Right to Water: Human Rights, State Legislation, and Civil Society Initiatives in India (Bangalore: Centre for Interdisciplinary Studies in Environment and Development, Technical Report, 2007).

Sivasubramanian, K. Irrigation Institutions in Two Large Multi-village Tanks of Tamil Nadu (Chennai: Madras Institute of Development Studies, PhD Thesis, 1995).
Smith, L. et al. Testing the Limits of Market-Based Solutions to the Delivery of Essential Services: The Nelspruit Water Concession (Johannesburg: Centre for Policy Studies, September 2003).
Srinivas, Sampath. Case of Public Interventions, Industrialisation and Urbanisation: Tirupur in Tamil Nadu, India (New Delhi: World Bank, 2000).
Suresh, V. and Pradip Prabhu. Democratisation of Water Management as a Way to Reclaiming Public Water: The Tamil Nadu Experience (Amsterdam: Transnational Institute, 2007).
Trewin, Dennis ed. 2007 Year Book Australia (Canberra, Australian Bureau of Statistics, Yearbook No. 89, ABS Catalogue No. 1301.0, 2007).
United Nations Development Programme. Safe Water 2000 (New York: UNDP, 1990).
United Nations. A Strategy for the Implementation of the Mar del Plata Plan for the 1990s (New York: UN Department of Technical Cooperation, 1991).
Vaidya, Chetan. The Tiruppur Area Development Program – Focus on Urban Infrastructure and Private Sector Participation (Indo-US Financial Institutions Reform and Expansion Project – Debt Market Component FIRE (D), Note No.13, 1999).
Vaidyanathan, A. and J. Saravanan. Household Water Consumption in Chennai City: A Sample Survey (Centre for Science and Environment, 2005).
Vermillion, Douglas. Impacts of Irrigation Management Transfer: A Review of the Evidence (Colombo: International Irrigation Management Institute, Research Report 11, 1997).
WaterAid. Drinking Water and Sanitation Status in India: Coverage, Financing and Emerging Concerns (New Delhi: WaterAid, 2005).
Wickeri, Elisabeth. Grootboom's Legacy: Securing the Right to Access to Adequate Housing in South Africa (New York: Center for Human Rights and Global Justice, Working Paper No. 5/2004, 2004).
Williams, Sandy and Sarah Carriger. Water and Sustainable Development: Lessons from Chile (Chile: Global Water Partnership (GWP), Technical Committee (TEC), Policy Brief No. 2, 2006).
Wong, Senator, Penny. Minister for Climate Change and Water, Murray Darling Deal Delivered, (Media Release PW 41/08, 26 March 2008).
World Bank/OED. Water Supply and Sanitation Projects: The Bank's Experience 1967–1989 (Washington: World Bank, Report 10789, June 1992).
World Bank. Water Resources Management (Washington: World Bank, Policy Paper, 1993).

World Bank. Staff Appraisal Report – Uttar Pradesh Rural Water Supply and Environmental Sanitation Project (Report No. 15516-IN, 1996).

World Bank/OED. Argentina: Country Assistance Review (Washington: World Bank, Report 15844, 1996).

World Bank. India Water Resources Management Sector Review – Initiating and Sustaining Water Sector Reforms (Report No. 18356-IN in 6 volumes, 1998).

World Bank. Water Resources Sector Strategy: Strategic Directions for World Bank Engagement (Washington, DC: World Bank, Report No. 28114, 2004).

World Bank/Operations Policy and Country Services. Review of World Bank Conditionality: Modalities of Conditionality (Background Paper SecM2005-0390/1, June 2005).

World Bank/Operations Policy and Country Services. Policy Conditions in World Bank Investment Lending: A Stocktaking, Board Report 36924, July 2006.

World Commission on Dams. Dams and Development: A New Framework for Decision-Making, Report of the World Commission on Dams (Earthscan: World Commission on Dams, 2000).

World Customs Organisation (WCO). Harmonised Commodity Description and Coding System: Explanatory Notes (Brussels: WCO, 3rd ed., 2002).

World Health Organization and United Nations Children's Fund. Global Water Supply and Sanitation Assessment 2000 Report (Geneva: WHO, 2000).

World Health Organisation. The Right to Water (Geneva: WHO, 2003).

World Water Council. A Water Secure World: Vision for Water, Life, and the Environment (Hague: World Water Council, Commission Report, 2000).

Wouters, Patricia. The Legal Response to International Water Scarcity and Water Conflicts: The UN Watercourses Convention and Beyond (Dundee, 2003).

Yuvajanavedi. Report on the Environmental and Social Problems Raised due to Coca Cola and Pepsi in Palakkad District (Thiruvananthapuram: Yuvajanavedi, November 2002).